AF659031

ENCYCLOPAEDIA OF HORTICULTURE AND ALLIED TERMS

Volume - 1

ENCYCLOPAEDIA OF HORTICULTURE AND ALLIED TERMS

Volume - 1

Amit Baran Sharangi

2024

Daya Publishing House®

A Division of

Astral International Pvt. Ltd.

New Delhi – 110 002

© EDITOR

First Published, 2009

Reprinted, 2024

ISBN: 978-81-7035-572-4 (HB)

Disclaimer:

Every possible effort has been made to ensure that the information contained in this book is accurate at the time of going to press, and the publisher and author cannot accept responsibility for any errors or omissions, however caused. No responsibility for loss or damage occasioned to any person acting, or refraining from action, as a result of the material in this publication can be accepted by the editor, the publisher or the author. The Publisher is not associated with any product or vendor mentioned in the book. The contents of this work are intended to further general scientific research, understanding and discussion only. Readers should consult with a specialist where appropriate.

Every effort has been made to trace the owners of copyright material used in this book, if any. The author and the publisher will be grateful for any omission brought to their notice for acknowledgment in the future editions of the book.

All Rights reserved under International Copyright Conventions. No part of this publication may be reproduced, stored in a retrieval system, or transmitted in any form or by any means, electronic, mechanical, photocopying, recording or otherwise without the prior written consent of the publisher and the copyright owner.

Published by : **Daya Publishing House®**
A Division of
Astral International Pvt. Ltd.
– ISO 9001:2015 Certified Company –
4736/23, Ansari Road, Darya Ganj,
New Delhi-110 002
Ph. 011-43549197, 23278134
E-mail: info@astralint.com
Website: www.astralint.com

Printed at : **Replika Press Pvt. Ltd.**

Dedicated to

—×—×—

Sri Girindranath Sharangi

&

Smt. Snehalata Sharangi

—×—×—

My beloved parents

Foreword

Contrary to the popular belief as a subject of mere aesthetic gardening, Horticulture emerged to be a very promising multidisciplinary field with unbound potentialities assuring nutritional as well as economic security throughout the globe. With the advancement of science, its area and scope has been widened both in terms of concept and dimension of newer areas with newer terminologies to be understood for better comprehension. The present book *Encyclopaedia of Horticulture and Allied Terms* by Dr. A.B. Sharangi, Associate Professor, Bidhan Chandra Krishi Viswavidyalaya, is a welcome effort in this direction.

There are many revered books available with this theme in various names and forms, but very few books like the present one are there which cover almost the entire range of terms associated with Horticulture and allied sciences. I like to congratulate Dr. Sharangi for his maiden attempt in compiling such a valuable book.

I am confident that the book as such can meet the long-felt need of a well balanced encyclopaedia of this kind for a considerable period of time. I hope the book will be accepted widely by the students, teachers, researchers, amateur gardeners, policy makers, corporate personnel and all individual readers having keen interest on the diverse fields of Horticulture as theory as well as in practice.

June 5, 2008

Prof. (Dr.) M.G. Som
Former Vice Chancellor,
Bidhan Chandra Krishi Viswavidyalaya
(State Agriculture University),
West Bengal, India

Contents

Prologue

Agriculture gave the primitive man a settled life on earth towards growth and development of civilization. The search for food with the creation of green life slowly made him satisfied, rational and social. Agriculture, by its nature, involves self-conscious attempts by humans to change or "manage" natural ecosystems. Horticulture initially started as aesthetic agriculture in a confined garden, slowly changing its stand towards high value high return crops per unit area, time and input with the ultimate aim for food and nutrition security of the ever-increasing population of the entire globe particularly in the developing countries. India is no exception to this pragmatic trend to assimilate and welcome horticulture in her vision and mission. Harnessing the compelling strength in biodiversity, India is on its way to becoming the global hub for fresh horticultural produce outsourcing with diverse agro climatic zones, and a wealth of native knowledge contributing to the sector's development. Our country also shares a warm relationship with several gross importers of horticultural produce.

Horticulture, itself is a unique subject which includes the broad and diverse areas of Botany, Plant Pathology, Entomology, Genetics, Breeding, Biotechnology, Genetic Engineering, Soil Science, Chemistry, Farm management, Economics, Sociology, Statistics, Medicine, Agricultural Engineering, Agrometeorology, etc in general and Pomology, Olericulture, Spices, Tea, Coffee, Rose, Wine, Post Harvest Technology, Processing, Bonsai, Flower arrangement, Pollen/Spore, Pesticides, Manures/Fertilizers, Natural/Synthetic chemicals, Instruments/Implements/Tools, Quality management, Water, etc in particular. The understanding of the entire area is more or less impossible without the understanding of the specific terminologies associated with each subject concerned. Few chapters have been placed seperately (*e.g.* Chapters 13, 14 and 21) with a due consideration of their relative importance and not placed in any common chapter like the others.

In spite of a good number of publications in horticultural terms, no concerted attempt has been made to satisfy the need of the interested readers to provide a comprehensive output of the entire areas related to horticulture. The present book "*Encyclopedia of Horticulture and Allied Terms*" is a humble attempt to bring all the important and related terms together. Each term has been defined with utmost care, accuracy and completeness to provide a rational understanding and which are readily accessible for the students, teachers, amateur gardeners, corporate sectors,and all professionals associated with Horticulture.

An encyclopedia can never be original in its strictest sense, but a collection of works and compilation of others. Many terms and their definitions have been derived from books, journals, websites collected from different sources. The author is grateful to them for rendering such indirect help and acknowledging each and everyone. My special thanks are due to Seema, my wife and Avijoy, my son for their silent help during the wee hours of manuscript preparation. I am realy indebted to Dr. P.K. Sahu, Reader, Agricultural Statistics, BCKV for kindly correcting the statistics section. There is every possibility of printing, typographical or other errors as the entire volume came into reality within a very short spell of time. Inspite of every effort to avoid duplications of terms in different chapters, there may be many such instances present in the book. Constructive criticism, valuable comments and all sorts of suggestions can only make the publication complete in its future appearances.

Dr. A.B. Sharangi

Chapter 1
Horticultural Terms

Accent plant–Attention-getting plant due to its color or form.

Acre–A measure of land totaling 43,560 square feet or 4000 square metre.

Aerate-Loosening or puncturing the soil to increase water penetration and aeration.

Aeration–To increase the amount of air space in the soil by tilling or otherwise loosening the soil.

Air drainage–The flow of cold air down a hill. Planting of crops that are sensitive to late spring and early winter frost on slopes is incessary, preferably south-facing, so that cold air will drain downhill and settle in the low spots below the planted area.

Air layering–1. A propagation method comprising a cut is made three-quartets of the way through the stem at an angle of 45 degrees. The cut is wrapped in sphagnum moss that has been soaked in water and wrung out. This is then wrapped in plastic sleeve and tied at the top and bottom with ties.

2. A specialized method of plant propagation accomplished by cutting into the bark of the plant to induce new roots to form; a process of producing a new plant by forming soil around a stem.

Alkaline Soil–Soil with a pH greater than 7.0, also called sweet soil, opposite of acid. Alkaline soils slow the growth of many plants when their pH is above 8.0.

Alpina/alpinus/alpinum–Epithet meaning from an alpine area or just a very dwarf plant. It is also used to denote smaller species in a large genus.

Alpine–Any plant that is normally found in wild on mountains or in mountainous areas that are above the tree line.

Alternate leaf arrangement–Leaves are arranged at one leaf per node on different sides along the stem.

Alluvial soil–Zonal soil formed on flood plains of river valleys and at river mouths of deposition of new material successively on the surface during flooding.

Annuals–Plants that sprout from seed, grow, flower, set seeds, and die all in one growing season ;Plants that complete their life cycles in one year or less.

Anther–1. The terminal part of a stamen, which contains pollen.

2. The part of the flower (on the upper part of the stamen) in which pollen is produced.

Anthracnose–1. A type of leaf or fruit spot disease caused by a fungus.

2. A fungus disease that causes spots and often death of foliage.

Aphid–A small green or white insect that sucks juices from plant parts.

Aquatic plants–Plants which grow in, live in, or live on the water.

Arboreal–Tree-like or pertaining to trees.

Arborescent—ree-like; defined arbitrarily as pertaining to a woody plant at least 20 ft. high as maturity with a single stem and more or less definite crown.

Arboretum–A garden with a large collection of trees and shrubs cultivated for scientific or educational purposes.

Arbor–Tree.

Arborcide–A type of herbicide chemical that kills trees.

Arboriculture–The science of the culture or care of trees.

Arborist–A certified/licensed tree doctor specialist.

Arching–Curving gracefully.

Arms–In a grape plant, all main branches two years and older.

Asexual reproduction–Duplicating a plant from any cell, tissue or organ of that plant; multiplying plants without use of seeds

Axillary buds–Buds that are borne laterally on the stem in the axils of the leaves.

B and B–Balled and burlapped, a method of digging and holding woody plants.

Bedding plant--Plants (mainly annuals), nursery grown and suitable for growing in beds. Quick, colorful flowers are usually produced.

Biennial plants–1. A plant that grows, flowers, produces seeds or fruit, and dies in two years. Some herbacous flowers and vegetables are biennial. Most biennial plants produce foliage the first year and bloom the second year.

2. Plants with a two year life cycle, that produce leaves in the first season, then flower and set seed in the second season.

3. Plants that require 2 years, or 2 growing seasons with a dormant period in between, to complete their life cycles.

Biological control–Reducing pests by utilizing other organisms. For example, controlling gypsy moths by using *Bacillus thuringensis*.

Biomass–Moss of living tissue, often a measure of weight or volume.

Blade–The flattened, green portion of the leaf.

Bolting–1. Going to seed. Plants will produce a flower stalk and seeds, and then die before the end of the season. Lettuce and other crops will bolt during very hot temperatures.

2. Vegetables which quickly go to flower rather than producing the food crop. Usually caused by late planting and too warm temperatures.

Bone Meal–Finely ground white or light gray bone. Adds nitrogen and phosphorus to the soil.

Bonsai–The art of growing carefully trained, dwarf plants in containers.

Botanical name–The Latin or "scientific" name of a plant, usually composed of two words,the genus and the species (Binomial nomenclature).

Botrytis blight–A gray, felt-type mold that covers parts of a plant and causes stunting, dieback, and distorted growth.

Bottom heat–The undersurface heat provided in the soil by electric cables or hot water pipes.

Bract–A modified and often scale like leaf, usually located at the base of a flower, a fruit, or a cluster of flowers or fruits.; Modified leaves growing just below a flower (often confused with the flower itself).

Branch–The part of the tree that arises from the trunk and supports shoots, twigs and leaves.

Budding–A type of grafting in which a non-fruitful (vegetative) scion is placed in a stock plant.

Bud–Early stages of development of a flower or plant growth. An undeveloped stem, consisting of a tiny bundle of cells, from which leaves, lateral buds, flower parts or all three will arise.

Builder's Sand–This horticultural sand has coarse, large grains that improve drainage and aeration of soil.

Bulb–The thickened underground storage organ of the group of perennials which includes daffodils and tulips;. A budlike structure that consists of a small stem with closely crowded fleshy or papery leaves or leaf bases.

Bulb Planter–Used to dig holes for bulbs. This cylindrical, hand held tool minimizes disturbance to nearby tree or shrub roots.

Bulbil–An aerial bulblet.

Bulblet–Miniature bulbs that develop from meristems in the axils of scaly bulbs.

Butterfly garden–Planting to attract, feed and support Lepidoptera.

Calcium nitrogen–A quick-acting fertilizer with about 48 per cent calcium and 16 per cent nitrogen.

Cambium–A thin ring of tissue within the stem, branch, and trunk that continually forms nutrient and water-conducting vessels. The thin membrane located just beneath the bark of a plant.

Canes–Previous season's growth from the arms or trunk of a grape plant; or, the supporting part of the raspberry plant, similar in function to the trunk of a tree.

Canker–Bacterial or fungal disease causing bark lesions.

Capsule–Simple, dry, dehiscent fruit with two or more locules (the cavities of the ovary of the pistil of a flower) which split in various ways.

Catkin–1. A type of spike arrangement of flowers on the floral axis that has unisexual flowers made up of sepals and petals.

2. A slender, spike like, drooping flower cluster.

Cellar gardening–Using a basement to grow plants with artificial light.

Certified seed–The progeny of registered seed stock. It is the final stage in the expansion program and is certified with a metal seal and blue tag.

Chalk–white limestone product or soil.

Charcoal–A soil additive to increase moisture intake.

Chilling injury–Damage to certain horticultural products, such as banana, papaya, cucumber and sweet potato, which results from exposure to cold but above-freezing temperatures.

Chilling requirement–A cold period required by certain plants and plant parts to break physiological dormancy or rest. The chilling requirement is expressed in terms of the required number of hours at 7 degrees centigrade or less.

Chlorophyll–1. The green pigment found in chloroplast and that is necessary for photosynthesis.

2. The green pigment in leaves. When present, it usually dominates all other pigments.

Chlorosis–Yellowing of foliage due to a loss or breakdown of chlorophyll. Chlorosis may result from disease or infestation, poor growing conditions, or mineral deficiency.

Clay Soil–Soil with very small, flat particles that tend to pack together tightly, hardens when dry, and drains poorly without organic amendments.

Clean cultivation–Gardening practice of removing all weeds at all times.

Colchicine–A toxic chemical used to double chromosome numbers in seeds.

Cold frame–A low frame with clear top used to acclimatize plants to, or protect them from, cold.

Cold storage–A type of insulated storage utilizing mechanical refrigeration to maintain a stable, cold temperature for long-term storage.

Commercial floriculture–Area of horticulture that includes the commercial production and distribution of cut flowers, flowering pot plants, foliage plants and bedding plants.

Complete fertilizer–1. A fertilizer containing nitrogen, phosphorus, and potassium, the three nutrients in which plants are most commonly deficient.

2. A plant food with all three of the primary elements: nitrogen, phosphorus and potassium.

Complete flower–Flower composed of a short axis or receptacle from which arise four sets of floral parts-sepals, petals, stamens, and pistils.

Compost–1. An organic soil amendment resulting from the decomposition of organic matter.

2. Organic material composed of decaying plant and animal matter for use in soil.

Compound leaf–A leaf whose blade divided into several distinct leaflets or sections.

Compound umbel–A type of flower structure in which a series of simple umbels arise from the same point on the main axis.

Conifer–Woody trees and shrubs that produce cones. Common conifers include pines, firs, spruce, juniper, redwood and hemlocks.

Cool-season grass–A grass that greens up and grows during the spring, sets seed in early summer, then goes dormant until fall, when it begins growing again. A second dormancy may occur during cold winter weather.

Corm–1. A short, solid, enlarged, underground stem from which roots grow. Corms are food-storage organs. They contain one bud that will produce a new plant.

2. Short, fleshy, underground stems with few nodes and very short spaces between nodes. Gladiolus and crocus are cormous plants.

Corona–A crown like structure on some corollas, as in daffodils and the milkweed family.

Corymb–Type of raceme flower cluster in which the pedicels of the lower flowers are longer than the pedicels of the upper flowers, resulting in a flat-topped flower cluster.

Creosote–An oil-based preservative used to treat wood used in gardens.

Cross-pollination–The process in which pollen is transferred from an anther (the upper part of the stamen in which pollen is produced) of one flower to the stigma (the pollen-receiving site of the pistil) of a second flower of a different cultivar.

Crown–The point where roots join the aboveground parts of a plant.

Cultivar–1. A plant derived from a cultivated variety that has originated and persisted under cultivation, not necessarily referable to a botanical species, and of botanical or horticultural importance, requiring a name.

2. A manmade plant variety, produced and maintained by vegetative propagation instead of seed.

Cuticle–The waxy covering or layer on leaves or fruit, which protects the tissue against excess moisture loss.

Cuttage–The process of rooting cuttings.

Cuttings–1. Detached vegetative plant parts which when placed under conditions favorable for regeneration will develop into a complete plant with characteristics identical to the parent plant.

2. Any stem or vegetative part of a plant used for propagation.

3. Stems, roots, or leaves taken from plants for propagation.

Cyme–A broad, more-or-less flat-topped determinate (having a fixed number) flower structure in which the central flowers bloom first.

Damping off–1. A fungus, usually affecting seedlings and causes the stem to rot off at soil level. Sterilized potting soil and careful sanitation practices usually prevent this.

2. Decay of young seedlings at ground level following fungal attack. Basically the disease is the result of soil borne diseases or over watering.

Day-neutral plant–A plant that will flower under any daylength.

Deadheading–1. The process of removing old or spent flowers to prevent them from setting seed. This will promote continued bloom.

2. Removing blooms that have faded.

3. The removal of old blossoms to encourage continued blooming or to improve the appearance of the plant.; The process of pinching off used or spent blooms to keep the plants well groomed and to prevent them from setting seed. This will promote continued bloom.

Deciduous–Trees and shrub that shed all their leaves annually, usually in the winter seasons.

Dehiscent–Type of dry fruit in which the carpel (a simple pistil) splits along definite seams at maturity. Determinate growth. Limited growth.

Desiccation–Dehydration or loss of water.

Dethatch–Process of removing dead stems that build up beneath lawn grasses

Dibble–A stick or tool that makes planting holes in soil or media.

Dibble stick–A pointed tool used to make holes in the soil for seeds, bulbs, or young plants.

Dioecious–Species that produce male and female gametes on separate plants.

Direct seeding–Germination of seed in their final spot as opposed to in a nursery.

Disbudding–The removal of vegetative or flower buds.

Disease–A fungus or virus infection.

Diseased plant–A plant that is abnormal because of a disease-causing organism or virus.

Dish garden–A miniature planting in a shallow tray.

Dividing–The process of splitting up plants, roots and all that have began to get bound together. This will make several plants from one plant, and usually should be done to mature perennials every 3 to 4 years.

Dividing/division–A method of propagating plants by separating each one into two or more sections and then replanting.

Dormancy–1. A state of suspended growth or the lack of outwardly visible activity caused by environmental or internal factors.

2. The yearly cycle in a plants life when growth slows and the plant rests. Fertilizing should be withheld when a plant is in dormancy.

Dormant–1. A state of rest and reduced metabolic activity in which plant tissues remain alive but do not grow.

2. Plants which are in a resting state, most often during low or high temperatures. They may die down to the roots.

3. Being in a state of reduced physiological activity such as occurs in seeds buds etc.

Dormant oil–Oil sprayed on deciduous trees while they are dormant. Dormant oils are used to kill over wintering insects or insect eggs on plant bark.

Double digging–Soil bed preparation done by two or more spading sessions, preparing the soil by systematically digging an area to the depth of two shovels.

Double flower–A flower with many overlapping petals which gives it a very full appearance.

Drip irrigation–1. The application of small quantities of water directly to the root zone through various types of delivery systems on a daily basis.

2. Watering plants by small droplets over a long period of time each day.

Drip line–1. On the ground, the line that marks the outer edge to which a tree's (or other plant's) branches spread. The drip line usually signals a change in microclimate, where the area under the tree, which sees less precipitation, sunlight, and wind and may also be subject to competition from the tree's roots, meets an area that is not sheltered by the tree.

2. The circle which would exist if a line below the tips of the outer most branches of a tree or plant is drawn.

Drupe–Type of fruit with a thin outer skinlike region, a center portion that is thick and fleshy, and an inner area (endocarp) that is hard and stony.

Ecosystem–Stable, though not necessarily permanent, community of plants that have developed interrelationships with each other and with native wildlife to form a distinct, self-sustaining system. A few examples of ecosystems are tallgrass prairie, boreal forest, estuary, and oak savannah. Though ecosystems are a useful concept, in real life a "pure" ecosystem is unusual; more common are areas in which several ecosystems overlap to various degrees.

Edging–A line that creates visual interest and separation, for example, between the lawn and an annual border

Edging plant–A compact, short plant used to line or trim borders.

Edible landscape–A landscape planted with vegetables and fruit plants.

Endangered species–A species likely to become extinct.

Endocarp–The inner area of the fruit wall.

Entomology–The study of insects.

Epidermis–The outer layer of cells of all parts of a young plant and of some parts of older plants, such as leaves and fruits. These cells are usually covered with a waxy substance, called the cuticle, that minimizes water loss.

Epiphyte–A plant which grows on another plant but gets its nourishment from the air and rainfall. They do no damage to the host plant.

Erosion–The wearing away, washing away, or removal of soil by wind, water or human activity.

Escape–A garden plant that has been seeded or spread into the wild.

Espalier–1. Any plant trained lattice fashion in one plane; woody plant trained in a lattice-like or fence-like pattern on some support.

2. Process of training a tree or shrub so that its branches grow in a flat pattern.

Ethylene–Naturally occurring gas to ripen fruit.

Etiolation–Stretching of a plant and loss of color due to a absence of needed light or in a greatly reduced amount of light.

Evaporation–Process by which water returns to the air. Higher temperatures speed the process of evaporation.

Evapotranspiration–Process of loss of water from a plant's tissue and soil.

Evergreen–A plant that retains all or most of it's foliage throughout the year.

Exocarp–The outer skinlike region of the fruit wall.

Eye–An undeveloped bud growth which will ultimately produce new growth.

Fallow–A system in which land is left without a crop for an extended period to accumulate moisture and restore nutrients.

Fertilizer–1. Organic or inorganic plant foods which may be either liquid or granular used to amend the soil in order to improve the quality or quantity of plant growth.

2. Chemicals, inorganic and organic, which are added to the soil to aid soil fertility are called fertilizers.

Flat–A shallow box or tray used to start cuttings or seedlings.

Fleshy fruits–Classification of fruits that includes the berry, drupe, and pome. They have a fruit wall that is soft and fleshy at maturity.

Floriculture–The study of growing, marketing, and arranging flowers and foliage plants.

Flower–A shoot of determinate (limited in number) growth with modified leaves that is supported by a short stem; the structure involved in the reproductive processes of plants that bear enclosed seeds in their fruits.

Focal plant–A plant with form or color calling attention or focus to an area.

Foliar–Concerning foliage or leaves

Foliar feeding–Fertilizer applied in liquid form to the plants foliage in a fine spray.

Forcing–1. The manipulation of environmental factors which makes it possible to produce a marketable pot plant or cut flower out of season.

2. The process of hastening a plants growth to maturity or bloom, stimulation of flowers or growth by controlling light and temperature.

Foundation plant–A plant suitable for planting around houses and below windows.

Foundation seed–Seed stock handled to most nearly maintain specific genetic identity and purity under supervised or approved production methods certified by the agency.

Frond–The term used to describe the branch and leaf structure of a fern or members of the palm family.

Frost–The condensation and freezing of moisture in the air. Tender plants will suffer extensive damage or die when exposed to frost.

Frost pocket–A depression in the ground into which cold air drains but from which it cannot escape, thus causing it to be an area very subject to freeze injury.

Fruit–An expanded and ripened ovary with attached and subtending reproductive structures.

Fumigation–Use of gas or vapors that sterilize soils or containers

Fungicide–1. A chemical used to kill fungus, ills, diseases or viruses.

2. A chemical that kills fungi or prevents them from infecting healthy plant tissue.

Genes–Basic units of hereditary material that dictate the characteristics of individuals.

Geocarpy–Ripening of fruits underground.

Germinate–The process of the sprouting of a seed.

Germination–1. The initiation of active growth by the embryo, resulting in the rupture of seed coverings and the emergence of a new seedling plant capable of independent existence.

2. Development of seeds into seedlings or leafy young plants; the first stage in the development of a plant from seed.

Gibberellin–A group of compounds that naturally controls stem elongation. Induction of parthenocary is another function.

Girdle–1. Removal of bark for grafting or killing of a plant. It is a process of choking of a branch by a wire or other material, most often in the stems of woody plants that have been tied tightly to a stake or support.

2. Encircling of plant roots, stems, trunks or branches resulting in a constriction of the plant part, or a reduction of water and nutrient flow through the girdled plant part.

Girdling–The choking of a branch by a wire or other material, most often in the stems of woody plants that have been tied to tightly to a stake or support.

Graft–1. To unite a stem or bud of one plant to stem or root of another plant.

2. Inserting a bud or stem into another plant for purposes of propagation.

Grafting–1. The joining of two separate structures, such as a root and stem or two stems, so that by tissue regeneration they form a union and grow as one plant.

2. The uniting of a short length of stem of one plant onto the root stock of a different plant. This is often done to produce a hardier or more disease resistant plant.

Grafting wax–A wax used to seal a bud or graft from water and air.

Green fertilizer/manure–A fast-growing crop used to add organic matter when ploughed.

Ground cover–A group of plants usually used to cover bare earth and create a uniform appearance; a spreading short plant used to cover soil; a living mulch.

Growing media/growing medium–A material used to culture plants.

Growth habit–General appearance or mode of growth.

Growth regulator–A chemical used to increase growth or shorter stems.

Guano–Bat or bird dropping used as a fertilizer high with about 13 per cent nitrogen.

Habit–Same as growth habit. The entire aboveground shape characteristic of a plant (such as a vase-shaped or columnar tree or a mounded shrub) and aspects of its pattern of growth that might influence the three-dimensional shape of it in a landscape (whether it forms colonies, for instance, or loses its lower branches with age, or tends to be flattened by a hard rain).

Habitat–Type of region in which a plant is native.

Ha-ha–Hidden ditch used to keep out people or animals.

Halophyte–Salt-loving or salt-tolerant plant.

Hardened–A plant condition created by various factors enabling it to withstand environmental stresses; contrast with succulent growth which is very vulnerable to environmental stress and damage.

Hardscape–Constructed elements of a landscape such as decks, walls fences and driveways. Structures and features of a landscape constructed from materials other than plants, such as walks, walls, and trellises made of wood, stone, or other materials, pathways, pools and ponds.

Hardening–The result of many changes that occur in a plant as it develops resistance to adverse conditions, especially cold.

Hardening off–1. The process of gradually acclimatizing greenhouse or indoor grown plants to outdoor growing conditions.

2. The process of plant adjustment to cold temperatures

3. The treatment of tender plants to enable them to survive a more adverse environment. Treatments involve withholding nutrients, lowering temperatures, allowing temporary wilting, and other methods to slow growth rate.

Hardiness–The ability of a plant to withstand low temperatures or frost, without artificial protection.

Hardpan–1. A layer in the soil that cannot be penetrated and which restricts root penetration as well as movement of air and water.

2. The impervious layer of soil or clay lying beneath the topsoil.

Hardwood cuttings–Cuttings made from woody deciduous species and narrow-leaved evergreen species, such as grape and hemlocks.

Hardy Plants–Resistant to pests and diseases, often more vigorous and productive. Can withstand cold winters.

Heading back–1. Type of pruning cut where the end portion of the shoot is removed but the portion at the base is not.

2. Cutting an older branch or stem back to a stub or twig.

Heeling in–Temporarily setting a plant into a shallow trench and covering the roots with soil to provide protection until it is ready to be permanently planted.

Herbaceous–Plants that are mainly soft and succulent, forming little or no woody tissue.

Herbaceous cuttings–Cuttings made from herbaceous plants (see definition below), such as chrysanthemums, coleus, and geraniums.

Herbaceous perennials–Plants with soft, succulent stems whose tops are killed back by frost in many temperate and colder climates, but whose roots and crowns remain alive and send out top growth when favorable growing conditions return.

Herbarium–1. A collection of plant specimens that have been taxonomically classified, pressed, dried, and mounted on a sheet of herbarium paper.

2. A collection of pressed and dry specimens for research purposes.

Herbicide–1. A chemical that kills or retards plant growth. Herbicides may kill the entire plant; or they may kill only the aboveground plant parts, leaving the roots alive.

2. A material that will kill plants. Herbicides may kill essentially all plants or be quite selective in their activity.

High-pressure sprayer–A sprayer using a high-pressure pump to force the spray through nozzles for both creating a fine spray and delivery to the plant.

Homeoclimactic–Regions that may be miles or continents apart but share similar climates.

Honeydew–The sticky secretion produced by sucking insects such as aphids.

Horticulture–1. The cultivation of plants for ornament or nourishment.

2. The science and art of growing fruit, flowers, ornamental plants, and vegetables in gardens.

Hortus–Latin word for garden.

Hotbed–A heated soil bed used for propagation as with cuttings in winter.

Humus–1. A dark colored organic material that is the product of organic matter breakdown by both microorganisms and chemical reactions.

2. The brown or black organic part of the soil resulting from the partial decay of leaves and other matter.

Husk–Outer covering of a fruit or seed.

Hybrid–1. In its simplest form, a first-generation cross between two genetically diverse parents. Imperfect flower. Flower lacking either the stamen or the pistil.

2. A plant created from a cross between two parent plants of different species, subspecies, or varieties of plant.

3. The offspring of two plants of different species or varieties of plants. Hybrids are created when the pollen from one kind of plant is used to pollinate and entirely different variety, resulting in a new plant altogether.

Hydroculture–Long-term plant growing in water-based nutrient solutions.

Hydroponics–The science of growing plants in mineral solutions or liquid, instead of soil.

IAA–Indole-3-acetic (natural auxin) acid; used to promote rooting in cuttings.

IBA–Indolebutyric acid; used to promote rooting in cuttings.

Inarching–A type of grafting that attaches a parent plant using an arched stem.

Incised–Cut by sharp and irregular incisions more or less deeply, but intermediate between toothed and lobed.

Incomplete flower–Flower lacking one or more of the four sets of floral parts.

Indehiscent–Type of dry fruit in which the fruit wall does not split at any certain point or seam at maturity.

Indeterminate growth–Growth that is potentially limitless.

Inflorescence–The arrangement of the flowers on the floral axis-a flower cluster.

Internode–The section of stem between two nodes.

Interstock–Intermediate stem piece that is grafted between the upper and lower pars of the graft.

Introduced–A plant that is exotic or brought from another region; a plant brought and sold in the nursery trade for the first time.

Landscape design–The profession concerned with the planning and planting of outdoor space to secure the most desirable relationship between land forms, architecture, and plans to best meet human needs for function and beauty.

Larva–An immature stage through which some types of insects must pass before developing into adults. Caterpillars are the larvae of moths and butterflies, and grubs are the larvae of beetles. Larvae are typically wormlike in appearance.

Lateral bud–A bud forming along the side of a stem or branch rather than at the end.

Lath–In gardening, an overhead structure of evenly spaced slats of wood or other materials used to create shade.

Lathhouse–A shaded environment created using slats of wood.

Lawn–Any expansive planting of grass, mowed to keep it short.

Leaf mold–Partially decomposed leaf matter, used as a soil amendment.

Layering–1. A vegetative method of propagation that produces new individuals by producing adventitious roots (roots growing from unusual places) before the new plant is separated from the parent plant.

2. A method of propagation, by which a branch of a plant is rooted while still attached to the plant by securing it to the soil with a piece of wire or other means..

3. Treating a shoot so it will form roots while still attached to the parent plant. A common method used for propagating roses.

Leaching–1. The downward movement of nutrients or salts through the soil profile in soil water. Leaching accounts for nutrient losses but can also be beneficial in ridding a soil of excess salts.

2. The removal or loss of excess salts or nutrients from soil. The soil around over fertilized plants can be leached clean by large quantities of fresh water used to 'wash' the soil. Areas of extremely high rainfall sometimes lose the nutrients from the soil by natural leaching.

Leaf arrangement–The arrangement of leaves on the stem.

Leaf mold–Partially decomposed leaf matter, used as a soil amendment.

Leaf-bud cuttings–Cuttings consisting of a leaf blade, leaf stalk, and a short piece of the stem. Leaf cuttings.

Entire leaves with or without the leaf stalks.

Leaflets–Individual parts of a compound leaf.

Leaves–Vegetative plant parts that are lateral outgrowths of stems that have developed special structures for photosynthesis.

Legume–A simple, dry, dehiscent fruit with one locale that splits along two seams.

Liquid manure–A fertilizer made by running water through or into solid manure.

Limb up–Prune off lower branches of *woody plants* to (further) expose the main trunk or trunks.

Lime–Ground limestone that is used to raise soil pH.

Limy soil–A soil with traces of calcium/ lime.

Loam–1. A type of soil consisting of sand, clay, and organic matter.

2. A humus(or organic matter)rich soil with up to 25 percent clay, 50 percent silt,and less than 50 percent sand.

Lobe–A segment of a cleft leaf or petal.

Locales–The cavities of the ovary of the pistil of a flower.

Long-day (short-night) plant–A plant that requires a day longer than its critical daylength to flower.

Macronutrients–Nutrients required by plants for normal growth. Macronutrients such as nitrogen, phosphorus, and potassium are needed in large quantities by most plants.

Malling series–A group of apple rootstocks that originated at the East Malling Research Station in England.

Malling-Merton series–A group of apple rootstocks that were the result of the crossing the Malling series with the 'Northern Spy' to add resistance to woolly aphids.

Manure–Organic matter, excreted by animals, which is used as a soil amendment and fertilizer. Green manures are plant cover crops which are tilled into the soil.

Market gardening–The growing of an assortment of vegetables for local or roadside markets. Mesocarp. The center portion of the fruit wall.

Mature–A later phase of growth characterized by flowering, fruiting, and a reduced rate of size increase.

Media/soil media–Any substrate (organic or not) used to hold plant roots.

Metamorphosed–Changed from one state to a different one.

Metamorphosis–Changing from one state to a different one.

Micronutrients–Mineral elements which are needed by some plants in very small quantities for normal growth. It is also called a trace or minor element. If the plants require specific 'trace elements' and they are not available in the soil, they must be added. *e.g;* calcium, copper, iron, zinc, and manganese.

Microbudding–Method of budding in citrus. A bud is removed form the bud stick by giving a flat cut just underneath the bud with a sharp knife. An inverted T-cut is made, in the stock wherein the microbud is slipped keeping the right side up. The entire inverted T-cut including the bud is covered and exposed only after 2-3 weeks when healing is completed.

Microclimate–1. The enviroment immediately surrounding a plant; very localized climate conditions. Many different microclimates may occur at the same time in different areas of a graden.

2. Variations of the climate within a given area, usually influenced by hills, hollows, structures or proximity to bodies of water. *(i.e. when it's raining at a place, and the sun is shining on the other side of it).*

Mites–A group of tiny animals related to spiders, many of which feed on plants.

Mold–A type of fungus that grows on decaying plant tissue. *e.g.* Mucor.

Monoecious–Plants that produce male and female flowers on the same plant.

Mother block–A certified group of plants used to derive propagation tissues.

Mound or stool layering–A layering procedure in which the new shoots develop in spring and soil is mounded around their bases, excluding light and enhancing root formation.

Mowing strip–Edging placed between a lawn and a planting bed or patio, set just below ground level to provide a flat surface for one wheel so the lawnmower blades can cut to the edge of the lawn, saving the later step of snipping the edge grass with scissors or a weed-whacker. Mowing strips are commonly made of bricks, pavers, or poured cement.

Mulch–1. A layer of organic or inorganic material on the soil surface. Mulches help to moderate the temperature of the soil surface, reduce loss of moisture from the soil surface, suppress weed growth, and reduce run-off.

2. Any loose material placed over the soil to control weeds and conserve soil moisture. Usually this is a coarse organic matter, such as leaves, clippings or bark, but plastic sheeting and other commercial products can also be used.

3. A material applied to the surface of a soil for a variety of purposes such as conservation of moisture, stabilization of soil temperature, and suppression of weed growth.

4. Any material spread over the soil surface to retain soil moisture, moderate soil temperature, and suppress the growth of weeds.

Mutation–A sudden heritable change in the expected genetic or tissue makeup of a plant. A sudden change in genetic material resulting in an altered individual. Generally disadvantageous to the mutated plant's survival.

Mycelium/mycelia–1. Thread-like growth parts of many fungi, often white or gray.

2. The vegetative body of a true fungus composed of branching hyphae.

Nascent–In the act of being formed.

Native–Original to an area.

Native plant–Any plant that occurs and grows naturally in a specific region or locality.

Naturalism–Planting to create the effect that plants grew in that space without human intervention; gardening to keep or enhance existing natural features.

Naturalize–To plant randomly, without a pattern. The idea is to create the effect that the plants grew in that space without man's help, such as to find wild flowers growing naturally.

Naturalized–Thoroughly established, but originally from a foreign area.

Neutral Soil–A soil having a pH value of 7.0, that is neither acid nor alkaline.

Nematode–Mostly microscopic worm-like organisms that damage plant roots.

Nitrate of soda–A strong alkaline fertilizer with about 15 per cent nitrogen.

Nitrification–The process of oxidizing ammonium salts into nitrogen for plants.

Nitrogen fixer–Some plants (notably many legumes such as clover, peas, beans, and alfalfa) have the ability to transform nitrogen from the air into a form of nitrogen that plants can use. As nitrogen is a key nutrient that often limits growth, nitrogen fixers can influence the fertility of the surrounding soil and the growth of neighboring plants. The actual "fixing" or chemical transformation is performed by bacteria that live in nodules on the roots of the host plant.

NPK–Nitrogen, Phosphorus and Potash; the three main food elements necessary for plant life. Nitrogen is required for foliage and growth, phosphorus for flowers and potash for roots.

Node–1. The location on the stem where buds, leaves and branches are attached.

2. The part of a stem from which a leaf or new branch starts to grow.

Novae-angliae–Epithet meaning from New England.

Nursery grower–A person who produces or distributes ornamental plants.

Nut–An indehiscent dry fruit similar to the achene except that the fruit wall is hard throughout.

Nutrient–A chemical required by a plant for growth.

Offshoots–Short, horizontal stems that occur in whorls or near whorls at the crown of stems.

Olericulture–The study of vegetable production.

Opposite leaves–Leaves arising from opposite sides of the same node.

Organic gardening–The method of gardening utilizing only materials derived from living or once-alive things. *(i.e. composts and manures)*

Organic–Something that was or is alive; containing carbon.

Ornamental–Showy or visually pleasing.

Ornamentals–Popular abbreviation for a division of Horticultural Science dealing with plant cultivated for landscape or display value.

Organic material–Any material which originated as a living organism. *(i.e. peat moss, compost, manure)*; A substance derived from plant or animal material(*Syn.* Organic Matter).

Organic matter–1. Materials rich in carbon of either plant or animal origin, which exist in all stages of decomposition of soils.

2. Mulch or other material derived from decomposed plant or animal products.

Ovary–Part of the pistil that contains one or many small bodies known as ovules.

Ovule–The immature seed in the ovary.

Pad and fan cooling–The use of a moistened surface and fans to cool a greenhouse.

Palmate leaf–A compound leaf with all the leaflets arising from one point at the end of the leaf stalk

Palmately veined leaves–Leaves with netted veins having several large veins radiating into the blade from the leaf stalk at the level where the leaf stalk and blade join.

Panicle–Type of flower cluster with either a cluster of racemes or corymbs, distinctly branched.

Parallel venation–Leaves with large veins that are essentially parallel to one another and are not connected by lateral veins.

Parasitic plant–A plant which lives on, and acquires it's nutrients from another plant. This often results in declined vigor or death of the host plant.

Pathogen–An organism capable of causing a disease. (Such as fungus, bacterium, or virus).

Peat–Partially degraded vegetable matter found in marshy areas. Peat is commonly used as a soil amendment.

Peat Moss–1. Partly decomposed moss, rich in nutrients and with a high water retention. Commonly used as a soil amendment.

2. The partially decomposed remains of various mosses. This is a good, water retentive addition to the soil, but tends to add the acidity of the soil pH.

3. An organic soil additive from Sphagnum and related mosses. The partially decomposed remains of various mosses. This is a good, water retentive addition to the soil, but tends to add the acidity of the soil pH.

Peduncle–The short stem of the flower cluster or the main axis of an inflorescence.

Pepos–Berries that have a hard rind around the fruit.

Perennial plant–1. A plant that lives for more than 2 years, often living for many years. Almost all woody plants and many herbaceous plants are perennials.

2. A *nonwoody plant* which grows and lives for more than two years. Perennials usually produce one flower crop each year, lasting anywhere from a week to a month or longer.

Perennials–1. Plants that do not die after flowering but live from year to year.

2. Plant species that can live for more than two years.

Perfect flower–A flower that has both a pistil (or pistils) and stamens but may lack sepals, petals, or both. Perianth. Structure consisting of the sepals and petals.

Pergola–An arbor with an open roof of rafters supported by posts or columns; an overhead covering for an outdoor structure. An open framework over a walkway or path, which is usually designed to be covered in climbing plants.

Pericarp–The fruit wall, consisting of three distinct layers; the exocarp, the mesocarp, and the endocarp.

Perlite–1. A mineral, which when expanded by a heating process forms light granuals. Perlite is a good addition to container potting mixes, to promote moisture retention while allowing good drainage.

2. White granules of a treated volcanic mineral, usually mixed with potting soil to improve drainage.

Pest–Any insect or animal which is detrimental to the health and well being of plants or other animals.

Pest management–The control of a pest or group of pests by a broad range of techniques from biological control to pesticides. The goal is to keep damage below economic levels without completely eliminating the pest.

Pesticide–A chemical used to kill an organism considered a pest.

Petals–Structures collectively making the corolla, which protect the inner reproductive structures and often attract insects by either their color or their nectar and thus facilitate pollination.

Petiole–A stalk that attaches the leaf to the stem; a leafstalk.

pH–A measure of acidity or alkalinity, expressed as the negative log of the hydrogen ion concentration on a scale of 1.0 to 14.0. A pH of 7 is neutral; less than 7 is acidic more than 7 is basic

Basically, pH is a measure of the amount of lime (calcium) contained in soil. Soil pH can be tested with an inexpensive test kit; A measure of the acidity or alkalinity of a substance; a measure of the relative concentration of hydrogen ions and hydroxyl ions.

The scale used to denote the acidity or alkalinity (lime content) of a soil or growing compost. 6.5 pH is considered neutral and supports most plant life.

Phenotype–1. The visual appearance of a plant regardless of genetics.

2. The bodily characteristics of an organism.

Phloem–Nutrient-conducting vessels found throughout the plant. Phloem vessels transport nutrients produced in the foliage down through the stems, branches, or trunk to the roots.

Photosynthesis–The internal process by which a plant turns sunlight into growing energy. The formation of carbohydrates in plants from water and carbon dioxide, by the action of sunlight on the Chlorophyll in the leaves.

Physiology–The study of chemical function of any organism.

Phyto–Prefix meaning plant.

Phytogenetics–The study of plant genetics and breeding.

Phytopathology–The study of plant diseases.

Pinching–1. Breaking off the uppermost growing point, thus allowing the willary buds to start to grow. This promotes outward growth rather than upward growth.

2. Removing a shoot tip with thumb and forefinger, resulting in a bushier growth.

Pinching back–Utilizing the thumb and forefinger to nip back the very tip of a branch or stem. Pinching promotes branching, and a bushier, fuller plant

Pinnate leaf–A compound leaf with the leaflets arranged along both sides of the midrib.

Pinnately veined leaves–Leaves with netted veins where secondary veins extend laterally from a single midrib.

Pistil–1. The female reproductive organ, consisting of the stigma, style, and ovary. Plant growth. A permanent increase in volume, dry weight, or both

2. The seed-bearing organ of a flower, consisting of the ovary, stigma, and style.

Plant pathogen–A microorganism that causes a plant disease.

Plant propagation–Increase in numbers or perpetuation of a species by reproduction.

Pollination–The transfer of pollen from the stamen *(male part of the flower)* to the pistil *(female part of the flower)*, which results in the formation of a seed. Hybrids are created when the pollen from one kind of plant is used to pollinate and entirely different variety, resulting in a new plant altogether.

Pome–Type of fruit in which the portions produced by the pericarp are enclosed within fleshy parts that are derived from parts of the flower other than the ovary. Typical of Rosaceae. An apple is a pome.

Pomology–The study of fruit production.

Post-harvest physiology–A science relating to fruit and vegetable storage and quality.

Potting Compost–Medium for potting plants or for sowing seeds.

Potting on–Placing a plant into a larger pot to promote growth etc.

Potting soil–A growth substrate suitable for container gardening. A soil mixture designed for use in container gardens and potted plants. Potting mixes should be loose, light, and sterile.

PPAF–Acronym for Plant Patent Applied For.

Pot-bound–Having restricted root growth and a circular pattern of root growth caused by a too-small container.

Potting soil–A soil mixture designed for use in container gardens and potted plants. Potting mixes should be loose, light, and sterile.

Pricking out–Transplanting small seedlings into larger trays or containers.

Propagation–1. Various methods of starting new plants ranging from starting seeds to identical clones created by cuttings or layering.

2. Means of reproducing plants, such as by seeds, cutting, budding or grafting.

Pruning–1. Removal of plant parts such as buds, developed shoots, and roots to maintain a desirable form by controlling the direction and amount of growth.

2. The cutting and trimming of plants to remove dead or injured wood, or to control and direct the new growth of a plant.

Pungent–With a sharp, hard point; sharp and acid to taste or smell.

Purity–The relative stability and uniformity of a breeding line.

Quick dip–Treating cuttings with a liquid hormone.

Raceme–Type of flower cluster in which stalked flowers are on pedicels approximately equal in length on a single floral axis.

Regionally appropriate–Plants that are winter- and summer-hardy in your area (check the USDA plant hardiness zone rating, a number between 1 and 11) and are not considered ***invasive*** by state or federal authorities.

Registered–A cultivar cataloged with an International Registrar (IRA).

Registrar–An international authority that catalogues cultivar names (IRA).

Relative humidity–The measurement of the amount of moisture in the atmosphere.

Rest–A state of suspended growth or outwardly visible activity due to internal physiological factors. Also referred to as ***physiological dormancy***.

Rhizome–1. Horizontal stems that grow underground. They serve as storage organs and a means of vegetative reproduction.

2. A modified plant stem which grows horizontally, under the surface of the soil. New growth then emerges from different points of the rhizome. Irises and some lawn grasses are rhizome plants.

Root–Vegetative plant part that anchors the plant, absorb and conduct water and minerals in solution, and often stores food.

Root ball–The network of roots(or clump of roots) along with the attached soil, of any given plant.

Root cuttings–Cuttings 5 to 15 cm long made from root sections in the fall or winter.

Rootbound–A condition which exists when a potted plant has outgrown its container. The roots become entangled and matted together, and the growth of the plant becomes stunted. When repotting, loosen the roots on the outer edges of the root ball, to induce them to once again grow outward.

Rooting hormone–1. A powder of liquid growth hormone, used to promote the development of roots on a cutting.

2. A growth hormone in powder or liquid form, which promotes the formation of roots at the base of a cutting.

Runner–1. A slender stolon with elongated internodes. These root at the nodes that touch the ground. Example: strawberries.

2. A slender stem growing out from the base of some plants, which terminates with a new offset plant. The new plant may be severed from the parent after it has developed sufficient roots.

Runners–Aboveground, trailing stems that form roots at their nodes when they make contact with moist soil.

Rust–A type of fungal disease that requires two hosts. For example, White Pine Blister Rust alternates between white pine and Ribes sp. Gives a rusty appearance to the plant.

Samara–An indehiscent dry fruit with either one or two seeds, in which pericarp bears a flattened wing-like outgrowth. The "helicopters" of the maple trees are double-samaras.

Sanitation–The process of keeping a clean and disease-free growing area.

Scales–Modified leaves that protect structures.

Scaling–A type of propagation that separates bulb scales from the mother bulb to induce the formation of bulblets.

Scarification–1. The chemical or physical treatment given to some seeds to break or weaken the seed coat sufficiently for germination to occur.

2. Deliberate wounding of a seed to allow water to enter

Scion (cion)–1. The upper part of the union of a graft.

2. A short length of stem, taken from one plant which is then grafted onto the rootstock of another plant; the top "slip" or desirable propagated tissue in a grafted plant.

Scree–Rock garden planting bed made of crushed stone and soil.

Screen–A tall hedge of many plants used to block wind or a bad view.

Scurfy–Describes a surface covered with bran-like particles.

Seascape–A landscape planted near the seashore or beach.

Seed Heads–Pods or clusters of seeds on mature plants at the end of the growing season.

Seed–Plant embryo with associated stored food encased in a protective seed coat.

Self Seeding–An annual's ability to drop seeds that germinate successfully and grow into seedlings.

Self-pollination–The process by which pollen is transferred from the pollen producing section of the plant to the pollen receiving part of the plant of the same flower or another flower of the same cultivar.

Semihardwood cuttings–Cuttings made from woody, broad-leaved evergreen species such as ligustrum and holly.

Sepals–Structures that usually form the outermost whorl of the flower; collectively, the calyx.

Separation–The use of bulbs and corms in propagation that utilizes the naturally detachable parts.

Serrate leaf–A leaf with serrations or "teeth" along the edge of the leaf.

Sessile–Without a petiole (as in some leaves), or without a pedicel (as in some flowers and fruits).

Sexual reproduction–The reproduction of plants through a sexual process involving cell division.

Shade–A low light planting environment.

Sheath–An elongated, more or less tubular structure enveloping an organ or plant part; a tube-shaped part containing a leaf of a grass or similar plant.

Shoot–Stem, 1 year old or less, that possesses leaves.

Short-season crop–A plant capable of producing parts in a minimum of days.

Simple bud–Bud containing either leaf or flower primordia, but not both.

Simple layering–A method similar to tip layering, except that the stem behind the end of the branch is covered with soil and the tip remains above ground.

Simple leaves–Leaf blades consisting of one unit.

Single flower–A flower having only a minimum number of petals for that variety of plant.

Sinus–The indentation of a leaf. Sinuses alternate with lobes.

Slimeflux–Exuded mass from a cut or wound woody plant trunk.

Slip–A plant cutting.

Sludge–Processed sewage, high in nitrogen, which is used as fertilizer.

Smut–A fungal disease that often has gray or black spores.

Sod-forming grass-A grass that spreads by creating new plants from existing plants by running roots across or under the ground (***stolons*** or ***rhizomes***, respectively), creating a dense turf. Also called ***spreading grass***. Contrast with ***clump-forming grass*** (also called ***bunchgrass***), in which single plants get larger over time but do not create new plants.

Softwood cutting–A herbaceous (often green) cutting of young stems.

Soil drench–A media treatment to kill fungi.

Soilless mix–A substrate for plant roots with manmade materials.

Slow-release fertilizer–1. Fertilizers that release their nutrients slowly and evenly, over a long period of time

2. A fertilizer that is made by coating the particles with a wax or other insoluble or very slowly soluble material to provide a predictable, slow release of the encapsulated materials.

Softwood cuttings–Cuttings taken from soft, succulent, new spring growth of deciduous or evergreen species of woody plants.

Soil–The outer or upper most weather layer of the earth's crust that has the potential to support plant life. Soil is made up of inorganic particles, organic matter, microorganisms, water, and air.

Soil management–The practices used in treating a soil, which may include various types of tillage and production systems.

Soil pH–Basically, pH is a measure of the amount of lime (calcium) contained in the soil. A soil with a pH lower than 7.0 is an acid soil, a soil pH higher than 7.0 is alkaline soil. Soil pH can be tested with an inexpensive test kit.

Soil structure–The arrangement of individual soil particles.

Soluble fertilizers–Fertilizers that dissolve easily in water and are immediately available for plant use.

Species–Plants which are genetically similar and which breed true to type from seed.

Sphagnum–A bog moss which is collected and composted. Most peat moss is composed primarily of sphagnum moss. This moss is also packaged and sold in a fresh state, and used for lining hanging baskets and air layering.

Spike–Flower cluster of an indeterminate (unlimited number) raceme of sessile flowers attached to the floral axis with the oldest flowers at the base.

Spikelet–Diminutive type of spike inflorescence found in grasses.

Spines–Sharp-pointed woody structures, usually modified from a leaf or part of a leaf.

Spore–The reproductive cell of ferns, fungi and mosses. *(These plants do not produce seeds)*

Spreader–A mechanical device to spray chemicals; a chemical used to disperse pesticides uniformly over a plant surface.

Staking–The practice of driving a stake into the ground next to, and as a support for a plant. When attaching the plant to the stake, be sure that it is tied loosely so it doesn't strangle the stem. When staking a potted plant, the stake should be set into the planter before the plant is added.

Stamen–1. Part of the flower consisting of the anther, in which pollen is produced, and a slender filament that holds the anther in a position favorable for pollen dispersal. Male reproductive organ.

2. The male reproductive organ of a flower, consisting of a filament and a pollen containing anther.

Staminate flower–Flower in which only the stamens (male reproductive parts) are present; there are no pistils (female reproductive parts).

Sterile–Lacking fertility or reproductive potential; said of a soil heated or treated to remove harmful organisms. Barren, not able to produce seed.

Sterilization–1. The process of killing insects and disease in soil

2. Process of killing harmful organisms on cultured plant tissue.

Starch–A stored carbohydrate in plant parts.

Stick–Verb used when one places cuttings in a rooting medium or bed.

Stigma–The pollen-adheres site of the pistil.

Stipules–Leaflike appendages often found on either side of the base of the petiole; they may subtend the leaf, but they are not actual parts of the leaf.

Stock (rootstock)–The lower part of a graft.

Stolon–An aboveground stem that reclines or becomes prostrate and may form roots at the nodes that may come into contact with the ground.

Stomates–Tiny pores located mainly on the underside of leaves. Oxygen, carbon dioxide, water vapor, and other gases move in and out of the leaf through these pores.

Stratification–1. The storing of seeds at low temperatures under moist conditions to break physiological dormancy or rest.

2. A temperature treatment of seed used to break dormancy.

Stratified–Arranged in horizontal layers.

Succulent leaves–Water storage structures characteristic of plants in and semiarid regions.

Sucker–A growth originating from the rootstock of a grafted plant, rather than the desired part of the plant. Sucker growth should be removed, so it doesn't draw energy from the plant.

Summer annual–A one year plant that sows in spring and flowers in summer or fall.

Sunscald–A burning of tissue due to strong light on foliage or bark.

Systemic–A chemical which is absorbed directly into a plants system to either kill feeding insects on the plant, or to kill the plant itself.

Systemic pesticide/systemic herbicide–A chemical that is absorbed by a plant and is translocated in its vessels, either to kill feeding insects on the plant, or to kill the plant itself.

Tap Root–1. The main, central root of a plant.

2. The main, thick root growing straight down from a plant. *(not all plants have tap roots)*

3. Primary root that persists and maintains its dominance.

Taxonomy–The area of botany dealing with the classifying and naming of plants.

Taxonomythe–The science of plant classification and nomenclature.

T-budding–A type of grafting that places a bud into a T-shaped cut on a stock.

Temperate zone–Northern hemisphere of the world.

Temperate fruit–A fruit plant that requires a cool period and is deciduous. Examples are apple, pear, and peach.

Tender plants–Plants which are unable to endure frost or freezing temperatures.

Tendrils–1. Slender, twining leaf modifications used for support, as in the terminal leaflets of the grape.

2. The twisting, clinging, slender growth on many vines, which allows the plant to attach themselves to a support or trellis.

Terminal buds–1. A bud at the end of a stem or branch

2. Large and vigorous buds at the tips of stems, responsible for terminal growth.

Terminal Stem-Growing tip on a plant or shrub.

Terrarium–A glass or plastic-enclosed growing environment with high humidity

Texture–The relative coarseness or fineness of foliage in the landscape

Thatch–The layer of dead stems that builds up under many lawn grasses. Thatch should be removed periodically to promote better water and nutrient penetration into the soil.

Thinning–Removing excess seedlings, to allow sufficient room for the remaining plants to grow. Thinning also refers to removing entire branches from a tree or shrub, to give the plant a more open structure.

Thorn–A short, sharp-pointed branch.

Tip layering–Layering in which rooting takes place near the tip of the current season's shoot, which naturally falls to the ground.

Tissue culture–The growing of masses of unorganized cells on agar or in liquid suspension. Useful for the rapid asexual multiplication of plants.

Topiary–A method of pruning and training certain plants into formal shapes such as animals.

Topsoil–The top layer of *native* soil. This term may also apply to good quality soil sold at nurseries and garden centers.

Top working–1. Grafting only the ends of a plant, leaving branches intact.

2. Grafting procedure by which branches of trees are changed to a more desirable cultivar.

Top-dressing–A fertilizer or compost applied at the soil level.

Topiary–The art of pruning and training plants into specific shapes.

Torsion-Twisting

Torulose–Twisted or knobby, irregularly swollen at close intervals.

Trailing–Prostrate, running down or along the ground, or not rooting.

Translocation–The movement of water and nutrients through a plant's vessels.

Transpiration–1. The release of moisture through the leaves of a plant.

2. Evaporation of water from plant tissue to the atmosphere. Transpiration occurs mainly through the stomates in the leaves.

Transplant–A seedling that grows its first true leaves, which resemble the adult plant leaves rather than the seed leaves that first appear.

Transplanting–1. The process of digging up a plant and moving it to another location.

2. The process of moving a plant from one soil mass to another.

Trench layering–A method of propagation using long branches rooted in a manmade ditch.

Trenching–Digging a shallow ditch around the border of a planting.

Trickle irrigation–The application of small quantities of water directly to the root zone through various types of delivery systems on a daily basis.

True leaves–The first leaves to emerge after the germination of a seed.

Tuber–A flat underground stem which stores food and plant energy and from which a plant grows. *(e.g. Dahlias).*

Tubers–Enlarged underground stems serving as storage organs of starch or related materials. Potatoes are tubers.

Twig–Stem, 1 year old or less, without leaves.

Umbel–Type of flower cluster in which the pedicels arise from a common point and are about equal in length.

Urban horticulture–The study of stresses affecting city trees, gardens and landscapes.

USDA–United States Department of Agriculture.

USNA–United States National Arboretum.

Variegated–1. Leaves with green and white striped pattern instead of the typical all-green look.

2. Leaves which are marked with multiple colors.

Vegetable–The edible portion of an herbaceous garden plant.

Vermiculite–1. A light weight, flaky, treated amendment that improves soil drainage.

2. The mineral 'mica' which has been heated to the point of expansion. A good addition to container potting mixes, vermiculite retains moisture and air within the soil.

3. The mineral mica, heated to the point of expansion. A good addition to container potting mixes, vermiculite retains moisture and air within the soil.

Wall garden–A garden against a man-made structure to protect plants.

Weed–A plant growing where it is not wanted or hindering the growth of more desirable plants.

Wet feet–A condition where plants are exposed to excess soil moisture caused by flooding or a high water table.

Whorled arrangement–Three or more leaves at a node.

Wildlife garden–Planting with plants to attract, feed and house wild animals, insects and birds.

Willow water–A solution of willow bark that aids in rooting cuttings.

Winter annual–A one year plant that sows in fall/winter and flowers early spring.

Woody plant–1. A tree or shrub; any plant with lignified tissues.

2. As opposed to ***herbaceous*** plants, woody plants have a year-round structure of live stems and branches, and these woody parts increase in diameter each year. Includes both ***evergreen*** and ***deciduous*** trees and shrubs, even small ones that may look like herbaceous plants.

Wilt–Loss of rigidity and drooping of plant parts caused by many different plant diseases. Generally, it involves an interference with water conduction inside the stems.

Woody perennials–Plants with woody fiber.

Wounding–Deliberately cutting plant stems to increase root formation.

Xeriscape–A landscape made of plants that require very little water or rain.

Xylem–Water-conducting vessels found throughout the plant. Xylem vessels transport water and minerals from the roots upward through the plant.

Yellows–A type of disease, common among some flowers, that is caused by a virus or mycoplasma-tike organism. This disease causes stunting and a yellowish appearance.

Zone–1. An area distinguished by a range of annual average minimum temperatures, used in describing plant hardiness.

2. Regions that share similar climatic and rainfall conditions producing similar growing seasons.

Chapter 2
Terms of Botany

Abortive–Undeveloped; rudimentary, only partially developed. Example: incomplete seed or frost-nipped bud.

Abscission–1. The falling of leaf, twig-tip, etc, from a clean-cut scar, by a self-healing wound;

2. A natural dropping of leaves, flowers and other plant parts.

Accessory buds–Buds found beside or above the true bud at a node.

Achene–A dry, indehiscent, one-seeded fruit. This develops from mono-carpellary superior ovary.

Acorn–The fruit of oaks, a thick walled nut with a woody cup-like base.

Aculeate–Prickly.

Acuminate–Coming to a point with concave, curved sides.; Having a sharp, gradually tapered point; the tip of a leaf blade that narrows gradually to a point.

Acute–The tip or base of a leaf blade whose margins come to a broad

Acute–With sides forming an angle of less than 90°; shape pointed.

Adpressed–Lying close to another organ, but not fused to it

Adult phase–The stage of plants capable of producing fruit and flowers.

Adventious bud–An extra or special bud not found in the axil of a leaf.

Adventitious–Not in the usual place (buds that have remained undeveloped so that they are no longer evidently axillary, or that really originate elsewhere, as on a root).

Aerial roots–Roots produced above ground, often used for climbing.

Aggregate flower–A single flower heaped or crowded into a dense flower cluster.

Aggregate fruit–A fruit formed by the coherence or the connation of pistils that were distinct in the flower.

Aggregated–Joined together, confluent, as applied to bundle traces.

Alba/albus/album–White or whitish, usually applied to flowers

Albo-maculata–Spotted in white.

Albo-marginata–White edged or margined.

Albo-striata–Striped or striated in white. Usually a variegated plant.

Allelopathic–Plant that produces chemicals affecting other nearby plants' growth. Usually used to indicate a negative effect, inhibiting the growth of nearby plants. Black Walnut (*Juglans nigra*) trees inhibit the growth of many other plants.

Alt–Latin prefix meaning tall.

Alternate–The arrangement of leaves or buds one at a time along a stem; leaf scars or leaves occurring one at each node.

Altissima–Tall.

Amplexicaul-Encircling the stem.

Anatomy–The internal structure of plant.

Anchor root–A large root that holds a plant in a physical soil mass.

Androecium–Stamens of a flower, as a unit.

Annual–A plant that germinates, grows, flowers, produces seeds in one growing season.

Annual ring–A concentric circle of wood produced by one year's growth in a tree trunk.

Annuals–Plants whose life cycle lasts only one year, from seed to blooms to seed.

Annular–Ring-like; leaf scars which encircle the bud, or bundle scars which are circular with an opening in the center.

Anterior–On the front side, away from the axis, toward the subtending bract.

Anther–1. The pollen bearing part of a stamen, borne at the top of a filament or sessile.

2. Any fleshy plant. Used mostly to denote edible aromatic ones.

Apetalous–Without petals. Ex. Grass flowers

Apex–The tip or terminal end; (pl. apices) the top or tip of a structure, such as a leaf or stem

Apical–Describes the apex or tip.

Apiculate–Ending in an abrupt pointed tip.

Apomixis–Process of producing viable seed without fertilization

Appressed–Flattened against the twig; not spreading.

Arcuate–Arched, bent like a bow.

Aril–A fleshy appendage of the seed, usually a fleshy seedcoat.

Armed–With pines (barberry, Japanese quince) or prickles (brambles). Leaves that are pungent at the tip or around the margin (holly), are not included in this limited definition.

Aromatic–Fragrantly scented, especially when broken or crushed.

Articular membrane–A membrane consisting of the thin, enlarged base of the petiole, on which the leaf scar occurs.

Articulate–Having nodes or joints where separation may naturally occur.

Ascending–A bud that is between spreading and appressed. Curving indirectly or obliquely upward.

Atro–Botanical name prefix meaning dark or intensely colored.

Atrosanguinea–Dark or blood red or maroon.

Atrovirens–Dark or blackish green.

Attenuate–Showing a long gradual slender taper; usually applied to apices, but equally appropriate for bases of leaves, petals, etc.

Aurescent–Becoming yellow or yellow part of the season.

Auxin–A chemical that stimulates plant growth.

Awl-shaped–Tapering to a slender stiff point.

Awn–A bristle-like appendage.

Axil–The location on a stem between the upper surface of a leaf or leafstalk and the stem from which it grows.

Axilary buds–Buds that form in leaf axils.

Axillary–In the axil.

Axis–The main stem or central support of a plant.

Azurea/azureus/azureum–Azure or deep blue.

Baccate–Pulpy, fleshy.

Backcross–A hybrid of two plants crossed once again back to one parent.

Barbed Bristles–Awns, etc. provided with terminal or lateral spine-like hooks that are bent sharply backward.

Bareroot–Plants offered for sale which have had all of the soil removed from their roots; (BR) a plant dug as bare root for transplanting.

Bark–The rough outer part of the cortex; loosely used for the entire cortex. It includes primary and secondary phloem, cortex and periderm.

Basal rosette–Ground level ring of leaves (single or multiple layers) around the plant's central stem where it joins the roots.

Basal–Pertaining to the extremity of an organ by which it is attached to its support; said of leaves when located at the base of the plant only. *See rosette.*

Base–The bottom of a structure, such as a leaf.

Beaked–Ending in a point, especially on fruits.

Bearded–Having long hairs; Patch of hairs on a flower or other part

Berry–A fleshy, indehiscent, pulpy, multi-seeded fruit resulting from a single pistil.

Bi–Prefix meaning twice or doubly.

Bicolor–A plant or flower in two colors.

Biennial–Ordinarily applied to plants that live only two seasons; during the first season only leaves and stems are produced above ground, while the flowers and seeds are borne the second summer. Here used in a special sense in separating the biennial canes of the raspberries and blackberries from the stems of other woody plants. In these species, the canes themselves are biennial from underground perennial stems.

Bifida/bifidum–Divided into two parts.

Bifurcate–Forked, as some Y-shaped hairs.

Bigeneric hybrid–Species formed by crossing two different genera.

Bipinnate–Twice compound with the leaflets along each side of a common axis. The leaflets are further divided into pinnules.

Bisexual–Said of a flower with both stamens and pistils.

Biternate–Twice ternate; structure basically ternate, but whose primary divisions are again each ternate.

Blade–The expanded part of a leaf.

Bloom–A waxy coating found on stems, leaves, flowers and fruits, usually of a grayish cast and easily removed.

Blush–Pink or reddish tint to a plant part

Bole–The stem of a tree.

Bolting–Vegetables quickly flowering rather than producing the food crop. Usually caused by late planting and too warm temperatures.

Bonsai–A living woody plant dwarfed by careful pruning and container culture.

Botanical name–The Latin or scientific name of a plant, usually composed of two words, the genus and the species.

Bract–1. A modified leaf, often small and associated with leaves.

2. A modified leaf of an inflorescence. Leaf-like or petal-like structure, which may appear with or below flowers, or have a bright flower-like appearance. Several bracts form an involucre.

Bracteate–With bracts. Bracted.

Branchlet–A smaller division of a branch.

Branch–One of the coarser divisions of a trunk or main stem; loosely, any division of the stem.

Bristly–With stiff hairs.

Broad-elliptic–Wider than elliptic.

Broad-ovate–Wider than ovate.

Broken–Not continuous, as applied to bundle-traces.

Bronzing–A metallic bronze or coppery color, especially of foliage after a winter.

Bud count–Number of early buds.

Bud eye–A potential growth point that is used for budding.

Bud scale–A modified leaf or stipule (there may be one, a few, or many) protective of the embryonic tissue of the bud.

Bud union–A swollen and distinct node when a bud has been grafted.

Bud–1. The rudimentary or resting end or branch of a stem; usually referring to the stage in which the growing tips pass the winter or dry season; also applied to undeveloped flowers or flower clusters. Scaly buds are protected by modified leaves or stipules. Naked buds lack such special protection. Usually one bud occurs in each axil or angle above a leaf, but these often branch and collateral buds, standing side by side, are thus produced; in some plants (walnuts, honeysuckle) several buds occur one above the other (superposed) in an axil.

2. A structure of compact embryonic tissues

Bulb–The thickened underground storage organ of the group of perennials which includes daffodils and tulips. A modified underground stem comprised of shortened central axis surrounded by fleshy scale-like leaves.

Bulbil–Small bulbs arising around the parent bulb.

Bulblet–Small bulbs arising in the leaf axils.

Bunchgrass–Grass that grows as distinct plants that get larger over time. Also called ***clump-forming grass***. Contrast with ***spreading grass*** (also called ***sod-forming grass***), which expands using running roots that create new plants from existing ones.

Bundle scar–Small dots or lines on the surface of the leaf scar marking the point of original departure of the vascular conducting strands into the leaf. Also called bundle trace.

Bur–Any rough or prickly seed envelope.

Burl–A knot or woody growth of very irregular grain.

Button center–A double rose flower with unexpanded central petaloid parts.

Buttressed–With supporting wings.

Caducous–Falling away early or prematurely.

Caerulea/caeruleum/caeruleus–Intensely blue.

Calcarate–Having a spur.

Callus–A hard protuberance, A thickened tissue that develops at the base of stem cuttings; or the new tissues formed in response to a wound.

Calyx–Sepals and often greenish bracts underneath a flower; the outer floral envelope. The outer set of perianth segments, usually green in color and smaller than the inner set.

Cambium–The thin membrane located just beneath the bark of a plant giving rise to secondary tissues, resulting in growth in diameter.

Campanulata–Latin name for a plant having campanulate or bell-like flowers.

Campanulate–Bell-shaped

Canes–Stem from a thin multi-stem plant; A long woody pliable stem rising from the ground. The long shoots of blackberry, grape, etc.

Canescent–Ash-colored, with gray hairs.

Capillary–Hair-like; very slender.

Capitata–Latin name for a plant with a head-like or large clustered feature.

Capitate–Headlike, in a dense rounded cluster.

Capsule–A dry fruit derived from a compound pistil, and opening in one of a number of ways.

Cardinalis/cardinale–Bright glowing red.

Carpel–A single pistil in a female flower part containing several pistils

Caryopsis–The fruit of members of the grass family; not basically distinct from an achene.

Castaneous–Dark brown.

Catkin–A spike-like inflorescence comprised of scaly bracts subtending unisexual flowers; often somewhat flexuous and pendulous but not necessarily so. Ex.–the flowers of willows or poplars.

Caudiform–A plant with a fleshy swollen base that is not a true bulb.

Caulescent–Having an evident leaf-bearing stem above ground.

Cauliflorous–Flowering on the trunk or on specialized spurs from it or from the larger branches (redbud).

Ceriferous–Waxy.

Cernuous–Drooping or nodding.

Certified/certification–Seed or stock verified to be true or lacking viruses.

Chambered–Pith when divided into small, empty compartments separated by transverse partitions. With cavities separated by plates or disks, as applied to pith; discoid.

Channeled–Grooved lengthwise.

Chartreuse–Variable green-yellow in coloration, often considered ineffective

Chimera/chimaera–A fusion of unrelated tissues such as in all variegated leaves; a man-made fusion of tissue caused by grafting.

Chlorophyll–The green pigment found in chloroplast used to convert light to energy.

Chlorosis–The yellowing of plant tissue due to nutrient deficiencies or disease.

Chromosome–A unit of heredity material in all living cells.

Ciliate–Hairy on the margin, like the eyelids.

Clasping–A stalkless leaf, with the base partly surrounding the stem. Growing around; amplexicaul.

Claw–The constricted petiole-like base of petals and sepals of some flowers.

Cleft–Divided to or about the middle into divisions.

Cleistogamous–A small, closed self-fertilized flower, usually near the ground.

Climax ecosystem–The ecosystem that will eventually overtake a certain area of the globe, perhaps after the landscape has moved through a succession of other ecosystems, and will persist indefinitely thereafter, ending the process of succession. Also called the dominant ecosystem.

Climber–A vine-like or rambling plant that raises its foliage by supporting itself on surrounding objects, either by twining or coiling around them, by the aid of tendrils or aerial roots, or by scrambling over them.

Clinal variation/cline–With a spectrum such as blue to green in spruces,

Clone–A line of plants propagated from a single mother plant, identical to each other and to the parent.

Club root–A fungal disease causing swollen roots.

Clump-forming grass–Grass that grows as distinct plants that get larger over time. Also called ***bunchgrass***. Contrast with ***sod-forming grass*** (also called ***spreading grass***), which expands using running roots that create new plants from existing ones.

Clustered–1. Multiple leaves, flowers or fruits seemingly arising from a point in the plants.

2. Leaves crowded in such a way that they are not to be clearly opposite or alternate, also said of the whorled condition.

Coccinea/coccineus/coccineum–Scarlet red or bright red.

Coherent–Two or more similar parts or organs touching one another in very close proximity by the tissues not fused.

Collateral–Extra or supernumerary buds that are inserted on either side of a normal axillary bud. Usually they grow side by side.

Comose–Tufted with hairs.

Companion plant–A plant making an showy, contrasting display or providing pollination.

Complete flower–One that has corolla, calyx, stamens and one or more pistils.

Composite–Compound; the common name of a member of the Compositae.

Compost–An organic soil amendment resulting from the decomposition of organic matter.

Compound–A leaf composed of two or more distinct leaf blades (leaflets).

Compound leaf–A leaf of several leaflets, in some cases (Citrus) the lateral leaflets may have been lost and only the terminal leaflet remain.

Compressed–Flattened from the sides, as applied to nodes or buds.

Concave–Curved like the inner surface of a sphere.

Cone–A coniferous fruit, having a number of woody, leathery, or fleshy scales, each bearing one or more seeds, and attached to a central axis; the seed-holding structure of trees like pines and spruce.

Conelet–A young, immature first season cone, in the pines.

Confluent–Blending together, not easily distinguishable apart, as applied to bundle-traces.

Conical–Cone shaped, as he young form of many spruces.

Conifer–A cone-bearing plant, with tiny needlelike leaves, that may be evergreen or deciduous.

Coniferous–Cone bearing.

Connate–Like parts fused together into one, fused in to a tube.

Continuous–Pith which is solid, not spongy, chambered or interrupted by cavities.

Convex–Curved like the outer surface of a sphere.

Coppice–Growth arising from sprouts at the stump, bushy.

Coppice growth–Arising from sprouts at the stump; bushy.

Cordate–A leaf base that extends in a curving, dual lobe-like fashion

Cordiform–Heart-shaped or formed.

Cordoned–Having been espaliered.

Corky–Soft and springy.

Corky-ridged–With elongated warts or ridges on the bark.

Corm–A solid bulb-like underground stem not differentiated into scales, often depressed-globose in form, bearing scale-like buds on the surface, usually tunicated. Ex. Gladiolus, crocus.

Cormel–A tiny corm (fleshy stem base) around a large mother corm.

Cornuta/cornutum/cornutus–Epithet meaning horned.

Corolla–1. The combined term for floral petals

2. The petals and normally colored portion of a flower; the inner floral envelope, consisting of petals.

Corona–A crown; an appendage or extrusion that stands between the corolla and stamens; an out gowth of perianth tissue in the "cup" of Narcissus, or of the androecium in milkweeds.

Cortex–The rind or outer covering of a stem; technically it lies outside of the vascular bundles, which constitute the stele.

Corymb–A more or less flat-topped indeterminate inflorescence whose outer flowers open first. Ex. Viburnum

Cottony–Much the same as woolly, with white hairs.

Cotyledon–The primary leaves of the embryo, present in the seed.

Cover crop–A crop that is planted in the absence of the normal crop to control weeds and add humus to the soil when it is plowed in prior to regular planting.

Creeping–Prostrate and spreading over the ground.

Crenate–1. Having rounded teeth at the leaf margins

2. Scalloped with rounded teeth.

Crenate-serrate–Having a mixture of blunt and sharp teeth.

Crenulate–Finely crenate.

Crisped–Wavy on the margin; short and curly when applied to pubescence.

Crown–On a clumping plant, the point at which the main stem (or stems) and the roots join together.

Crucifer–A member of the Mustard family.

Cruciform–Cross shaped.

Cryptogam–A plant reproducing by spores such as most ferns and algae.

Cucullate–Hooded.

Culm–An upright flowering stem from a rhizome; stem of grasses and sedges or Gramineae and Cyperaceae plants.

Cultigen–Any plant arising from cultivation and not the wild.

Cultivar group–An association of related cultivars with a species, subspecies, or variety due to common origin or groups of traits.

Cultivar–A cultivated variety; does not occur naturally.

Cuneate–1. Having a very narrow leaf base

2. Wedge-shaped with an essentially straight side, the structure attached at the narrow end.

Cupped–A flower with depressed or concave center.

Cuspidate–With an apex somewhat abruptly and concavely constricted into an elongated sharp-pointed tip.

Cuticle–The outermost waxy layer of epidermal cell walls.

Cymbiform–Boat shaped.

Cyme–A more or less, flat-topped determinate inflorescence whose outer flowers open last. Ex. Sambucus

Cytology–The study of cells and often chromosomes.

Decidua–A deciduous species in a normally evergreen genus.

Deciduous–Perennial plants whose leaves die all at once (and usually fall) at the end of each growing season, to be replaced by new leaves at the next growing season. Most deciduous plants are broad-leaved, though a few, such as *Larix laricina* (Tamarack), have needles. Plants whose leaves live year-round are *evergreen*.

Decompound–Repeatedly compound.

Decumbent–Reclining on the ground with the tip ascending; trailing stems on the ground with lateral shoots upright.

Decurrent–Continued down the stem in a ridge or wing, as applied to leaf-bases.

Decussate–Situation where opposite leaves at any two consecutive nodes on a stem are oriented at 90 degree angles to one another

Defoliate/defoliation–Loss of leaves such as premature summer leaf drop; casting off or falling off of leaves.

Dehiscent–Opening to discharge the seeds or pollen.

Deliquescent–Breaking up into fine branches.

Deltoid–Triangular, with equal sides.

Dense–Crowded together, thick, compact.

Dentate–Leaf margins that are toothed with outward pointing notches.

Denticulate–Finely dentate.

Denuded–Naked through the loss of covering.

Depressed–Flattened, as if compressed somewhat.

Determinate–Said of an inflorescence when the terminal flower opens first and the prolongation of the axis is thereby arrested.

Di–A prefix indicating two.

Diaphragmed–The pith which is solid with transverse bars of denser tissue at intervals between the nodes (tulip tree) or at the nodes (grape).

Dichotomous–Forked in pairs.

Dicot–A plant of the angiosperm group having two cotyledons.

Dieback–Death of shoot tips such as by winter cold or chemical damage.

Diffuse–Loosely or widely spreading, an open form.

Digitate–Spreading like the fingers, the leaflets leaving the petiole at one point as applied to compound leaves. The same as palmate.

Dimorphic–Having two forms.

Dingy–Neither white nor brighly colored, as applied to pubescence.

Dioecious–Species in which male and female reproductive organs are found on separate plants. Most hollies (genus *Ilex*) are dioecious; female plants will only bear fruit when a male grows nearby, and male plants do not bear fruit.

Diploid–Having two set of chromosomes in each cell. Represented by 2n.

Disarticulating–Falling away by abscission, leaving a clean cut scar, as with most leaves, many flowers, some twig tips, etc.

Disbudding–The processing of removing extra buds to promote larger flowers.

Discoid–The same as chambered, when applied to pith.

Disk–An enlarged tip, as applied to tendrils.

Disk flower–The tubular flower in the center of the usual Compositae inflorescence.

Display index–Percentage of peak display for flowers or other showy part.

Dissected–Cut, lacerated or divided in narrow, slender segments, said of fringed petals or leaves.

Dissectum–Deeply cut, as in foliage.

Distal–Toward the apex, away from the base.

Distichous–Two ranked.

Diurnal–Blossoms opening only during the day.

Divaricate–Spreading very wide apart.

Divergent–The same as spreading.

Divided–Deeply lobed when applied to leaves; separated to the base into divisions.

Domestica–A plant having been domesticated or cultivated near residences.

Dormancy–A state of suspended growth; the yearly cycle in a plant's life when growth slows and the plant rests.

Dorsal–Relating to the back or outer portion of a plant part; of or on the back or outer surface of a leaf.

Double–A flower with numerous and full petals.

Double nose–Bulbs with two growing apices.

Doubly serrate–Serrations bearing minute teeth on the margins.

Downy–Pubescent with soft short straight hairs.

Drupe–1. A fleshy fruit whose seed is enclosed in a stony structure

2. A stone fruit (cherry, plum, peach). Many berry-like fruits are technically small drupes, like huckleberry.

Dwarf–A genetically smaller form of variety of any plant, often 1/4 or less normal size.

Dwarf varieties–Varieties bred to grow smaller than their parent plants. Dwarf plants may lose the ability of the parents to set fruit (Bailey's dwarf highbush cranberry, *Viburnum trilobum 'Bailey's Compact'*). They may not resemble miniatures of their parents (dwarf Alberta spruce, *Picea glauca var. albertiana 'Conica'*).

Ellipsoid–Elliptical in section, like a football.

Elliptical–Having a relatively broad middle and tapering at both ends

Endemic–Native or local to an area.

Ensis–Botanical name suffix meaning from a region or country.

Entire–1. Having a smooth margin (without teeth); may or may not be lobed

2. Neither toothed nor lobed, as applied to leaves.

Ephemerals–Plants that emerge and bloom during one season, then die back for the remainder of the year. Many spring ephemerals bloom in woodlands, including trillium and ladyslipper.

Epidermis–Outer skin of the twig; sometimes peeling or cracking during the first winter; ultimately destroyed by growth in diameter of the twig.

Epiphyte–A plant growing on another without being a parasite (orchid) or on a rock or tree trunk (moss, lichen). Contrasted usually with plants rooted in the soil; sometimes parasites.

Erect–Upright habit of growth.

Even-pinnate–If no terminal leaflet is present and the total number of leaflets is an even number.

Evergreen–Holding or having green foliage through the winter.

Everlasting–A flowering plant that retains colors when cut and dried.

Excavated–Hollowed out, as applied to pith, making the stem fistulous.

Excurrent habit–The cone-shaped form typical of many conifers and a few angiosperms with a clearly identified central leader and growing point at the top of the plant

Exfoliate–Peeling bark from a branch or trunk

Exfoliating–Peeling away; bark that peels and sheds off.

Exotic–Said of a plant not native; rare or tropical plants.

Explant–Any removed portion of a plant used for tissue culture.

Exserted–Projecting beyond, as stamens beyond a corolla.

Extra-axillary–Above rather than in the axil. Same as supra-axillary.

Eye–An undeveloped bud growth which will ultimately produce new growth; a vegetative or leafy bud.

F_1 hybrid/F_1 hybrid–A first generation of a cross of different plants.

Falcate–Sickle shaped.

Falls–The outer whorl of petals of an iris flower, often broader than the inner petals and often drooping or flexuous.

Fasciated–Abnormally much flattened or cockscomb-like growth, and seemingly several units fused together.

Fascicle–A tight bundle or cluster, as in the needle leaves of pines

Fascicled–Clustered, like the leaves on a spur of barberry.

Fastigiate–With upright branches; having a narrowly columnar or pillar-like growth form.

Feathering–Spreading out roots before planting a potted plant.

Fertile–Capable of producing seed.

Fibrous–Having long narrow shreds or flakes.

Fiddlehead–An unfurling fern frond that resembles the end of a violin.

Filament–The slender stalk or stem of the anther or pollen sac.

Filiform–Thread-like, long and very slender.

Filamentous–Thread-like.

Fireblight–A serious bacterial disease that kills members of the rose family.

Fissure–Bark that is furrowed and ridged, or splitting lengthwise

Fissured–Torn lengthwise, in vertical furrows, as applied to bark, or to pith, for which the more general term spongy is used.

Fistulous–Hollow, with excavated pith.

Flaking–Shredding, but with short segments.

Fleshy–For a stem, not hard and woody; for a fruit or bud scales, not dry. Succulent.

Floret–Technically a minute flower; applied to the flowers of grasses and Composites.

Florida–With abundant flowers.

Floriferous–Having many flowers compared to most cultivars or species.

Fluted–Grooved.

Foliage–Mass of leaves; leaves collectively are termed so.

Follicle–A small dry fruit opening down one edge. Ex. Columbine, milkweed

Forbs–Flowering ***herbaceous plants*** that are not grasses and sedges. As grasses and sedges do produce (relatively inconspicuous) flowers, the term "forbs" is often used (instead of "flowering plants") to specify the plants with conspicuous flowers the grow among the grasses in a meadow or prairie. Technically, this use of the term excludes small shrubs (such as leadplant, *Amorpha canescens*) that may grow among the grasses and forbs, produce flowers, and have a form similar to herbaceous plants.

Form–A botanical variation of a variety differing in only one trait, such as color. A subdivision of species that occurs occasionally in the wild, seldom breeds true, and does not develop a natural population or distribution.

Fragmented–Not continuous, as applied to bundle scars.

Fringed–Ciliate with glands or scales rather than fine hairs.

Frond–A leaf, once applied only to the leaves of ferns but now also said of the leaves of palms too.

Fruticose/fruticosa/fruticosum/fruticosus–Shrubby or shrub-like, in the sense of the stems being woody. Often in a genus of smaller form.

Fusiform–Spindle-shaped; tapering to each end from a smaller mid-section.

Gall–A swelling on a leaf or stem caused by fungi or insects like wasps.

Genotype–The true genetic makeup of a plant.

Genus–A group of species possessing fundamental traits in common but differing in other lesser characteristics.

Girdling Root–Roots that strangle a plant by encircling it tightly.

Glabrate–Nearly glabrous.

Glabrescent–Becoming smooth.

Glabrous–Smooth.

Glands–Secreting organs. Leaf teeth and stipules often end in minute glands.

Glandular-bristly–With stiff gland-tipped hairs.

Glandular-ciliate–Fringed with small glands.

Glandular-pubescent–With gland-tipped hairs.

Glauca/glaucum/Glaucous–Covered with a white or bluish bloom which can usually be rubbed off easily.

Globose–Shaped like a globe; spherical.

Glossy–1. Very shiny

2. Shining, reflecting more light than if lustrous.

Glumes–Pair of bract at the base of grass inflorescence or spikelet.

Graft union–The region where rootstock and scion come together; there can be slightly deformed growth at the union that is noticeable, but does not effect the function of the tree.

Grandiflora/grandiflorus/grandiflorum–Epithet meaning larger flowers than normal for a species; a class of roses with large and clustering blooms.

Half- or sub-shrub–A plant with stems that are woody at the base, usually dying back to the woody stems or even back to the ground after severe winters. Suffrutescent.

Hardwood–A deciduous tree.

Hardwood cutting–A cutting taking from mature or lignified stems.

Hardy–Used in the horticultural sense, enduring winter conditions.

Harvesting–The removal of a desirable plant part.

Haustorium–A projection that acts as a penetrating and abosrbing organ.

Heading back–Cutting an older branch or stem back to a stub or twig.

Healing–The process of closing of a plant wound or graft union.

Heaving–Lifting of soil and plants due to frost and ice.

Hedge–A linear planting of plants as a barrier.

Heeling in–Temporarily setting a plant into a shallow trench and covering the roots with soil to provide protection until it is ready to be permanently planted.

Heirloom–Any open-pollinated plant that has a history of domesticity and that has been grown for at least 50 years.

Heliotropic–A plant that responds to sun or light.

Herb–A plant dying to the ground at the end of the season; one whose aerial stems are soft and succulent without appreciable parenchymatous xylem tissue, a plant not woody in texture.

Herbaceous–Plants that die down in winter and sprout forth the following spring; a plant with soft or succulent tissue.

Herbaceous plants–Unlike ***woody plants***, herbaceous plant stems do not add new growth each year. Many herbaceous plants die to the ground in winter, then grow new stems and leaves in Spring, but others stay ***evergreen***. These may steadily increase in size by adding leaves and stems, but the stems do not gain hardness or grow in circumference with each new growing season. Herbs are not all herbaceous plants; some common herbs, such as lavender (*Lavandula sp.*) and sage (*Salvia sp.*), are technically shrubs because their woody stems increase in size each year when grown in climates in which they are winter-hardy.

Herbs–This term has two different (and incompatible) definitions.

(1) In common usage, herbs are plants used by humans for cooking, medicine, or ritual. Herbs can be ***herbaceous plants*** or ***woody plants***, annual or perennial, ***deciduous*** or ***evergreen***. Some trees (such as willow) can be considered herbs under this definition.

(2) In the field of botany, the terms "herb" and "herbaceous plant" mean the same thing, and that meaning is listed under the term "herbaceous plants" in this glossary.

Hesperidium–A fleshy berry-like fruit with a hard rind and definite longitudinal.

High centered–Having petals taller or more pointed than outer petals.

Hip–Fruit of a rose.

Hippocrepiform–Horseshoe-shaped.

Hirsute–Pubescent with coarse or stiff hairs.

Hoary–With a close white or whitish pubescence.

Horizontalis/horizontale–Having habit or branches very horizontal; branches spreading outward rather than being erect.

Hybrid–The offspring of two plants of different species or varieties of plants. Hybrids are created when the pollen from one kind of plant is used to pollinate and entirely different variety, resulting in a new plant altogether. Hybrid seed will not be true therefore it isn't suitable for seed saving.

Iana–1. Suffix meaning commemorative, or in memory of.

2. Suffix on plant names honoring the person who found it.

Imbricate–Scales which overlap like shingles; the opposite of valvate in which the scales meet along a line without overlapping.

Imperfect flower–A flower that lacks either stamens (pollen producing structures) or pistils (seed producing structures).

Imposter clone–A clone that has incorrectly replaced the true clone in the trade.

Impressed–Having sunken veins as viewed from the upper leaf surface

Inbred/inbreeding–An undesirable plant made by crossing two related plants.

Incised–Intermediate between toothed (dentate or serrate) and lobed, being a sharply inward cut leaf (the inward cuts are called incisions)

Included–Not protruding as stamens not projecting beyond a corolla; as opposed to exserted.

Incomplete flower–One which lacks any one or more of these parts; calyx, corolla, stamens, and pistils.

Indehiscent–Not opening, as applied to fruits.

Indeterminate–Those kinds of inflorescence whose terminal flowers open last, hence the growth or elongation of the main axis is not arrested by the opening of the first flowers.

Indicator plant–A sensitive plant that alerts to some condition such as drought.

Indurate–Hardened.

Inequilateral–With unequal sides.

Inermis–Without thorns or less thorny.

Inflorescence–1. A grouping of flowers attached to a specialized stem (peduncle)

2. A flower cluster, but sometimes large and showy

Infra–Below.

Inorganic–Something that was never alive; of human manufacture.

Internode–The portion of a stem between two nodes

Interspecific–Between species; hybrid cross.

Interstem/interstock–A piece of stem tissue grafted between a rootstock and a scion.

Invasive–An introduced (non-native) plant that disrupts the local ecosystem and isn't checked by climate, grazing, or other natural means. Extreme examples include ***kudzu vine*** (*Pueraria lobata*), which can grow 35 feet in one season.

Involucrate–With an involucre or cluster of bracts.

Involucre–A cluster of modified leaves about a flower cluster.

Japonica/japonicus/japonicus–Epithet meaning from Japan; such plants are widely Asian.

Jointed–Having nodes or points of real or apparent articulation.

juvenile–1. Chronologically speaking, it is an organ or tissue that is not fully developed; developmentally speaking, a plant that is unable to produce flowers (reproduce); contrast with mature which is able to reproduce

2. An early phase of plant growth, usually characterized by non-flowering, vigorous increase in size, and often thorniness.

Keel–A ridge on the back of a leaf or bud scale.

Key–A small indehiscent fruit with a wing.

Knees–Bent projection of trunk; pointed or dome-like outgrowths such as those at the base of a bald cypress from cypress roots rising above the water.

Labiate–Having lips as in the two lips in flowers in the mint family

Lacebark–With exfoliating or bi-colored bark,usually with a fine texture.

Lacinate–Cut or dissected as in leaf margins or some lacy flower petals.

Lactiferous–Milky.

Lactiflora/lactiflorus/lactiflorum–White or milky-colored.

Lanceolate–1. Much longer than wide, widest at the base and tapering to the tip; less wide than in ovate.

2. Shaped like a lance-head, as applied to leaves.

Lateral bud–A bud borne in the axil of a previous season's leaf.

Lateral–Buds which appear along the sides of the twig; at or along the side.

Latex–A milky waste product. It mainly contains alkaloids, starch, grains, sugars, minerals salts etc.

Latifolia–Epithet meaning wider leaves than normal for a genus or species.

Latiloba/latilobus/latilobum–Epithet meaning wider lobes or tissues sections in a leaf or flower.

Leader–The primary or terminal shoot; the trunk of a tree.

Leaf scar–1. Scars from which leaves have fallen. They usually occur characteristically either singly (alternate) or paired (opposite) or in groups of more than 2 (whorled) at each node. Leaf scars differ greatly in size and shape, and offer some of the best winter characteristics. The points where woody strands of vascular tissue passed up into the leaf are usually evident, and are called bundle scars or bundle traces. Typical leaf scars are essentially at the level of the stem; but they are raised o a pronounced base or leaf cushion in some cases, or the buds are covered by an articular membrane in other.

2. The healed structure which was once a point of leaf attachment

Leaflet–Flat division of a leaf; one of the divisions of a compound leaf.

Leaflets–A division of a compound leaf

Leafspot–One of many types of fungal or bacteria diseases causing round marks.

Leaf–The foliage appendages of the stem, though not always serving as foliage; sometimes metamorphosed into a spine (barberry), or tendril (clematis), or reduced to a scale (juniper). Leaves originate at and mark the node or joints of the stem. Buds normally occur in the angles or axils above leaves and are correspondingly alternate, opposite or whorled on the stem.

Leaves or stipules which do not persist in a green condition throughout the winter; usually falling in the autumn.

Leggy–Abnormally tall and spindly growth; a lot of stem space between leaves; condition where a shrub or multi-stemmed small tree does not have significant foliage in its lower portion

Legume–A pod; the characteristic fruit of the pea family.

Lenticel–A dot or corky spot on stems or branches, usually of different color than the stem

Lenticels–Very small wart-like structures, breaking through the bark of most young twigs. Corky in texture and made of loosely packed cells, providing gaseous exchange between the inner tissues of the stem and the atmosphere.

Lifting–A form of transplantation usually involving little root disturbance.

Lignified–Woody, hardened.

Ligule–1. A strap shaped organ.

2. A minute projection from the top of the leaf sheath in grasses.

3. The strap shaped corolla in the ray flower of Composites.

Liliflorum/liliflora/liliflorus–Epithet meaning flowers like a lily; with big white trumpets.

Line–A group of plants used for breeding.

Linear–Parallel sides, long and very narrow

Lineate–Lined; bearing thin parallel lines.

Lined–Lightly ridged or ribbed.

Lobe–A portion of a leaf that projects outward and divides the leaf into distinct parts, but not enough to make them separate leaflets; lobes may be rounded or pointed

Lobed–Divided rather deeply, as applied to leaves.

Lustrous–Having a slight metallic gloss, less reflective than glossy.

Lutea/luteum/luteus–Epithet meaning yellow, often a lighter shade than aurea (golden).

Macrantha/macranthus/macranthum–Epithet meaning larger flowers.

Maculata/maculatum/maculatus–Epithet meaning spotted or mottled in color.

Malodorous–Emitting a foul smell or odor

Margin–The edge of a leaf or leaf blade.

Maritima/maritimus/maritimum–From a seashore or maritime area.

Mature–Chronologically speaking, it is an organ or tissue that is fully developed; for example, ripe; developmentally speaking, a plant that is able to produce flowers (reproduce); contrast with juvenile which is unable to reproduce

Meristem–The growing tissue area in a plant; all buds contain a meristem. Areas of actively dividing cells and plant growth; capable of developing into specialized tissues.

Merous–A suffix preceded by an Arabic number or numeric prefix such as tri-, indicating how many of each part a flower possesses; 5-merous would be a flower with 5 sepals, 5 petals, etc., while trimerous would have three of each part.

Midrib/midvein–The rib or central vein running down the center of a leaf.

Milky sap–Whitish in color, often thicker than water.

Mollis–A covering of soft hairs.

Moniliform–Repeatedly constricted laterally and appearing bead-like.

Monocarpic–Fruiting once and then dying.

Monoclonal cultivar–A cultivar that correctly represents a single clone.

Monocot–A plant of the angiosperm group having one cotyledon.

Monoculture–Growing only one particular type of plant.

Monoecious–A species with unisexual flowers, having male and female reproductive structures on the same plant; a species that has separate male and female reproductive structures (flowers or cones) on the same plant.

Monopodial–Continuing growth from a terminal bud each year.

Morifolium–Epithet meaning leaves like mulberry, such as mums.

Morphology–The study of structure or forms.

Mucilaginous–Slimy.

Mucronate–Ending in a fine, slender tip.

Multiple fruit–A fruit formed from several flowers included in a single structure having a common axis. Ex. Pineapple, mulberry.

Multiple–Applied to several buds in or over an axil, instead of the customary single or solitary bud.

Multi-stemmed–Having more than one stem or trunk.

Mummy–A dried, shriveled fruit.

Naked–A bud which is not covered by scales; the outer pair of leaves serves the same purpose. Or of a flower having no floral envelopes (perianth).

Nana/nanus/nanum–Epithet meaning dwarf or smaller.

Needle–A slender, leaf usually in the shape of a needle or awl commonly found in some conifers, such as Junipers; contrast with scale-like

Net veined/netted venation–With veins reticulated and resembling a fish net; the interstices close.

Netted venation–The veins reticulated and resembling a fish net; the interstices close.

Neutral flower–A sterile flower consisting of perianth without any essential organs.

Niger/nigrum/nigrus/nigra–Dark or blackish in color.

Night-blooming/night-flowering–Plants that bloom only in dark or shade.

Nocturnal–Opening at night and closing during the day.

Nodal–At or pertaining to a node.

Node–The point where a leaf or other structure meets a stem node; the part of a stem from which a leaf or new branch starts to grow; the point of attachment of leaves or buds on a stem

Nodule–Round bacteria-filled swellings on the roots of legume plants.

Nomenclature–The study of naming plants including spelling and format.

Notched–With V-shaped indentations.

Nut–A dry, indehiscent, 1-celled, 1-seeded fruit having a hard and bony mesocarp; the outermost endocarp may be fibrous or slightly fleshy.

Obcordate–A leaf with rounded lobes at the tip.

Oblique–Lop-sided, as one side of a leaf base being larger, wider or more rounded than the other.

Oblong–longer than broad, with the margins parallel except at the extreme basal and apical ends

Obovate–Inversely ovate, broadest above the middle.

Obscure–Not easily made out, as applied to buds, bundle scars, etc.

Obtuse–With sides forming an angle of more than 90 degrees; having a rounded and blunt apex.

Ob–Used as a prefix to indicate inversion; obcordate in contrast with cordate, oblanceolate in contrast to lanceolate, etc. A cordate leaf is heart-shaped with rounded lobes at the base of the leaf. An obcordate leaf has rounded lobes at the tip.

Occidentalis/occidentale–Globally western in origin.

Odd pinnate–A compound pinnate leaf with a terminal or odd numbered leaflet at the tip.

Oedema–Corky portions of leaves caused by excessive humidity.

Offset–Small bulb or plantlet used to propagate new plants.

Oides–Plant name suffix meaning like or similar to.

Open-pollinated–Seeds produced from plants that are allowed to pollinate naturally, primarily through insects, wind and water.

Opposite–Leaves or leaf scars which are paired on opposite sides at each node; the arrangement of leaves and buds that arise from a single node on either side of a stem

Orbicular–Rounded in two-dimensional outline as many plant leaves

Orbiculate–Circular or disc-shaped.

Oval–Twice as long as broad, widest at the middle, both ends rounded.

Ovary–The ovule-bearing part of a pistil; the female part of a flower containing immature seeds (ovules).

Ovate–Egg shaped, broadest below the middle.

Ovoid–A bud which is egg-shaped, with the broadest portion near the base.

Ovule–The egg-containing unit of an ovary, which after fertilization becomes the seed.

Pacifica/pacificum/pacificus–From near the Pacific ocean or the Western US.

Palmate–Radiating,fan-like from a common point; as of leaflets of a palmately compound leaf or veins of a palmately-veined leaf. Digitate.

Pandurate–Fiddle shaped.

Panicle–A branched cluster of stalked flowers.

Parallel–Running side by side from the base to tip; especially of veins.

Parasitic plant–A plant which lives on, and acquires its nutrients from another plant without benefitting the other organism.

Parasitoid wasp–Many species of wasp prey on other insects. They deposit their eggs into the larvae or pupae of the host insect, and the wasp larvae then consume the host. These wasps are correctly termed parasitoid rather than parasitic, since they kill their hosts. Parasites or parasitic animals live off a host without killing it, in a relationship that could be viewed as a type of symbiosis.

Parterre–An ornamental flower garden where beds and paths are arranged to form a pattern. A formal garden construction on a level surface consisting of planting beds, edged in stone or tightly clipped hedging and gravel paths arranged to form a pleasing pattern.

Parvi–Botanical name prefix meaning small or reduced.

Parviflora–Having smaller flowers.

Parvifolia–Having smaller or shorter leaves.

Pedate–A palmately divided or compound leaf whose two lateral lobes are again cleft or divided.

Pedicel–The stalk of a flower or fruit when in a cluster or when solitary; the stalk which directly supports a flower or fruit.

Peduncle–The stalk of a flower cluster or a single flower when the flower is solitary, or the remaining member of a reduced inflorescence; the main supportive stalk of a flower or fruit cluster.

Pendula/pendulous/pendulum/pendulus–Weeping or drooping with branches hanging down from a central stem; a weeping or spreading cultivar.

Pendulous–Hanging, drooping or declined; can refer to plant form or parts, like branchlets or inflorescences.

Perennial–A plant that dies down to the ground during winter and survives to grow again each spring.

Perfect flower–A flower having both functional stamens and pistils; a plant with both functioning male and female parts.

Perianth–The two floral envelopes of a flower; a collective term embracing both corolla and calyx as a unit; often used when it is not possible to distinguish one series from the other (as in most monocots) and the parts then called tepals.

Perlite–A light weight white-colored soil additive from volcanic materials.

Persicifolia–Having leaves shaped like the peach.

Persistent–Not deciduous, as applied to leaves; not disappearing, as applied to pith, pubescence, epidermis, etc. Adhering to a position instead of falling, whether dead or alive.

Pest–Any insect or animal which is detrimental to the health and well being of plants or other animals.

Petals–Whorl of flower parts, usually colored, encircling the androecium.

Petiole–Leaf stalk; the attachment structure between a leaf blade and the stem.

Petiolule–Leaflet stalk.

Picta/pictum/pictus–Painted; often a variegated or multicolored plant.

Pilose–Shaggy with soft hairs.

Pinching back–Nipping off the very tip of a branch or stem. Pinching promotes branching, and a bushier, fuller plant

Pinetum–Woody plant collection emphasizing conifers.

Pinked–Notched.

Pinna–A leaflet of a compound leaf; when applied to ferns, the primary division attached to the main rachis; feather-like.

Pinnata/pinnatum/pinnatus–Having pinnate or feather-patterned veins or leaflets. Pinnatio means many small, lateral branches growing off a rib or axis.

Pinnate–A compound leaf with leaflets or segments along each side of a common axis or rachis; feather-like.

Pinnately compound–When the leaflets are arranged along a common rachis or if only three are present at least the terminal leaflet is petioled;

Pinnule–The leaflet of a pinna; a secondary leaflet of a pinnately decompound leaf.

Pisifera/pisiferus–Literally, pea-bearing.

Pistil–The female part of a flower composed of ovary, style and stigma.

Pistillate–An imperfect flower with a pistil, or seed organ, but having no functional stamens (male pollen producing organs).

Pit–A thin area of secondary cell-wall.

Pith–The central part of a twig, usually lighter or darker than the surrounding wood. similar to marrow.

Pithy–Sometimes used in the sense of having a large pith and little wood.

Plicate–Folded, as in a folding fan, or approaching this condition.

Plugs–Young seedlings grown in roughly one-inch-across and four-inch-deep tubular holes, generally grown in a "plug pack" with six such holes or a "plug tray" with three or four dozen. Growers can sell the entire plug pack or plug tray, or they can transfer the seedlings to larger individual pots.

Plume–Feather-like erect inflorescence.

Plumose–Feather-like, plumy.

Pod–A dry dehiscent fruit; capsule, legume; fruit from a member of the legume (bean) family.

Pollen–The male cells or microspores produced by the stamens.

Pollination–The transfer of pollen from the stamen (male part of the flower) to the pistil (female part of the flower), which results in the formation of a seed. Hybrids are created when the pollen from one kind of plant is used to pollinate and entirely different variety, resulting in a new plant altogether.

Pollinator–Cultivar, often a male clone, proven to pollinate other cultivars.

Polycarpic–Flowering and fruiting many times.

Polyclonal cultivar–A cultivar representing two or more distinct clones.

Polygamous–Bearing unisexual and bisexual flowers on the same plant.

Pome–A type of fleshy fruit represented by the apple, pear or related genera. A fruit with a papery or bony core at the center and with sepals or scars from which the sepals have fallen at the blossom end. Many "berry-like" fruits are really small pomes (cotoneaster).

Posterior–At or toward the back; opposite the front; nearest the axis; away from the subtending bract.

Precocity–Tendency of a given species/ cultivar to mature early.

Preformed–Already with definite shape or structure, as with leaves within a bud.

Prehensile–Clasping, coiling in response to touch.

Prickles–Slender, sharp outgrowths of the stem tissues beneath the epidermis.

Primocane–The first year's shoot or cane of a biennial woody stem, ex. Rubus.

Procumbens–An epithet meaning trailing low along the ground.

Procumbent–Lying flat on the ground but the stem not rooting at nodes or tip.

Progeny testing–The evaluation of seedlings of a particular cross or source.

Prominent–Projecting outward, conspicuous.

Propagation–Various methods of starting new plants ranging from starting seeds to identical clones created by cuttings or layering.

Prostrate–Lying flat on the ground; a general term.

Protogynous–Having the stigma receptive to pollen before the pollen is released from the anthers of the same flower.

Provenance–The geographic wild origin in terms of site and population.

Proximal–Toward the base, away from the apex.

Pruning–Reducing the growth on a shrub / tree to promote new flowers, stems or foliage. Also for cosmetic reasons. The cutting and trimming of plants to remove dead or injured wood, or to control and direct the new growth of a plant.

Ptero–Botanical name prefix meaning winged.

Puberulent–Minutely hairy,when viewed with a hand lens.

Pubescent–Fine hairiness of a leaf, petiole, pedicel, peduncle, or stem.

Punctate–With translucent or covered dots,depressions or pits.

Quadrangular–Four angled, or pith or a twig.

Raceme–A simple indeterminate inflorescence with pedicel led flowers. An elongated, unbranched cluster of stalked flowers.

Rachis–The stem or petiole-like central axis of a pinnately compound leaf, connecting the petiolules to the petiole;the axis along which the leaflets of a pinnate leaf are arranged (walnut).

Radial–Arranged around and spreading from a common center.

Radical–Of or pertaining to the root.

Radicle–The embryonic root of a seed *i.e.* first organ to emerge from seed during seed germination.

Ranked Foliage–Arranged in longitudinal planes around the stem.

Ray flower–A flower with a strap-shaped corolla, usually found on the outer edge of a Compositae inflorescence.

Rebloomer–A plant with two or more flowering periods.

Reblooming–Typically said of a plant with a spring and later fall blooming

Recurrent–A plant more or less blooming in two or more sessions or months.

Recurved–Bent or curved backward, usually roundly or obtusely so.

Reflexed–Bent abruptly backward or downward.

Reniformis/reniforme–Epithet meaning kidney shaped.

Reticulate–Net-like appearance; for example, referring to a leaf having veins that branch many times in a dense pattern, often more prominently noticed when the veins are impressed.

Rhizome–An underground stem distinguishable from a root by the presence of nodes, buds or scale-like leaves; and may give rise to new plantlets.

Rhizomes–Roots that travel underground laterally and send up new plants. An example is the perennial sweet woodruff (*Galium odorata*), which travels laterally an inch or so underground, then breaks the surface to grow a new stem.

Rhombic–A four-margined leaf that is diamond-shaped, having three prominent tips, two on the side and one at the top; with four nearly equal sides, but unequal angles.

Rib–A conspicuous vein of a leaf; or a prominent ridge.

Root anchor–A large root that holds a plant in a physical soil mass feeder; a small surface root which easily absorbs nutrients and water.

Root ball–The clump consisting of the main roots of a plant and the soil (or other growing medium) clinging to them. The network of roots, with attached soil, of any given plant. The root ball should be kept intact when transplanting.

Root zone–The area of ground under which a given tree's (or other plant's) roots spread. Often the root zone covers the area encircled by the *drip line*; that is, the roots often spread underground to the same distance that the tree's branches extend above.

Rootbound–A condition which exists when a potted plant has outgrown its container. The roots become entangled and matted together, and the growth of the plant becomes retarded. When repotting, loosen the roots on the outer edges of the root ball, to induce them to once again grow outward.

Rootstock–The portion of a grafted plant that provides the root; grafted plants typically consist of a scion, which develops into the shoot or crown and a rootstock that provides the root system; rootstocks may include a signicant length of stem, called standards, commonly used for weeping trees or shrubs

Rosea/roseus/roseum–Pink.

Rosette–A crown of prostrate leaves radiating from a stem, and at or close to the surface of the ground.

Rudiment–The beginning of an undeveloped plant part.

Rufous–Reddish brown.

Rugose–Wrinkled, usually covered with wrinkles.

Runner–A slender trailing shoot that usually roots at the tip and some nodes; a long offshoot of a plant. A slender stem growing out from the base of some plants, which terminates with a new offset plant. The new plant may be severed from the parent after it has developed sufficient roots.

Saccate–With a sac-like shape.

Saggitate–Shaped like an arrowhead.

Samara–A dry indehiscent fruit bearing a wing (Maple, Ash).

Sanguinea/sanguineus/sanguineum–Bright or blood red.

Saprophyte–A plant that lives on decaying organic matter, such as mushrooms.

Savannah or savanna–Grassland with scattered trees, fairly common in the central plains of the U.S. before development.

Scab–A fungal scale disease that leaves spots on foliage and fruit.

Scabrous–Rough or gritty to the touch; rough-pubescent.

Scald–Burning due to strong sun or light

Scale–A small and usually dry bract or vestigial leaf or a structure resembling such; a small bract or leaf-like covering of a flower or leaf bud. A small round insect.

Scale-like–Small adpressed leaf or bract; contrast with needle-like scales and becoming obvious in winter during plant dormancy

Scandent–Climbing, usually without tendrils.

Scape–A leafless peduncle arising from the basal rosette of a few or not basal leaves; sometimes a few scale-like leaves or bracts may be borne on it; a scape may be one or many-flowered.

Schizocarp–A dry dehiscent fruit that splits into two halves.

Scion–The shoot portion of a rootstock-scion graft.

Scorch–Injury to plant parts due to burning by wind or water loss; sometimes used as a synonym of scald, such as burning by the sun.

Scrambler–A plant that climbs without twining or grasping in some way.

Secund–One sided.

Sedges–Plants of the genus *Carex*, which includes over 1,500 species; sedges look similar to grasses.

Seed–A fertilized ripened ovule that contains an embryo; a ripened plant ovary capable of germinating to produce another plant.

Seed coat–The covering or protective layer around a seed.

Seed Pods–The hardened protective shell that contains the maturing seeds of a plant and which functions to prevent mechanical injury to and water loss from the embryo, it may alos regulate germination in several ways.

Seedling–A young plant grown from seed.

Self–To cross a plant with its own flowers or one of identical type; said of flowers of one color as opposed to a bicolor.

Self-fertile–A plant capable of producing viable seed with its own pollen.

Self-pollination–A plant capable of pollinating its own flowers.

Self-sterile–A plant not capable of reproducing with its own pollen.

Semi-double–A flower with only a few stamens converted to petals; a flower with less than twice the normal number of petals.

Semi-evergreen–1. Having some leaves deciduous and some leaves persistent throughout the winter, with the degree of persistent leaves determined by the harshness of the winter season.

2. A plant whose foliage is retained all year in warm climates but not in colder ones; some of the foliage is retained in most climates where hardy.

Semp–A common abbreviation for Sempervivum and related genera.

Sepal–Modified leaves forming the outer floral envelope.

Septum–A portion of a fruit.

Serrate–Having marginal teeth that lean toward the tip of the leaf; margins can be singly or doubly serrated; saw-toothed, the teeth pointing forward.

Serrulate–Minutely serrate.

Sessile–Without an obvious attachment structure, such as a leaf attached to a stem without an obvious petiole ; without a stalk.

Setaceous–Bristle-like, very narrow; resembling or having a bristle.

Shrub–A woody plant that is usually under 15 feet tall and has many stems.

Shrubby–Applied to woody plants which at maturity are less than 20 feet high, have more than one stem, and no definite crown shape (a purely arbitrary definition); See also "Arborescent".

Sib–Shorthand for sibling or sister/brother in a seedling population.

Simple–A leaf with a single leaf blade (when not compound), or of an inflorescence when unbranched.

Sinensis/sinense–Epithet meaning from China; from Asia generally.

Single flower–A flower having only a minimum number of petals for that variety of plant.

Sinuate–With a strongly wavy margin.

Sinus–The recess between lobes of a leaf or petal; the space between two lobes, segments, or divisions; as of leaves or flower parts.

Solitary–Single, one.

Sooty mold–A fungus that develops on insect secretions.

Spaghetti irrigation/spaghetti tube–The popular name for drip irrigation because of long narrow tubes used.

Spathe–A large hood-like bract enclosing flowers.

Spatulate–Spoon shaped.

Speciation–The evolutionary process of a wild population becoming a new distinct species.

Species–A natural group of plants composed of similar individuals which can produce similar offspring; usually including several minor variations.

Specimen or specimen plant–A plant that is grown in relatively open ground with little competition and therefore develops an unnaturally (in most cases) broad spread and dramatic form. Contrast with masses, drifts, thickets, or groves of plants, in which individual specimens intermingle with each other and may even be hard to distinguish from each other.

Sphagnum–Dried bog moss. Most peat moss is composed primarily of sphagnum moss. This moss is also packaged and sold in a fresh state, and used for lining hanging baskets and air layering.

Spicate–With spikes; having the shape or form of a spike.

Spike–A usually unbranched, elongated, simple, indeterminate inflorescence in which the main axis is elongated and the flowers are sessile.

Spine–A modified very sharp leaf, similar to thorn and prickle; sharp outgrowths of the twig, sometimes but not always paired at the nodes.

Spinosa–Epithet meaning with spines, prickles or thorns.

Spinose/spinosissima–With many spines.

Spittle bug–Frog hoppers that suck sap from plant stems and produce saliva.

Spongy–Irregularly interrupted by small, sometimes scarcely distinguishable cavities; porous.

Spore–A minute reproductive body comprised of a single gametophytic cell; a simple reproductive cell of ferns, mosses or fungi.

Sporocarp–A body containing sporangia or spores.

Sporophyll–A spore bearing leaf.

Sport–A shoot that differs from a plant and can be reproduced as such.

Spreading grass–Grass that spreads by creating new plants from existing plants by running roots across or under the ground (***stolons*** or ***rhizomes***, respectively), creating a dense turf. Also called ***sod-forming grass***. Contrast with ***bunchgrass*** (also called ***clump-forming grass***), in which single plants get larger over time but do not create new plants.

Spreading–Growing outward or horizontally.

Spur–A projection of a flower; a short thick stem that produces flowers and fruit.

Spur shoots–Short stubby branches with greatly crowded leaf scars and very slow growth, often bearing flower buds. Also called spurs.

Stake–A piece of wood/plastic/metal used to support stems or flowers.

Stalk–A supporting structure of a leaf, flower or fruit.

Stalked–Buds with scales elongated perceptibly below the lowest scales (alder).

Stamen–The male part of a flower composed of the anther and filament; the pollen bearing organ of a seed plant.

Staminate–An imperfect flower with stamens or pollen producing structures, but with no pistil, or seed producing structure.

Standard–1. A small tree commonly produced by grafting a weeping or dwarf form on a trunk of the desired height.

2. The erect petals of an iris flower, as opposed to the broader and often drooping falls petals.

3. The upper, more or less erect petals of a pea-like flower which consists of three types of petals: standard, wings and keel.

Stellate–Star-like; with several arms radiating from a center. Often said of hairs that are once or twice forked.

Stem–The trunk and its branches; one of the three fundamental parts of a higher plants–root, stem and leaf.

Stigma–The receptive female portion, usually the distal end of a pistil, of a flower that receives pollen.

Stipe–The stalk of a fern frond or stalk of a flower pistil.

Stipel–A stipule of a leaflet.

Stipellate–Having stipules at the base of the leaflets.

Stipular–Having stipules at the base of the leaves.

Stipule–An appendage at the base of the petiole, often in pairs and sometimes called auricles (ears)

Stipule scars–A pair of marks left after the stipules fall off, to either side of the leaf scar.

Stipules–Small leaf-like organs occurring in pairs on either side of the leaves; occasionally each one extends half way around the twig, respectively.

Stock–The basal or rooted part of a grafted plant; a group of preserved nursery plants used for propagation.

Stolon–Along with runners, horizontal stems involved in vegetative propagation; a horizontal stem (usually just above, at, or below ground level) that roots at its tip and gives rise to a new plant; similar to runners

Stolon–A slender horizontal stem which produces new plants; a horizontal stem that roots at its tip and grows a new plant.

Stoloniferous–Producing stolons.

Stolons–Surface roots that travel laterally and root at the joints, growing new plants. An example is ground ivy (*Glechoma hederacea*, also called **creeping Charlie**).

Stomata–A minute pore in the epidermis, especially in the lower surface on the leaf. The ***"breathing pores"*** of a leaf.

Stomate–A breathing or water pore of a plant leaf or stem; sweat gland.

Stool–A clump of roots used for propagation, usually applied to layering

Strain–A race within a species reproduced by seed; many are now cultivars.

Striate–With fine longitudinal lines, channels or ridges;such as bark

Strike–To produce roots, as with a cutting.

Strobilus–A term for a conifer cone.

Style–The more or less elongated part of a pistil between the stigma and the ovary. The long female stalk or pistil between the stigma and ovary.

Sub–A prefix often used to mean nearly, as in subacute, subsessile, etc

Subcontinuous pith–With occasional but not regular gaps in the pith.

Suberization–A formation of corky bark on wounded cuttings or stems.

Submerged bud–A bud hidden by the petiole or embedded in the callus of the leaf scar.

Subopposite–Paired leaf scars that are close, but not at exactly the same height on the twig (staggered).

Subshrub–A partly woody plant.

Subtending–Standing below.

Succulent–A fleshy plant that hold water in itself, usually in the leaves or stem. Thickened, juicy, fleshy tissues that are more or less soft in texture.

Sucker–A vigorous shoot arising from a plant base or from below ground; also the adhering discs of a vine. A growth originating from the rootstock of a grafted plant

Suckers–Many stems arising from the base of a tree or shrub and gradually spreading the diameter of its basal area

Suffrutescent–A plant with stems that are woody at the base, usually dying back to the woody stems or even back to the ground after severe winters. Half–or sub-shrub.

Suffruticose–Shrubby; densely bushy plants.

Superficial–On the surface, not connected to inner tissues.

Superposed–Extra buds that appear above the true axillary buds; usually flower buds.

Supine–Lying flat, face upwards.

Suture–A line of dehiscence or groove marking a face of union.

Sylvestris/sylvestre–Epithet meaning of the woods.

Sympodial–Multiple upward branching from a common branch point, resembling an upward-turned hand

Syncarp–A rounded aggregation of fruits; a fleshy aggregate fruit.

Tap root–The primary or thick main root growing straight down from a plant.

Tapestry hedge–A row of mixed species of shrub.

Tap rooted–Having one main fleshy root that extends straight down into the soil. Examples include carrot and dandelion.

Tender plants–Plants that are unable to endure frost or freezing temperatures.

Tender–In the horticultural sense of not enduring winter conditions.

Tendril–The twisting, clinging, slender growth on many vines, which allows the plant to attach themselves to a support or trellis; a twisting thread-like shoot used to support the plant. A climbing organ metamorphosed from a leaf, stipule or stem.

Tepal–A segment of perianth not differentiated into calyx or corolla. Ex.–tulip, magnolia.

Terminal–Applied to the end bud beyond which no further growth takes place normally until the following season. At the tip or distal end.

Ternate–In threes.

Ternately compound–When the leaflets are in threes; palmately compound–when three or more leaflets arise from a common point.

Thallus–A flat leaf-shaped part in the early life cycle of a fern

Thorn–One of various sharp and pointed outgrowths of plants that can arise from stems, leaves, or fruits; botanically, a modified twig which has tiny leaf scars and buds and is usually pointed and sharp; other sharp structures are prickles and spines which are technically different from thorns and each other, but which have the same overall effect

Thorns–Sharp outgrowths of the twig that represent modified branches; usually bearing leaf scars, or branched.

Throat–The inner mouth-like opening of a tubular flower.

Tomentose–With short matted hairs; woolly.

Tomentulose–Microscopically tomentose or woolly.

Tomentum–Densely matted hairs.

Toothed–The condition of a margin broken into small projecting segments, either serrations, dentations, or crenations.

Transverse ridge–A ridge that runs across the stem from one leaf scar to its pair on t he opposite side of the twig.

Trichomes–A bristle or hair.

Tricolored–A plant with three colors.

Trifoiolate–A leaf with three leaflets. ex. Trillium

Triloba/trilobus/trilobum–Epithet meaning three major lobes in the leaf or other part.

Triploid–With three sets of chromosomes; many of these plants are seedless.

True Leaves–The first hardy leaves, usually the second pair, on a new plant.

Truncate–With a flat base or apex, perpendicular along the margin to the midrib; with a leaf or part abruptly cut.

Truss–A non-technical term for a domed flower mass.

Tuber–A flat underground stem that stores food and plant energy, with growing points from which a plant grows.

Tubercle–A miniature tuber, tuber-like structure or projection.

Tuberous–With tubers or part resembling them.

Tumid–Swollen

Tunicated–With concentric layers, often of fleshy scales.

Turf–A dense mat of lawn grasses cut very low.

Turgid–A plant part that is swollen and full of water. Swollen to firmness.

Twiggy–Having many divergent twigs.

Twining–A stem winding around a support.

Type–Identifies each plant according to various accepted groupings such as shrub, perennial or grass. A pressed herbarium specimen.

Umbel–Usually a flat-topped flower cluster, with pedicels and peduncles arising from a common point, resembling the supports of an umbrella.

Unarmed–Without either spines or prickles, though the leaves may have sharp teeth at the margins or tip like holly.

Uncinate–Hooked at the tip.

Understory trees–Small trees that grow well under larger ones. Common understory trees in a birch-poplar-spruce forest include ironwood, mountain maple, and serviceberry. Other common understory trees in a mixed maple-hardwood forest are witch hazel and black cherry.

Undulate–Wavy along a margin, as a leaf margin.

Unisexual flowers–Flowers having either pistils or anthers, but not both.

Unisexual–Withonly one sex (male or female) of flowers on a single plant.

Valvate–Applied to bud scales that meet along a definite, usually longitudinal, line without overlapping; the reverse of imbricate.

Variant–A variation between strains; a plant that is genetically different from the wild plant.

Variedad–Spanish word for variety; a true cultivar.

Variegated–Striped, margined or mottled with colors other than green, where green is normal; having markings or zones of tissue lacking green chlorophyll.

Variety–One of possibly many closely-related plant species. A naturally occurring variation of a wild species.

Vascular bundle–A discrete group of conducting vessels.

Váscular bundle scar–A minute spot within the leaf scar where the vessels were positioned.

Vascular–The vessels that conduct water or nutrients in plants.

Vegetative–The leafy or green portions of a plant; referring to non-reproductive structures or growth

Vein–1. A vascular tube in a leaf, typically referring to any within the blade except for the midrib

2. The rib or branch of vessels in a plant leaf or other flat part.

Venation–Arrangement or pattern of veins or vessels and their relative association.

Venulose–With very fine hairs.

Verdant–Characterized by abundance of verdure, green in tint or color.

Verdure–Lush greenness of flourishing vegetation.

Vernacular name–A name in a common local language.

Vernal–Related to spring.

Vernation–The pattern of leaves within a bud.

Verrucose–Having a wart-like surface.

Verticillate–Arranged in whorls or arranged in a radiating patterns.

Vesicle–A small bladdery sac or cavity filled with air or fluid.

Vestige–The remains of an exhausted or dead plant structure.

Viable/viability–Seed or spores capable of germination with suitable treatment.

Villous–Having long, soft, shaggy hairs that are not matted.

Vine–A slender stemmed plant that trails and climbs by means of attaching parts and stems.

Virescent–Becoming green.

Virginiana/virginianum/virginianus–From the Southeastern USA.

Virus–Small, infectious that conreplicate only in a living host celeb *i.e.*, deforming microorganism that causes disease and death.

Virus indexing–A procedure to test plant stock to determine if it has a virus.

Viscid–Sticky or with appreciable viscosity.

Vivipary/viviparous–With live plantlets on a mother plant.

Voluable–Readily flowing, twining.

Wardian case/Wardian–An ornate 19th century terrarium used for ferns and rare plants.

Warm-season grass–Grass that grows and "greens up" in summer, sets seed in fall, and goes dormant (loses its green color and stops growing) during winter and spring. According to Easy Lawns, most warm-season grasses are ***bunchgrasses*** and will not tolerate any shade.

Water sprout–A vigorous sucker from the base of a woody plant.

Waterlogged–Said of soil that is oversaturated with water.

Waxy–Having a thick cuticle

Weeping–Dropping conspicuously, pendent.

Whip–A long scion used for tongue and groove grafting.

Whorled–With parts arranged in whorls, as in a leaf arrangement with three or more leaves arising from a node

Woolly–Having long, soft, more or less matted or tangled hairs; like wool.

X–Indicates a hybrid.

Xeriphyte–A plant naturally adapted to drought or low moisture conditions.

Chapter 3
Manures and Fertilizer Terms

Acid forming fertilizers–Fertilizers having acidic effect on soil are known as acid forming fertilizers. Examples: Urea, Ammonium sulphate, etc.

Acidification–The reduction in pH of the soil brought about by removal of cations (mainly Ca^{2+} and Mg^{2+}) by leaching or removal in the crop.

Additive–Material incorporated in a fertilizer:

(*a*) to improve its physical performance, *e.g.* in storage or spreading characteristics,

(*b*) to provide a minor nutrient, *e.g.* boron,

(*c*) to provide a non-nutrient biological action, *e.g.* pesticides or growth regulators.

The material is technically called as filler or conditioner.

Amendment–*Search* Mineral amendment.

Ammonification–The process in which ammonium compounds are formed by decomposition of nitrogenous organic compounds is called ammonification.

Ammonia volatilization–The process of loss of ammonia in vapour form is known as ammonia volatilization. Usually it occurs in soils with low cation exchange capacity and high pH associated with high concentration of ammonia and high temperature.

Ammonium citrate–Reagent used to determine the P_2O_5 content of various fertilizers and to provide an assessment of its availability to plants. Three principal variants of the ammonium citrate method are used, depending on relevant legislation–(*a*) neutral ammonium citrate solution, (*b*) alkaline ammonium citrate (Peterman), (*c*) alkaline ammonium citrate (Joulie).

Analysis–*Search* Content, formula.

Anion–A negatively charged atom or group of atoms, such as chloride (Cl^-), nitrate (NO_3^-) or sulphate (SO_4^{2-}).

Annual application–A system in which the annual fertilizer requirements of the crop are given each year (rather than at longer intervals).

Anticaking agent–Substance incorporated in or applied as surface treatment to granular fertilizer to prevent caking.

Application–General term for putting fertilizer on fields, crops, grassland, forests etc.

Application rate–Weight of fertilizer (for liquids often volume) applied per unit area.

Approved fertilizer–Legal definition in the classification of fertilizers which defines the type and content of approved fertilizers.

Autumn fertilization–Application of fertilizers before the winter sowing, frequently as basic PK fertilization, also as N stater fertilization.

Availability–1. The extent to which the nutrients in a fertilizer can be taken up by crops, often measured approximately by solubility in water or dilute acids; also used of soil nutrients to describe the fraction that can be taken up by plants. *Search* Solubility.

2. The amount of nutrient in a fertilizer that is soluble in water or a specified solution. By extension, some solubilities are used as a measure of nutrient availability to plants.

Balanced fertilization–The application of nutrients in proportions best suited to the needs of the crop taking account of the amounts of each supplied by the soil.

Banded over seed–Placement in a band over the seed row.

Band placement–Placement in bands parallel to the lines of seeds or plants, often carried out at the same time as sowing the crop.

Basal dressing (base dressing)–Fertilizer (usually P and K) applied before establishment of a crop or orchard to provide for its needs over a season or period of years.

Basic slag–A by-product of a steel industry obtained from phosphate iron ore which contains about 6-18 per cent acid soluble phosphoric acid.

Big bags–Closed flexible packages, holding from 250 to 1500 kg of material, used for storage, transport and application of fertilizers.

Binary fertilizer–Any fertilizer supplying two major nutrients(N, P and K) to the plants is known as binary fertilizer. Example–Ammonium phosphate, Diammonium phosphate, Nitrophosphate etc.

Bioavailable–Available to living organisms.

Biodiversity–Diversity of animal and crop species, between species in different ecosystems.

Blended fertilizer (mixed fertilizer–US)–Compound fertilizer made by dry mixing of several fertilizer materials of similar granulometry, with no chemical reaction.

Bone meal–It is an organic phosphatic fertilizer made up of powdered bones of animals, either raw or steamed. The raw bone meal contains small quantity of nitrogen(about 2 per cent) but the steamed one has only phosphate.

Broadcasting–Application of fertilizer as uniformly as possible to the surface of a field (including grassland, forests, etc).

Buffer–Buffer is a mixture of two solutions that is the solution of weak acid and its salts formed by a strong base or a weak base and its salts formed by a strong acid *e.g.*, a mixed solution of acetic acid and sodium acetate. A buffer has a definite pH which remains more or less constant for a long time.

Bulk blend–A mixed fertilizer prepared without a chemical reaction from components that retain their physical form. Usually prepared shortly before use and supplied to farm in bulk.

Bulk fertilizer–Fertilizer in a stockpile or transported loose, without any form of packaging or closed container.

Bulky organic manures–The manures having small quantity of plant nutrients but more organic matter as bulk.

Caking–The agglomeration of fertilizer particles by adhesion at their points of contact to form a compact mass that is difficult to break up.

Cation–Positively charged atom or group of atoms; *e.g.* potassium (K^+), ammonium (NH_4^+), calcium(Ca^{+2}), magnesium(Mg^{+2}), hydrogen(H^+), etc.

Cation exchange capacity, CEC–The ability of soils to hold cations electrostatically.

Chelates–These are certain micro-organic complexes which are soluble,readily absorbed by plants and never converted to insoluble forms when applied into the soil. The metallic ions, commercially chelated are Fe, Cu, Zn and Mn. Examples-EDTA (Ethylene diamine tetra acetic acid),EDDHA (Ethylene diamine dihydroxyphenoxy acetic acid), etc.

Chelated fertilizer–Fertilizer in which one or more trace elements are held by complex organic molecules (chelating agents) and are gradually released, thus prolonging the period of plant availability.

Chemical fertilizer–*Search* Mineral fertilizer.

Chemical fixation of nutrients–Due to the soil reaction, the available plant nutrients get fixed and/or reversed into unavailable forms. This is known as chemical fixation of nutrients. In sodic or calcareous soils exchangeable sodium replaces Ca from colloid which reacts with monocalcium phosphate and converts the phosphorus into tri and tetra calcium phosphate, the unavailable forms. Similarly, in acid soils, high availability of Al, Fe and Zn turns the soil toxic.

Chemigation–Application of fertilizers and/or pesticides in irrigation water to fertilize crops and control pests.

Coated fertilizer–A fertilizer in which the particles are covered by a layer of a different substance; thin layers are used to prevent moisture pick-up or caking and thicker layers or less permeable materials to delay or prolong the release of nutrient.

Colloid–Colloids are organic or inorganic matter with very small particle size and a correspondingly large surface area per unit of mass.

Compacted fertilizer–Fertilizer of flattish and angular form made by fragmentation of thin sheets of material agglomerated by high pressure or by the fragmentation of thin sheets of dried material on a revolving drum.

Complete fertilizer–Fertilizers which contain all three essential elements namely, N, P and K.

Complex fertilizer–Fertilizers containing two of the primary elements (known as *incomplete complex fertilizers*) or more (known as *complete complex fertilizers*), then they are said to be as complex fertilizers.

Compost–It is a mass of rotten organic matter made from farm wastes.

Composting–The process of making compost is known as composting.

Concentrated organic manures–The manures having more plant nutrients and less organic matter. Examples–Oil cake, Bone meal etc.

Conservation tillage–It is the minimal mechanical soil cultivation with at least 30 per cent of the field surface covered by crop residues. The term is also known as *no-till* and *reduced-till*.

Content (analysis, formula)–Plant nutrient concentration of a fertilizer, usually expressed as a percentage.

Corrective application–Fertilizer applied to increase the content in the soil of a plant nutrient in which the soil is deficient.

Critical concentration–Concentration in the soil of a plant nutrient below which yield loss might be expected.

Crushing strength–The ability of a granular or other particulate product to resist breakdown by physical pressure.

Crystalline fertilizer–Crystals clearly visible, for example ammonium sulphate. Often called a powder fertilizer.

Decalcification–Elimination of free calcium carbonate and Ca^{2+} ions from the soil as part of the process of acidification.

Declaration–The analytical information shown on label and/or invoice to comply with relevant legislation.

Deep placement–Placement of fertilizer at some depth, so that the full depth of the root zone is enriched with nutrients.

Deficiency–An inadequate supply or non-availability of an essential nutrient, restricting plant growth. It is usual to distinguish true deficiency (due to an inadequate content of a nutrient in the soil) and induced deficiency (arising from a chemical interaction with other soil constituents).

Degradation–The process of breakdown of soil organic matter with evolution of energy, formation of humic and fulvic acids and conversion of plant nutrients to inorganic form.

Denitrification–Loss of nitrate from the soil by its conversion by anaerobic microorganisms to nitrogen oxides and gaseous nitrogen. It occurs in flooded, waterlogged or partly waterlogged soils where oxygen concentration falls below 0.1 ppm in the soil solution. Actually the anaerobic microorganisms use oxidized forms of nitrogen as electron acceptors in their respiration and use to reduce the nitrate form to N_2 or N_2O.

Distribution–The channels through which fertilizer passes from place of manufacture to place of use.

Dolomite–Dolomite is a naturally occurring material composed of calcium and magnesium carbonate. It is used both as a liming material and a fertilizer.

Dressing–An application method of fertilizer.

Drip line placement–Application in a band round the perimeter of spread of a bush or tree.

Dung–*Search* Farmyard manure.

Economic optimum–The most profitable rate of fertilizer application taking into account value of the crop and cost of fertilizer.

Equivalent acidity–This is the quantity of calcium carbonate required to neutralize the acid residue caused by application of acid forming fertilizers in the soil. The equivalent acidity of some important acid forming fertilizers are–Ammonium sulphate (110), Ammonium sulphate nitrate (95),Urea (80), Ammonium chloride (128), Ammonium phosphate (86), etc.

Equivalent basicity–The residual basicity due to the application of basic fertilizers is known as equivalent basicity. It is expressed in terms of calcium carbonate equivalent of basic residue left by a fertilizer material (in kg/100 kg of fertilizer salt). The equivalent basicity values of some fertilizers are–Sodium nitrate (29), Calcium nitrate (21), Calcium cyanamide (63), Dicalcium phosphate (25), etc.

Efficiency–*Search* Nutrient recovery.

Essential plant nutrients–The important nutrients without which plants fail to grow develop and complete their life cycle are said to be as essential plant nutrients.

Eutrophication–An increase in the rate of addition of nutrients to water bodies, leading to an excessive multiplication of algae and other undesirable aquatic plant species.

Farm yard manure (manure, dung)–Organic manure consisting essentially of faeces, urine and straw or other litter.

Fertilization–Any or all aspects of the use of fertilizers to improve crop growth and soil fertility.

Fertigation–Application of fertilizer dissolved in irrigation water.

Fertilizer–1. Solid, liquid or gaseous material containing one or more fertilizer nutrients more or less immediately available to plants and acting essentially on the chemical component of soil fertility is known as fertilizer.

2. Fertilizer is any chemically synthesized material capable of supplying nutrient to the plants when applied into the soil or sprayed over the foliage.

Fertilizer burn–Damage caused by applying commercial fertilizer directly to foliage or roots.

Fertilizer distributor–Equipment used to apply fertilizer in the desired way, including broadcasting, placement, deep application, etc.

Fertilizer grade–It is an indication of percentages of nitrogen, phosphorus and potash respectively in a fertilizer material.

Fertilizer legislation–The laws and regulations formulated by the Govt. of India (1957) to be enforced with a view to regulate the quality of fertilizers sold to the consumers. It ensures each bag to be labeled with fertilizer grade and the quantity of nutrients present in it.

Fertilizer nutrient–In most countries, N, P and K, but sometimes including other nutrients supplied in fertilizers, *e.g.* Mg, S, etc.

Fertilizer plan–The field-by-field and crop-by-crop fertilizer recommendations for a complete farm arrived at after consideration of all relevant crop, soil and climatic factors is said to be the fertilizer plan.

Fertilizer recommendation–Advice on specified type and amount of fertilizer to be applied taking account of soil analysis and/or other information on the crop and soil.

Fertilizer requirement–It is the amounts of plant nutrients required for the optimum growth and development of a crop during its growth period in addition to the amount supplied by the soil.

Fertilizer sample–Representative sample of fertilizer obtained by a prescribed procedure (a) for the purpose of quality control, (b) in the course of statutory checks on fertilizer composition.

Fertilizer spreader–*Search in* Fertilizer distributor.

Filler–Material, *e.g.* sand or limestone, not containing any nutrient which is recognized by relevant legislation or application of which is required by the crop, added to fertilizer to obtain a specified NPK content. These materials are basically neutral in reaction.

Fixation–*Search* Immobilization.

Fixed potassium–Potassium held by a solid which is neither water-soluble nor readily exchangeable.

Flotation applicator–A fertilizer applicator equipped with large, low-pressure tyres, which spread the weight of the vehicle over a larger soil surface area.

Fluid fertilizer–*Search* Liquid fertilizer.

Foliar application–Application (spraying) of nutrient solution dissolved in water to the foliage of a crop for stomatal and cuticular absorption with a view to get a quicker response.

Foliar fertilizer (foliar feed)–A product containing major, secondary and micro-nutrients, as required, designed for application to, and nutrient uptake by the foliage of the crop; usually applied at a low concentration.

Formula (analysis)–Nutrient content of a fertilizer expressed in the order N-P-K, and in terms elements or oxides of P and K according to relevant legislation.

Free acidity–The sum of acidic components present in an unneutralized state in a fertilizer.

Free flowing–Used to describe a fertilizer that flows easily.

Granular fertilizer–Fertilizer composed of spheroid particles, usually 1.5-5 mm in diameter, and formed by any of a number of granulation processes. This kind of fertilizer is superior in efficiency compared to the fine or powdered fertilizers due to ease in handling and less fixation in soil.

Granulometry–Term describing the size and shape attributes of individual granules and other particles or the particle size distribution within a product.

Greenhouse gas–A gas whose higher concentration in the atmosphere may change the average world temperature by absorbing solar radiation. The gases include carbon dioxide, methane and nitrous oxide.

Green manure–The biomass of a crop grown specifically for incorporation in the soil to improve fertility.

Green manuring–The process of turning green plants into the soil either by raising them in the same field(called as *green manuring in situ*) or plants grown elsewhere, cut at green stage, transported at manuring site and incorporated into soil(*green leaf manuring*) is known as green manuring.

Growth medium–Substance without significant chemical fertility having good capacity for water retention and root growth. May be used in combination with nutrient solutions or in admixture with soil.

Growth regulator–An organic compound applied to a crop to influence cell multiplication or cell growth.

Growth substance–*Search* Growth regulator.

Hardness–The ability of a granule or other fertilizer particle to resist deformation by physical pressure.

Heavy metal–Metallic element with a high specific weight, such as cadmium and lead, often toxic to mammals above critical levels.

Immobilization (fixation)–The conversion of plant nutrients to unavailable or sparingly available form in the soil, by physical adsorption or chemical reaction.

Impregnation–Mixing or spraying a small amount of a micronutrient, herbicide, fungicide or insecticide with or on a fertilizer. Normally carried out during bulk blending operations but also by using equipment that permits impregnation of the fertilizer during application.

Incorporation–Cultural operation by which fertilizer applied to the surface of the soil is mixed with the soil.

Inhibitor–A substance, usually synthetic, stopping or delaying the activity of specific groups of soil micro-organisms or enzymes produced by them, *e.g.* nitrification inhibitors, urease inhibitors.

Injection–Placement of fertilizer below the soil surface by injection; usually refers to application of anhydrous ammonia in this way to avoid gaseous loss.

Inorganic fertilizer–*Search* Mineral fertilizer.

Input–Any management operation or application with an economic cost designed to enhance crop production.

Integrated management–Integrated management takes all relevant factors into account and implements them as a whole on the farm. It includes Integrated Crop Management (ICM), Integrated Farm Management (IFM) and Integrated Nutrient Management (INM).

Intensive fertilization–The use of high rates of fertilizer supplying the nutrient needs of the crop when grown under optimum conditions.

Intermediate–Product of the chemical industry intended for use in a subsequent stage of fertilizer manufacture, *e.g.* anhydrous ammonia, ammonium phosphate, but often capable of direct use as a fertilizer.

Intermediate bulk containers–*Search* Big bags.

Inter-row placement–Placement in bands midway between the rows or between every other row.

Isotope–Atoms of the same element that have a different number of neutrons in their nuclei *e.g.* 15N, 32P. Used to label nutrients in order to trace their fate in soils and plants.

Kaolinite–It is an aluminosilicate mineral of 1-1 crystal lattice group.

Label–Printed material on or attached to the package identifying the product and providing any information required by law.

Labile–Used of soil nutrients to describe the fraction that is immediately available to plants, *e.g.* labile P as measured by various radioactive tracer techniques.

Late fertilization–Application of fertilizers during advanced growth of annual crops, mostly N for higher protein content in the grain.

Leaching–The loss of plant nutrients from soils in percolating water.

Lime–The materials rich in CaO [*viz.*, ground limestone, hydrated lime, burnt lime, etc.,], which are used to reclaim soil acidity are known as lime.

Lime requirement–The lime application rate needed to bring the soil to a specified pH.

Liming–The application of liming materials (usually Ca-based, but sometimes including Mg), usually to improve soil physical properties (pH, structure) but sometimes also improving chemical fertility.

Liming material–Material with a neutralizing action, usually calcium carbonate, hydrate lime [$Ca(OH)_2$] or quicklime (CaO), but including magnesium compounds.

Liquid feed–In horticulture, application of fertilizer in solution in water.

Liquid fertilizer–Fertilizer in solution and/or suspension form. *e.g.* Anhydrous ammonia, aqueous solution of nitrogen, etc.

Liquid manure–Crude solution/suspension arising from urine and litter juices or from a dung heap.

Localized application–*Search* Placement.

Lodging–The collapse of a crop into a recumbent or semi-recumbent position may be of nutritional, parasitic or climatic origin.

Low analysis fertilizers–These are the fertilizers which have lower quantities of nutrients in them; as for example, SSP (single super phosphate). Higher unit cost as well as the higher storage and handling costs make these fertilizers very much expensive.

Maintenance application–Fertilizer applied to maintain the chemical fertility of the soil.

Major nutrients (primary nutrients)–N, P and K; the three nutrients required by plants in large or moderate amounts and which it is the main purpose of fertilizers to supply.

Manure–Manures are the decomposed organic material, primarily the organic refuse from plants and animals.

Manuring–A general term for the use of fertilizer or organic manure.

Micronutrients (trace elements, oligo-elements)–B, Cl, Cu, Fe, Mn, Mo, Zn–An essential element required by plants in very small quantities.

Mineral amendment (Amendment)–Inorganic material applied to improve the physical state of the soil, *e.g.* gypsum.

Mineral fertilizer (chemical fertilizer, inorganic fertilizer)–Fertilizer containing one or more major nutrients in inorganic form or as urea or cyanamide (or rarely in the form of synthetic organic compounds).

Mineralization–Microbial breakdown of soil organic matter and crop residues releasing plant nutrients in available form.

Mixed fertilizer–*Search* Compound fertilizer.

Moisture content–Percentage of water contained by a fertilizer as measured by one of several accepted methods.

Municipal compost (town compost)–Organic fertilizer made from household wastes and/or sewage sludge, sometimes with the addition of inorganic nutrients.

Mycorrhizae–A class of fungi that associate with the roots of certain plant species, contributing to the solubilization and supply to the plant of phosphate and also to the availability of water and micronutrients

Neutralizing value–The quantified effect of a liming material in neutralizing the acidity of the soil, usually expressed in relation to the effect of $CaCO_3$.

Night soil–The solid and/or liquid human excrement used in the form of town compost, sewage, sludge, etc. for crop cultivation is known as night soil.

Nitrification–Bacterial conversion to the nitrate forms of ammonium nitrogen from fertilizer or formed during mineralization of organic matter.

Nonessential nutrient–Chemical element sometimes beneficial to, but not essential for, plant growth, *e.g.* Na, Si etc.

NPK compound (ternary compound)–Compound containing all three of the primary nutrients N, P and K.

NPK ratio–It is the relative proportions of NPK(per cent) in a fertilizer mixture, *e.g.* a fertilizer with analysis 12-12-18 has a 1-1-1.5 ratio. It is expressed in terms of N, P or P_2O_5 and K or K_2O depending on local legislation.

Nutrient–*Search* Plant nutrient.

Nutrient accounting–Calculation of the nutrient inputs (such as mineral fertilizers, manure, atmospheric deposition, biological nitrogen fixation and sedimentation) and nutrient outputs (in crops harvested, crop residues, leaching, gaseous losses and erosion). The results are used to prepare nutrient budgets and nutrient management plans.

Nutrient cycle–The series of transformations which a nutrient undergoes in the biosphere (soil, plant, animals, man, the environment) before returning to its original form.

Nutrient recovery (coefficient of utilization, efficiency)–Proportion of applied nutrient taken up by the crop, usually expressed as a percentage.

Nutrient removal–Amount of each nutrient removed from the field in the harvested crop.

Nutrient return–Quantity of fertilizer nutrients returned to the soil in crop residues, and sometimes also indirectly in the form of organic manure.

Nutrient uptake–Amount of a nutrient taken up by a crop during its growth period; usually refers to the maximum nutrient content.

Oilcakes–After extraction of oil from oilseeds, the remaining solid portion is known as oilcake. It may be of two types, edible which can be safely fed to the livestock and non-edible which are not fit for that. Oilcakes are added to the soil as concentrated organic manure to supply organic matter and all three major nutrients (N, P, K). However, mostly it implies nitrogen.

Oligo-elements–*Search* Micronutrients (trace elements).

Organic fertilizer–Material of animal or plant origin containing one or more fertilizer nutrients, usually not immediately available to plants (*e.g.* in the form of dried blood, guano, industrial wastes). Organic fertilizers usually act on their chemical component of soil fertility. *Search* Organic manure.

Organic manure (bulky organic manure)–A material usually composed mainly of vegetable matter (but often also containing animal faeces and urine) partially decomposed or capable of decomposition by soil micro-organisms. Organic manures (farmyard manure, composts etc.) can act on biological, physical and chemical components of soil fertility.

Organo-mineral fertilizer–Fertilizer containing nutrients in both organic and inorganic form.

Over-fertilization–Level of fertilization too high resulting in a yield depression.

Package–Any closed container used for the storage and transport of fertilizer, *e.g.* sack, drum.

Periodic application–A system of application in which the fertilizer requirements (usually P and K) for a number of years are given in a single dressing; or, the quantity so given.

pH–Notation based on hydrogen ion concentration in the soil solution which provides a measure of soil acidity or alkalinity (pH 4 to 5.5–very acidic; pH 5.5 to 6.5–acidic; pH 6.5 to 7.3–neutral ; pH above 7.3–alkaline).

Placement (localized application)–It is the method of application in which fertilizer is concentrated in specified parts of the soil which may be either on or below the surface; or, the location in which fertilizer is placed.

Plant food unit–*Search* Unit.

Plant nutrient (essential nutrient)–Chemical element essential for plant growth.

Plough sole placement of fertilizers–It is a method of placing fertilizers in a continuous band on the bottom of a furrow.

Powdered fertilizer–Fertilizer in the form of fine particles, formed by precipitation, crystallization or grinding of larger particles.

Precision farming–The modification of soil and crop management according to the different conditions found within the field. *Search* Variable rate application.

Pressurized solution–Solution of ammonia with or without other fertilizer, requiring storage under low or medium pressure.

Prilled fertilizer–Particulate fertilizer of nearly spherical form made by the solidification of free-falling droplets.

Primary nutrients–*Search* Major nutrients.

Profitability–Margin over cost incurred

Value/Cost ratio–The ratio of the value of additional crop to the cost of fertilizer applied.

Rate–*Search* Application rate.

Ratio–*Search* NPK ratio.

Raw material–Material extracted from mine or quarry to be processed into fertilizer (intermediates or final products), *e.g.* sylvinite, phosphate rock, sulphur, pyrites, naphtha etc.

Reorganization–Microbial activity utilizing available plant nutrients in the soil for the production of soil biomass; in some circumstances this may reduce nutrient levels critically, especially N.

Residual effect–The effect of a fertilizer or manure in continuing to provide plant nutrients usually when at least a year has passed since application.

Residual value–Quantification of residual plant nutrients derived from fertilizers and remaining in the soil after the crop has been harvested.

Retention zone–Soil zone where nutrients are concentrated following a fertilizer application.

Reversion–The chemical transformation of phosphate from water soluble to water insoluble form in fertilizers or in the soil.

Rock phosphate–The acid or citrate soluble phosphates present in a rock are known as rock phosphate.

Run-off–The lateral movement of excess water on the surface of the soil towards stream, rivers etc.

Salinization–Surface or near-surface accumulation of salts. Includes alkalization (mostly by sodium salts).

Secondary nutrient–Ca, Mg, S–nutrients other than N, P, K which are required in moderate or large amounts by plants. These are usually not applied directly but are supplied as by-product of commercial fertilizers used for NPK fertilization in the soil.

Segregation–The physical movement of fertilizer particles of different size, shape density within a mixture resulting in its composition becoming heterogeneous; segregation usually results from undue vibration or poor handling procedures before packaging.

Sewage sludge–Wet solid waste from sewage treatment works.

Side band placement–Placement in bands parallel to but separate from the line of Searchds or plants.

Side-dressing–Application at or alongside the base of the plants when these are grown in wide rows.

Slow release fertilizer–A fertilizer which releases nutrients over a period of time; slow release–gradual release over much or all of the growth period; controlled release–a delayed release of plant nutrients some time after application. (For example, sulphur coated urea, neem coated urea, IBDU etc.).

Sludge–The precipitate or solid deposit of sewage is known as sludge.

Slurry–1. Liquid effluent from livestock production units consisting of faeces and urine sometimes diluted with water.

2. U.S.A., Suspension fertilizer.

Soil analysis–Physical, chemical or biological assay of soil samples, usually in the laboratory, providing information on fertilizer requirement and its characteristics and, when repeated over a period of years, on changes in soil fertility.

Soil conditioner–Synthetic material having a beneficial effect on a physical aspect of soil fertility.

Soil fertility–The overall ability of a soil to support vigorous crop growth by ensuring adequate plant nutrients and suitable conditions for water uptake, and by providing favourable conditions for root growth and development. Fertility is a synthesis of chemical, physical and biological components and is also influenced by climatic and management factors.

Soil sample–Representative amount of soil taken from an area for assessment of soil fertility characteristics, usually chemical.

Soil solution–The liquid phase present in the soil, containing dissolved nutrients in plant available form.

Solubility–The amount of nutrient in a fertilizer that is soluble in water or a specified solution. By extension, some solubilities are used as a measure of nutrient availability to plants.

Solution fertilizer–Fertilizer dissolved in water to form a clear solution.

Sphericity–The degree to which the shape of a granule or other fertilizer particle approaches the form of a pure sphere.

Split application–A system of application in which fertilizer is applied at two or more growth stages.

Spot application–Application to individual plants or trees when these are widely spaced.

Spray application–Application of a fluid fertilizer to the soil or crop in the form of fine droplets.

Spreading–Usually the operation of broadcasting fertilizer on the soil surface, but sometimes refers to any fertilizer application operation.

Stabilized fertilizer–Fertilizers to which nitrification inhibitors or urease inhibitors (*Search* Inhibitors) have been added.

Starter application–Band placement of part of the annual fertilizer application near or with the Searchd to give rapid early growth.

Starter fertilizer–Fertilizer containing readily available plant nutrients in amounts suitable for rapid early growth of the crop and generally placed close to the Searchd.

Storage–The holding of a supply of fertilizer, in large or small quantities, until it is needed for application. The conditions of storage must meet safety requirements and guard against product deterioration.

Straight fertilizer–Fertilizer containing only one fertilizer nutrient (N, P or K).

Stubble fertilization–Application of fertilizers directly after grain harvest but before soil cultivation.

Super granules–Large (up to 11mm diameter) particles of fertilizer designed to reduce nutrient losses, usually with respect to urea applied to wetland rice.

Suspension fertilizer–Fertilizer partially dissolved in water and partially held in dispersed solid form, for example by the addition of a small quantity of clay.

Sustainable Development–Development that meets the needs of the present without compromising the ability of future generations to meet their own needs (Brundtland report, 1987).

Teast–Soils containing higher concentration of molybdenum are known as teast.

Tissue analysis–Analysis of plant tissues in order to estimate the crops' nutrient requirements.

Top-dressing–Application not followed by incorporation, usually to grassland or an established crop.

Town compost–*Search* Municipal compost.

Trace elements–*Search* Micronutrients.

Tramline–Parallel wheel tracks left unplanted at suitable spacing to facilitate mechanical operations in a standing crop.

Two nutrient fertilizer–*Search* Binary fertilizers.

Unit (plant food unit)–Expression of variable meaning referring according to local custom to (*a*) 1 kg of plant nutrient; used in stating fertilizer requirements, (*b*) 1 per cent of 1 cwt (1.12 lb = 0.507 kg) of plant nutrients; used decreasingly in stating fertilizer requirements, (*c*) 1 ton or 1 tonne of plant nutrient; used in fertilizer formulation and production planning, (*d*) 20 lb of plant nutrient or 1 per cent of a short ton of 2000 lb.

Urea condensate–The products of reaction between urea and aldehyde(s) that produce a slow release nitrogen fertilizer.*Search* Slow release fertilizer.

Urea formaldehyde–Slow release nitrogenous fertilizer produced by the reaction between urea and formaldehyde.*Search* Slow release fertilizer.

Value/Cost ratio (Profitability)–The ratio of the value of additional crop to the cost of fertilizer applied.

Variable-rate application–Variation of nutrient application rates as the applicator moves across the field, according to changes of the nutrient levels in the soil. It is also known as *site-specific application*.

Volatilization–It is the process of gaseous loss of a substance from the soil or from fertilizer or manure, *e.g.* as ammonia (ammonia volatilization) or as elemental nitrogen or oxides of nitrogen (*denitrification*).

Yield determinant–Any factor, natural or due to human intervention, influencing crop yield.

Chapter 4
Biotechnology and Genetic Engineering Terms

A–Adenine residue, in either DNA or RNA.

Ab–*Search in* antibody.

Abiotic stress–Abiotic stress is simply the effect of non-living factors which can harm living organisms. The examples of non-living factors are drought, extreme temperatures, pollutants, etc.

Abscisic acid(ABA)–It is a plant growth regulator which is involved in abscission, dormancy, stomatal opening/closure, and inhibition of seed germination. In some plant species ABA also affects the regulation of somatic cell embryogenesis. Its rate of synthesis dramatically increases when plant is under stress especially the water shortage.

Absciss; abscissa–The horizontal axis of a graph is known as absciss or abscissa. *The opposite word is* ordinate.

Absorb (L. *ab*, away + *sorbere*, to suck in)–It means *to suck up*, or *to take in*. In the cell, materials are taken in (absorbed) from a solution. *The opposite word is* adsorb.

Absorption–In general it is the process of absorbing/taking up of water and nutrients by assimilation or imbibition. The taking up by capillary, osmotic, chemical or solvent action, such as the taking up of a gas by a solid or liquid, or taking up of a liquid by a solid. *The opposite word is* adsorption.

In biology: the movement of a fluid or a dissolved substance across a cell membrane.

In plants: water and mineral salts are absorbed from the soil by roots.

In animals: solubulized food material is absorbed into the circulatory system through cells lining the alimentary canal.

Abzyme–*Search in* catalytic antibody.

Acaricide–It is a type of pesticide used for killing or controlling mites or ticks.

Accessory bud–It is a lateral bud occurring at the base of a terminal bud or at the side of an axillary bud.

Acclimatization–The adaptation of a living organism (plant, animal or micro-organism) to a changed environment that subjects it to physiological stress is known as acclimatization.It should not be confused with adaptation. *The opposite word is* acquired.

Acellular–Describing tissues or organisms that are not made up of separate cells but often have more than one nucleus. *The opposite word is* syncytium.

Acentric chromosome–The chromosome fragment which lacks a centromere.

Acetyl co-enzyme A acetyl CoA–A compound formed in the mitochondria when an acetyl group (CH_3CO-)–derived from breakdown of fats, proteins, or carbohydrates–combines with the thiol group (-SH) of co-enzyme A.

Acquired–Developed in response to the environment, not inherited, such as a character trait (acquired characteristic) resulting from environmental effect(s). *The opposite word is* acclimatization.

Acridine dyes–A class of positively charged polycyclic molecules that intercalate into DNA and induce frameshift mutations.

Acrocentric–A chromosome that has its centromere near the end is called acrocentric.

Acropetal–1. Developing or blooming in succession towards the apex, such as leaves or flowers developing acropetally.

2. Transport or movement of substances towards the apex, such as the movement of water through the plant.

The opposite tendency is termed basipetal.

Acrosome–An apical organelle in the head of a spermatozoan is known as acrosome.

Acrylamide gels–*Search in* polyacrylamide gels.

Actin–One of the two contractile proteins in muscle (the other being myosin). Actin is also found in the microfilaments that form part of the cytoskeleton of all cells.

Activated charcoal; activated carbon–Charcoal which has been treated to remove hydrocarbons and to increase its adsorptive properties. It acts by condensing and holding a gas or solute onto its surface; thus inhibitory substances in nutrient medium may be adsorbed to charcoal included in the medium. Rooting factors such as phenolamines present as contaminants in charcoal may stimulate growth *in vitro*. Its addition to rooting medium may stimulate root initiation in some plant species. Activated charcoal may differ in origin and in composition. *The opposite word is* charcoal; phenolic oxidation.

Active collection–The International Undertaking on Plant Genetic Resources (FAO, 1983) defined it as a collection which complements a base collection and is a collection from which seed samples are drawn for distribution, exchange and other purposes such as multiplication and evaluation.

Activator–1. A substance or physical agent that stimulates transcription of a specific gene or operon.

2. A compound that enables the active site of the enzyme to bind to the substrate by binding to an allosteric site on an enzyme. *Search in* gene expression.

Active site–1. A site on the surface of a catalyst at which activity occurs.

2. The site on the surface of an enzyme molecule that binds the substrate molecule.

Adaptation (L. *ad*, to + *aptare*, to fit)–Adjustment of a population to changed environment over generations, associated (at least in part) with genetic changes resulting from selection imposed by the changed environment. *Not* acclimatization, q.v.

Adaptation traits–The complex of traits related to reproduction and survival of the individual in a particular production environment is known as adaptation traits.They contribute to individual fitness. The traits are subjected to selection during the evolution of animal genetic resources. By definition, these traits are also important to the ability of the animal genetic resource to be sustained in the production environment. (FAO, 1999)

Adaptive radiation–The evolution of new forms, sub-species or species from one species of plant or animal in order to exploit new habitats or food sources.

Adaptor–1. A synthetic double-stranded oligonucleotide that has a blunt end, while the other end has a nucleotide extension that can base pair with a cohesive end created by cleavage of a DNA molecule with a specific type II restriction endonuclease. The blunt end of the adaptor can be ligated to the ends of a target DNA molecule and the construct can be cloned into a vector by using the cohesive ends of the adaptor.

2. A synthetic single-stranded oligonucleotide that, after self-hybridization, produces a molecule with cohesive ends and an internal restriction endonuclease site. When the adaptor is inserted into a cloning vector by means of the cohesive ends, the internal sequence provides a new restriction endonuclease site.

Addendum (pl: addenda)–In formulation of tissue culture media: an item or a constituent substance to be added.

Additive allelic effects–Effects of alleles at a locus, where the heterozygote is exactly intermediate between the two homozygotes.

Additive gene effects–Additive allelic effects summed across all the loci that contribute to genetic variation in a quantitative trait.

Adenilate cyclase–The enzyme that catalyses the formation of cyclic AMP.

Adenine ($C_5H_5N_5$; f.w. 135.14) (symbol: A)–Adenine is a white crystalline purine base. It is the constituent of DNA and RNA and nucleotides such as ADP and ATP. A B-group vitamin (B_4) generally available as $C_5H_5N_5.3H_2O$, m.w. 189.13. It is added to some tissue culture media, as adenine sulphate, to promote shoot formation and for its weak cytokinin effect. It is present in plant tissues combined with aminoamide, phosphoric acids and D-ribose.

Adenosine disphosphate (ADP)–*Search in* ATP.

Adenosine monophosphate (AMP)–*Search in* ATP.

Adenosine triphospate (ATP)–*Search in* ATP.

Adenovirus–A group of DNA viruses which cause diseases in animals. In man, they produce acute respiratory tract infections with symptoms resembling common cold. They are used in gene cloning, as vectors for expressing large amounts of recombinant proteins in animal cells. They are also used to make live-virus vaccines against more dangerous pathogens. *Search in* viral vaccines.

ADEPT (antibody-directed enzyme pro-drug therapy)–A way to target a drug to a specific tissue. The drug is administered as an inactive pro-drug, and then converted into an active drug by an enzyme administered with a second injection. The enzyme is coupled to an antibody that concentrates it in the target tissue. When the enzyme arrives at the target tissue, the pro-drug is activated to form the active drug, while elsewhere it remains inactive. *Search in* drug delivery; targeted drug delivery.

Adhesion (L. *adhaerere*, to stick to)–Adhesion is basically the attraction of dissimilar molecules for each other. It is a sticking together of unlike substances, such as soil and water.

A-DNA–A right-handed DNA double helix that has 11 base pairs per turn. DNA exists in this form when partially dehydrated.

Adsorb–*Search in* adsorption.

Adsorbent (noun)–A substance to which compounds adhere. In tissue culture, an adsorbent is added to the culture medium to adsorb compounds released by cultured cells or tissues, thus minimizing any adverse effect on the subsequent growth in culture. A common adsorbent in tissue culture is *activated charcoal*.

Adsorption–Adsorption is the formation of a layer of gas, liquid or solid on the surface of a solid. *The opposite word is* absorption.

Adult cloning–The creation of identical copies of an adult animal by nuclear transfer from differentiated adult tissue is known as adult cloning. *Search in* cloning; Dolly.

Advanced–It is applied to an organism or a part thereof, implying considerable development from the ancestral stage or from the explant stage.

Adventitious (L. *adventitius*, not properly belonging to)–A structure arising at sites other than the usual ones, *e.g.*, shoots from roots or leaves, and embryos from any cell other than a zygote.

Aerate–To supply with or mix with air or gas. The process is aeration.

Aerobe–A micro-organism that grows in the presence of oxygen is called as aerobe. Opposite: anaerobe.

Aerobic–Active in the presence of free oxygen.

Aerobic bacteria–Bacteria that can live in the presence of oxygen.

Aerobic respiration–It is a type of respiration in which foodstuffs are completely oxidized to carbon dioxide and water, with the release of chemical energy, in a process requiring atmospheric oxygen.

Affinity chromatography–A method for separating molecules by exploiting their ability to bind specifically to other molecules. There are several types of biological affinity chromatography. A biological molecule can be immobilized and a smaller molecule (ligand) to which it is to bind can be stuck to it, or the smaller ligand can be immobilized and the macromolecule stuck to it. A variant is to use an antibody as the immobilized molecule and use it to "capture" its antigen: this is often called **immuno–affinity chromatography**. A variation is **pseudo–affinity chromatography**, in which a compound which is like a biological ligand is immobilized on a solid material, and enzymes or other proteins are bound to it. Other techniques include **metal affinity chromatography**, where a metal ion is immobilized on a solid support: metal ions bind tightly and specifically to many biomolecules. The metal ion is bound to a chelator or chelating group, a chemical group that binds specifically and extremely tightly to that metal.

Affinity tag; purification tag–An amino acid sequence that has been engineered into a protein to make its purification easier. These can work in a number of ways. The tag could be another protein, which binds to some other material very tightly and thus allowing the protein to be purified by affinity chromatography. The tag could be a short amino acid sequence, which is recognized by an antibody. The antibody would then bind to the protein whereas it would not have done so before. One such short peptide, called FLAG, has been designed so that it is particularly easy to make antibodies against it. The tag could be a few amino acids, which are then used as a chemical tag on the protein. For example, a string of positively charged amino acids will bind very strongly to a negatively charged filter: this could be used as the basis of a separation system. Some amino acids bind metals very strongly, especially in pairs: this chemical property can be exploited by using a filter with metal atoms chemically linked onto it to pull a protein out of a mixture of proteins. *The opposite word is* affinity chromatography.

Aflatoxin–Toxic compounds, produced by moulds (fungi) of the *Aspergillus flavus* group, that bind to DNA and prevent replication and transcription. Aflatoxins can cause acute liver damage and cancer. They are natural carcinogens. Animals may be poisoned by eating stored food or feed contaminated with the mould.

AFLP–*Search in* amplified fragment length polymorphism.

Ag–*Search in* antigen.

Agar (Malay, *agar-agar*)–A polysaccharide solidifying agent used in nutrient media preparations and obtained from certain types of red algae (*Rhodophyta*). Both the type of agar and its concentration can affect the growth and appearance of cultured explants.

Agarose–The main constituent of agar. Extracted from seaweed.

Agarose gel electrophoresis–A process in which a matrix composed of a highly purified form of agar is used to separate larger DNA and RNA molecules. *Search in* electrophoresis.

Aggregate–1. A clump or mass formed by gathering or collecting units.

2. A body of loosely associated cells, such as a friable callus or cell suspension.

3. Course inert material, such as gravel, that is mixed with soil to increase its porosity.

4. A serological reaction (aggregation) in which the antibody and antigen react and precipitate out of solution.

Agonist–A drug, hormone or transmitter substance that forms a complex with a receptor site that is capable of triggering an active response from a cell.

Agricultural biological diversity–*Search in* agrobiodiversity.

Agrobacterium–A genus of bacteria that includes several plant pathogenic species, causing tumour-like symptoms. *Search in Agrobacterium tumefaciens;* crown gall; hairy root culture; Ri plasmid; Ti plasmid.

Agrobacterium tumefaciens–A bacterium that causes crown gall disease in some plants. The bacterium infects a wound, and injects a short stretch of DNA into some of the cells around the wound. The DNA comes from a large plasmid–the Ti (tumour induction) plasmid–a short region of which (called T-DNA, = transferred DNA) is transferred to the plant cell, where it causes the cell to grow into a tumour-like structure. The T-DNA contains genes which *inter alia* allows the infected plant cells to make two unusual compounds, nopaline and octopine, that are characteristic of transformed cells. The cells form a gall, which hosts the bacterium. This DNA-transfer mechanism is exploited in the genetic engineering of plants. The Ti plasmid is modified so that a foreign gene is transferred into the plant cell along with or instead of the nopaline synthesis genes. When the bacterium is cultured with isolated plant cells or with wounded plant tissues, the "new" gene is injected into the cells and ends up integrated into the chromosomes of the plant.

***Agrobacterium tumefaciens*–mediated transformation**–A naturally occurring process of DNA transfer from the bacterium *A. tumefaciens* to plants.

Agrobiodiversity; agricultural biological diversity–That component of biodiversity that is relevant to food and agriculture production. The term agrobiodiversity encompasses within-species, species and ecosystem diversity. (FAO, 1999)

AI–*Search in* artificial insemination.

AIDS (acquired immunodeficiency syndrome)–AIDS is a fatal human disease in which the immune system is destroyed by a retrovirus (Human Immunodeficiency Virus, HIV). The virus infects and destroys helper T-cells resulting in diminished immune responsiveness and increased susceptibility to a variety of infections and cancers, which are essential for combating infections.

Airlift fermenter–A cylindrical fermentation vessel in which the cells are mixed by air introduced at the base of the vessel and that rises through the column of culture medium. The cell suspension circulates around the column as a consequence of the gradient of air bubbles in different parts of the reactor.

Albinism–Hereditary absence of pigment in an organism. Albino animals have no colour in their skin, hair and eyes. The term is also used for absence of chlorophyll in plants.

Albino–1. An organism lacking pigmentation, due to genetic factors. The condition is *albinism*.

2. A conspicuous plastome (plastid) mutant involving loss of chlorophyll.

Aleurone–The outermost layer of the endosperm in a seed.

Algal biomass–Single-celled plants, such as *Chlorella* spp. and *Spirulina* spp., are grown commercially in ponds to make feed materials. *Chlorella* is grown commercially to make into fish food, it is fed to zooplankton, and these in turn are harvested as feed for fish farms. This is a means of converting sunlight into food in a way more convenient and controllable than normal farming.

Alginate–Polysaccharide gelling agent.

Alkylating agents–Chemicals that transfer alkyl (methyl, ethyl, etc.) groups to the bases in DNA.

Allele (Gr. *allelon*, of one another, mutually each other); **allelomorph** (adj: allelic, allelomorphic)–One of a pair, or series, of variant forms of a gene that occur at a given locus in a chromosome. Alleles are symbolized with the same basic symbol (*e.g.*, *B* for dominant and *b* for recessive); $B_1, B_2, ..., B_n$ *for n* additive alleles at a locus). In a normal diploid cell there are two alleles of any one gene (one from each parent), which occupy the same relative position (locus) on homologous chromosomes. Within a population there may be more than two alleles of a gene. *Search in* multiple alleles.

Allele frequency–The number of copies of an allele in a population, expressed as a proportion of the total number of copies of all alleles at a locus in a population.

Allele-specific amplification (ASA)–The use of polymerase chain reaction (PCR) at a sufficiently high stringency that only a primer with exactly the same sequence as the target DNA will be amplified. A powerful means of genotyping for single-locus disorders that have been characterized at the molecular level.

Allelic exclusion–A phenomenon whereby only one functional allele of an antibody gene can be assembled in a given B lymphocyte. The "allele" on the other homologous chromosome in a diploid mammalian cell cannot undergo a functional re-arrangement, which would result in the production of two different antibodies by a single plasma cell.

Allelomorph–*Search in* allele.

Allelopathy–The phenomenon by which the secretion of chemicals, such as phenolic and terpenoid compounds, by a plant inhibits the growth or reproduction of other plant species with which it is competing.

Allergen–An antigen that provokes an immune response.

Allogamy–Cross fertilization in plants. *Search in* fertilization.

Allometric–When the growth rate of one part of an organism differs from that of another part or of the rest of the body.

Allopatric speciation–Speciation occurring at least in part because of geographic isolation.

Allopolyploid (Gr. *allos*, other, + polyploidy)–A polyploid organism (usually a plant) having multiple sets of chromosomes derived from different species. Hybrids are usually sterile, because they do not have sets of homologous chromosomes and therefore pairing cannot take place. However, if doubling of the chromosome number occurs in a hybrid derived from two diploid (2n) species, the resulting tetraploid (4n) is a fertile plant, since it contains two sets of homologous chromosome and pairing may occur; this tetraploid is an allotetraploid.

Allosteric control–*Search in* allosteric regulation.

Allosteric enzyme–An enzyme that has two structurally distinct forms, one of which is active and the other inactive. Active forms of allosteric enzymes tend to catalyse the initial step in a pathway leading to the synthesis of molecules. The end product of this synthesis can act as a feedback inhibitor, converting the enzyme to the inactive form, thus controlling the amount of product synthesized.

Allosteric regulation–A catalysis-regulating process in which the binding of a small effector molecule to one site on an enzyme affects the activity at another site.

Allosteric transition–A reversible interaction of a small molecule with a protein molecule, resulting in a change in the shape of the protein and a consequent alteration of the interaction of that protein with a third molecule.

Allotetraploid–An organism with four genomes derived from hybridization of different species. Usually, in forms that become established, two of the four genomes are from one species and two are from another species. *Search in* allopolyploid.

Allozygote–A diploid individual that is homozygous at a locus in which the two genes are not identical by descent from a common ancestor.

Allozyme–*Search in* allosteric enzyme.

Alphalactalbumin–Protein component of milk.

Alternative mRNA splicing–The inclusion or exclusion of different exons to form different mRNA transcripts. *Search in* RNA.

Alu sequences–A family of 300-bp sequences occurring nearly a million times in the human genome.

Ambient temperature–Air temperature at a given time and place; not radiant temperature.

Amino acid (Gr. *Ammon*, from the Egyptian sun god, in M. L. used in connection with ammonium salts)–An acid containing the group NH_2. In particular, any of 20 basic building blocks of proteins with a free amino (NH_2) and a free carboxyl (COOH) group, and having the basic formula NH_2–C_R–COOH. According to the side group R, they are subdivided into: polar or hydrophilic (serine, threonine, tyrosine, asparagine and glutamine); non-polar or hydrophobic (glycine, alanine, valine, leucine, isoleucine, proline, phenylalanine, tryptophan and cysteine); acidic (aspartic acid and glutamic acid) and basics (lysine, arginine, hystidine). The sequence of amino acids determines the shape, properties and the biological role of a protein. Plants and many micro-organisms can synthesize amino acids from simple inorganic compounds, but animals are unable to synthesize some of them, called essential amino acids, so they must be present in the diet.

Aminoacyl site; A-site–One of two sites on ribosomes to which the incoming aminoacyl tRNA binds.

Aminoacyl tRNA synthetase–Enzyme that attaches each amino acid to its specific tRNA molecule.

Amitosis–Cell division (cytokinesis), including nuclear division through constriction of the nucleus, without chromosome differentiation as in mitosis. The maintenance of genetic integrity and diploidy during amitosis is uncertain. This process occurs in the endosperm of flowering plants.

Amniocentesis–A procedure for obtaining amniotic fluid from a pregnant mammal for the diagnosis of some diseases in the unborn foetus. Cells are cultured, and metaphase chromosomes are examined for irregularities (*e.g.*, Down syndrome, spina bifida, etc., in humans).

Amnion–The thin membrane that lines the fluid-filled sac in which the embryo develops in higher vertebrates, reptiles and birds.

Amniotic fluid–Liquid contents of the amniotic sac of higher vertebrates, containing cells of the embryo (not of the mother). Both fluid and cells are used for diagnosis of genetic abnormalities in the embryo or foetus.

Amorph; null mutation–A mutation that obliterates gene function.

AMP (adenosine monophosphate)–*Search in* ATP.

Amphidiploid–A species or type of plant derived from doubling the chromosomes in the F_1 hybrid of two species; an allopolyploid. In an amphidiploid the two species are known, whereas in other allopolyploids they may not be known.

Amphimixis–True sexual reproduction involving the fusion of male and female gametes and the formation of a zygote.

Ampicillin (?-lactamase)–A penicillin-derived antibiotic that prevents bacterial growth by interfering with synthesis of the cell wall.

Amplification–1. Treatment (*e.g.*, use of chloramphenicol) designed to increase the proportion of plasmid DNA relative to that of bacterial (host) DNA.

2. Replication of a gene library in bulk.

3. Duplication of gene(s) within a chromosomal segment.

4. Creation of many copies of a segment of DNA by the polymerase chain reaction (PCR)

Amplified fragment length polymorphism (AFLP)–A type of DNA marker, generated by digestion of genomic DNA with two restriction enzymes to create many DNA fragments, ligation of specific sequences of DNA (called adaptors) to the ends of these fragments, amplification of the fragments via PCR (using a set of primers with sequences corresponding to the adapters, plus various random combinations of three additional bases at the end), and visualization of fragments via gel electrophoresis. The PCR will amplify any fragment whose sequence happens to start with any of the three-base sequences in the set of primers. AFLPs have the important advantage that many markers can be generated with relatively little effort. They are a very useful means of quantifying the extent of genetic diversity within and between populations. Their major disadvantage is that they are not specific to a particular locus and, because they are scored as the presence or absence of a band, heterozygotes cannot be distinguished from homozygotes, *i.e.*, they are inherited in a dominant fashion.

Amplify–To increase the number of copies of a DNA sequence, either *in vivo* by inserting into a cloning vector that replicates within a host cell, or *in vitro* by polymerase chain reaction (PCR).

Ampometric–*Search in* enzyme electrode.

Amylase–A group of enzymes that degrade starch, glycogen and other polysaccharides, producing a mixture of glucose and maltose.

Amylolytic–The capability of breaking down starch into sugars.

Amylopectin–A polysaccharide comprising highly branched chains of glucose molecules. The water-insoluble portion of starch.

Amylose–A polysaccharide consisting of linear chains of 100 to 1000 glucose molecules. The water-soluble portion of starch.

Anabolic pathway–A pathway by which a metabolite is synthesized; a biosynthetic pathway.

Anaerobe–An organism that can grow in the absence of oxygen. Opposite: aerobe.

Anaerobic–An environment or condition in which molecular oxygen is not available for chemical, physical or biological processes.

Anaerobic digestion–Digestion of materials in the absence of oxygen. *Search in* anaerobic respiration.

Anaerobic respiration–Respiration in which foodstuffs are partially oxidized, with the release of chemical energy, in a process not involving atmospheric oxygen, such as alcoholic fermentation, in which one of the end products is ethanol.

Analogous–Features of organisms or molecules that are superficially or functionally similar but have evolved in a different way or contain different compounds.

Anaphase (Gr. *ana*, up + *phais*, appearance)–The stage of mitosis or meiosis during which the daughter chromosomes (sister chromatids) pass from the equatorial plate to opposite poles of the cell (toward the ends of the spindle). Anaphase follows metaphase and precedes telophase.

Anchor gene–A gene that has been positioned on both the physical map and the linkage map of a chromosome.

Androgen–Any hormone that stimulates the development of male secondary sexual characteristics, and contributes to the control of sexual activity in vertebrate animals. Usually synthesized by the testes.

Androgenesis–Male parthenogenesis, *i.e.*, the development of a haploid embryo from a male nucleus. The maternal nucleus is eliminated or inactivated subsequent to fertilization of the ovum, and the haploid individual (referred to as androgenetic) contains in its cells the genome of the male gamete only. Androgenesis is detected by cytological staining. *Search in* anther culture; gynogenesis; parthenogenesis.

Aneuploid (Gr. *aneu*, without + ploid)–An organism or cell having a chromosome number that is not an exact multiple of the monoploid *(x)* with one chromosome being present in greater (*e.g.*, trisomic $2n + 1$) or lesser (*e.g.*, monosomic $2n–1$) number than the normal diploid number.

Animal cell immobilization–Entrapment of animal cells in some solid material in order to produce some natural product or genetically engineered protein. Animal cells have the advantage that they already produce many proteins of pharmacological interest, and that genetically engineered proteins are produced by them with the post-translation modifications normal to animals. However, because animal cells are much more fragile than bacterial ones, they cannot tolerate a commercial fermentation process. Typical materials are hollow fibre membrane bioreactors, or porous carriers made of polysaccharide, protein, plastic or ceramic materials with microscopic holes inside which the cells grow.

Animal cloning–*Search in* cloning.

Animal genetic resources databank–A databank that contains inventories of farm animal genetic resources and their immediate wild relatives, including any information that helps to characterize these resources. (FAO, 1999)

Animal genome (gene) bank–A planned and managed repository containing animal genetic resources. Repositories include the environment in which the genetic resource has developed, or is now normally found *(in situ)* or facilities elsewhere *(ex situ–in vivo* or *in vitro)*. For *in vitro, ex situ* genome bank facilities, germplasm is stored in the form of one or more of the following: semen, ova, embryos and tissue samples. (FAO, 1999)

Anion–A negatively charged ion; opposite: cation.

Anneal–The pairing of complementary DNA or RNA sequences, via hydrogen bonding, to form a double-stranded polynucleotide. Most often used to describe the binding of a short primer or probe.

Annealing–The process of heating (denaturing step) and slowly cooling (re-naturing step) double-stranded DNA to allow the formation of hybrid DNA or complementary strands of DNA or of DNA and RNA.

Annual (L. *annualis*, within a year)– 1. (adj:) Taking one year, or occurring at intervals of one year.

2. Noun: In botany, a plant that completes its life cycle within one year. During this time the plant germinates, grows, flowers, produces seeds, and dies. *Search in* biennial, perennial.

Anonymous DNA marker–A DNA marker detectable by virtue of variation in its sequence, irrespective of whether or not it actually occurs in or near a coding sequence. Microsatellites are typical anonymous DNA markers.

Antagonism–An interaction between two organisms (*e.g.*, moulds or bacteria) in which the growth of one is inhibited by the other. *The opposite word is* synergism.

Antagonist–A compound that inhibits the effect of an agonist in such a way that the combined biological effect of the two becomes smaller than the sum of their individual effects.

Anther culture–The aseptic culture of anthers for the production of haploid plants from microspores. *Search in* androgenesis; gynogenesis; parthenogenesis.

Anther–Microsporangium bearing microspores which develop into pollen (microgametophytes). The upper part of a stamen, containing pollen sacs within which are numerous pollen grains.

Anthesis–The flowering period or efflorescence. Anthesis is the time of full bloom, which lasts till fruit set.

Anthocyanin–Water-soluble blue, purple and red flavonoid pigments found in vacuoles of cells.

Antiauxin–A chemical that interferes with the auxin response. Antiauxin may or may not involve prevention of auxin transport or movement in plants. Some antiauxins are said to promote morphogenesis *in vitro*, such as 2,3,5-tri-iodobenzoate (TIBA; M.W. 499.81), or 2,4,5-trichloro-phenoxyacetate (2,4,5-T; m.w. 255.49), which stimulate the growth of some cultures.

Antibiosis–The prevention of growth or development of an organism by a substance or another organism.

Antibiotic–A class of natural and synthetic compounds that inhibit the growth of or kill some micro-organisms. Antibiotics such as penicillin are often used to control (to some extent kill) contaminating organisms. However, resistance to particular antibiotics can be acquired through mutations. Some contaminating organisms are only suppressed or their metabolism slowed to an insignificant level. *Search in* antibiotic resistance; bactericide; bacteriostat.

Antibiotic resistance–The ability of a micro-organism to produce a protein that disables an antibiotic or prevents transport of the antibiotic into the cell.

Antibody (Gr. *anti*, against + body)–An immunological protein (called an immunoglobulin, Ig) produced by certain white blood cells (lymphocytes) of the immune system of an organism in response to a contact with a foreign substance (antigen). Such an immunological protein has the ability of specifically binding with the foreign substance and rendering it harmless. The basic immunoglobulin molecule consists of two identical heavy and two identical light chains. *Search in* monoclonal antibodies; polyclonal antibodies.

Antibody class–The class to which an antibody belongs, depending on the type of heavy chain present. In mammals, there are five classes of antibodies: IgA, IgD, IgE, IgG, and IgM.

Antibody structure–Antibodies have a well defined structure. Each antibody has two identical "light" chains and two identical "heavy" chains. Each chain comprises a constant region, *i.e.*, a region that is the same between antibodies of the same class and sub-class, and a variable region that differs between. The antigen-binding region or binding site–*complementarity determining region*–is in the variable region. The antibody can be cut by proteases into several fragments, known as Fab, Fab', and Fc.

Antibody-mediated (humoral) immune response–The synthesis of antibodies by B cells in response to an encounter of the cells of the immune system with a foreign immunogen.

Anticlinal–The plane of cell wall orientation or cell division perpendicular (at right angles) to the surface of an organ. *Search in* tunica; periclinal.

Anticoding strand–The strand of the DNA double helix that is actually transcribed. Also known as the antisense or template strand.

Anticodon–A triplet of nucleotides in a tRNA molecule that pairs with a complementary triplet of nucleotides, or codon, in an mRNA molecule during translation. *Search in* codon; mRNA; RNA; tRNA.

Antigen; immunogen–A compound that elicits an immune response by stimulating the production of antibodies. The antigen, usually a protein, when introduced into a vertebrate organism is bound by the antibody or a T cell receptor. *Search in* antigenic determinant; antigenic switching.

Antigenic determinant–A surface feature of a micro-organism or macromolecule, such as a glycoprotein, that elicits an immune response. *Search in* epitope.

Antigenic switching–The altering of a micro-organism's surface antigens through genetic re-arrangement, to elude detection by the host's immune system.

Antihaemophilic globulin–Blood globulin that reduces the clotting time of haemophilic blood.

Anti-idiotype antibodies–Antibodies which recognize the binding sites of other antibodies. Their binding sites are complementary to the binding sites of another immunoglobulin. When an animal becomes immune to something, it not only acquires antibodies against that something, it also acquires antibodies against those antibodies. This forms a network of antibodies which can all bind to each other to various degrees, helping to regulate the immune response. Some allergic responses are in part due to the breakdown of this sort of regulation.

Antimicrobial agent–Any chemical or biological agent that harms the growth of micro-organisms.

Antinutrients–Compounds that inhibit normal uptake of nutrients.

Anti-oncogene–A gene whose product prevents the normal growth of tissue. *The opposite word is* recessive oncogene.

Antioxidant solution–Pre-treatment solution (*e.g.*, Vitamin C; citric acid) that retards senescence and browning of tissue. It is employed to incubate explants prior to surface sterilization.

Antioxidant–Compound that slows the rate of oxidation reactions.

Antiparallel orientation–The normal arrangement of the two strands of a DNA molecule, and of other nucleic-acid duplexes (DNA-RNA, RNA-RNA), in which the two strands are oriented in opposite directions so that the 5′-phosphate end of one strand is aligned with the 3′-hydroxyl end of the complementary strand.

Antisense DNA–1. The strand of chromosomal DNA that is transcribed.

2. A DNA sequence that is complementary to all or part of an mRNA molecule.

Antisense gene–A gene that produces a transcript (mRNA) that is complementary to the pre-mRNA or mRNA of a normal gene (usually constructed by inverting the coding region relative to the promoter).

Antisense RNA–An RNA sequence that is complementary to all or part of a functional mRNA molecule, to which it binds, blocking its translation. *Search in* RNA.

Antisense therapy–The *in vivo* treatment of a genetic disease by blocking translation of a protein with a DNA or an RNA sequence that is complementary to a specific mRNA.

Antiseptic–Any substance that kills or inhibits the growth of disease-causing micro-organism (a micro-organism capable of causing sepsis), but is essentially non-toxic to cells of the body.

Antiserum–The fluid portion of the blood of an animal (after coagulation of the blood), containing antibodies

Anti-terminator–A type of protein which enables RNA polymerase to ignore certain transcriptional stop or termination signals and read through them to produce longer mRNA transcripts.

Antitranspirant–A compound designed to reduce transpiration when sprayed or painted on leaves of newly transplanted trees, shrubs or vines, or used as a dip for cuttings in lieu of misting; may interfere with photosynthesis and respiration if the coating is too thick or unbroken.

Apex (L. *apex*, a tip, point, or extremity; pl: apices)–The tip, point or angular summit. The tip of a leaf; the portion of a root or shoot containing apical and primary meristems. Usually used to designate the apical tip of the meristem.

Apical cell–A meristematic initial in the apical meristem of shoots or roots of plants. As this cell divides, new tissues are formed.

Apical dominance–The phenomenon of inhibition of growth of lateral (axillary) buds in a plant by the presence of the terminal (apical) bud on the branch, due to auxins produced by the apical bud.

Apical meristem–A region of the tip of each shoot and root of a plant in which cell division is continually occurring to produce new stem and root tissue, respectively. Two regions are visible in the apical meristem: (i) An outer 1-4-cells layered region (called the tunica), where cell divisions are anticlinal, *i.e.*, perpendicular to the surface; and below the tunica, (ii) the corpus, where the cells divide in all directions, giving them an increase in volume.

Apoenzyme–Inactive enzyme that has to be associated with a specific organic molecule called a co-enzyme in order to function. The apoenzyme/co-enzyme complex is called a holoenzyme.

Apomixis (Gr. *apo*, away from + *mixis*, a mingling; adj: apomictic)–The asexual production of diploid offspring without the fusion of gametes. The embryo develops by mitotic division of the maternal or paternal gamete, or, in the case of plants, by mitotic division of a diploid cell of the ovule.

The opposite word is androgenesis; gynogenesis; panmixis; parthenogenesis.

Apoptosis–The process of cell death, which occurs naturally as a part of normal development, maintenance and renewal of tissue in an organism. Apoptosis differs from necrosis, in which cell death is caused by a toxic substance.

Aquaculture–Aquaculture is the growing of water plants and animals, rather than harvesting them from wherever they happen to grow in rivers or seas. Usually aquaculture uses fresh water; when it uses sea water it can be called mariculture. It is considered to be a part of biotechnology because it is a new commercial development often involving growing organisms in large volumes of water, which has similarities to growing large volumes of yeast or bacteria. Biotechnology also provides clean, well-aerated water for the animals to grow in; food, such as krill or powdered synthetic food; and food additives, such as astaxanthins to ensure that fishes and prawns have the right colour. Aquaculture has also been used to mass-produce macro–and micro-algae for chemicals, vitamins and pigments. For both animals and plants, biotechnologists have been using genetic methods to produce triploid and tetraploid organisms, and hybrid algae through plant cell fusions. Triploid trout, for example, are sterile, and can be used for biocontrol of weeds without the threat of their being able to breed themselves.

Arabidopsis–A genus of flowering plants in the Cruciferae. *A. thaliana* is used in research as a model plant because it has a small genome (5 pairs of chromosomes; 2n = 10) and can be cultured easily, with a generation time of two months.

ARS (autonomous replicating sequence)–Any eukaryotic DNA sequence that initiates and supports chromosomal replication; they have been isolated in yeast cells. Also called autonomous(ly) replicating segment.

Artificial inembryonation–Non-surgical transfer of embryo(s) to a recipient female. As *in vitro* embryo technology develops, artificial inembryonation will gradually replace artificial insemination.

Artificial insemination–The deposition of semen, using a syringe, at the mouth of the uterus to make conception possible. It is used in the breeding of domestic animals.

Artificial medium–*Search in* culture medium.

Artificial seed–Encapsulated or coated somatic embryos (embryoids) that are planted and treated like seed.

Artificial selection–The practice of choosing individuals from a population for reproduction, usually because these individuals possess one or more desirable traits.

ASA–*Search in* allele-specific amplification.

Ascorbic acid; vitamin C ($C_6H_8O_6$; f.w. 176.12)–A water-soluble vitamin present naturally in some plants, and also synthetically produced. Aside from its role as a vitamin, it is used as an antioxidant in plant tissue culture; and included in disinfection solutions.

Ascospore–One of the spores contained in the ascus of certain fungi.

Ascus (pl: asci)–Reproductive sac in the sexual stage of a type of fungi (*Ascomycetes*) in which ascospores are produced.

Aseptic–Asepsis or sterile. The state of being free of contaminating organisms (bacteria, fungi, algae and all micro-organisms except viruses) but not necessarily free of internal symbionts. *Search in* axenic.

Asexual (Gr. *a*, without + L. *sexualis*, sexual)–Any type of reproduction not involving meiosis or the union of gametes.

Asexual embryogenesis–The sequence of events whereby embryos develop from somatic cells. a.k.a. somatic cell embryogenesis.

Asexual propagation–Vegetative, somatic, non-sexual reproduction of a plant without fertilization. *The opposite word is* apomixis.

Asexual reproduction–Reproduction that does not involve the formation and union of gametes from the different sexes or mating types. It occurs mainly in lower animals, micro-organisms and plants. In plants, asexual reproduction is by vegetative propagation (*e.g.*, bulbs, tubers, corms) and by formation of spores.

A-site–*Search in* aminoacyl site.

Asn–*Search in* asparagine.

Asparagine (abbr: Asn; $C_4H_8N_2O_3$; f.w. 132.12)–One of the 20 essential amino acids. It is occasionally included in plant tissue culture media as a source of reduced nitrogen.

Aspartic acid (abbr: Asp; $C_4H_7NO_4$; f.w. 132.12)–An amino acid necessary for nucleotide synthesis and occasionally included in plant tissue culture media.

Assay–1. To test or evaluate.

2. The procedure for measuring the quantity of a given substance in a sample (chemically or by other means).

3. The substance to be analysed.

Assortative mating–Mating in which the partners are chosen on the basis of phenotypic similarity.

Assortment–*Search in* segregation.

Asynapsis–The failure or partial failure in the pairing of homologous chromosomes during the meiotic prophase.

ATP (adenosine triphosphate)–A nucleotide of fundamental importance as a carrier of chemical energy in all living organisms. It consists of adenosine with three phosphate groups, linked together linearly. The phosphates are attached to adenosine through its ribose (sugar) portion. Upon hydrolysis, bonds yield either one molecule of ADP (adenosine diphosphate) and an inorganic phosphate, or one molecule of AMP (adenosine monophosphate) and pyrophosphate; in both cases releasing energy that is used to power biological processes. ATP is regenerated by rephosphorilation of AMP and ADP, using chemical energy derived from the oxidation of food.

ATP-ase–An enzyme that brings about the hydrolysis of ATP, by the cleavage of either one phosphate group with the formation of ADP and inorganic phosphate, or of two phosphate groups, with the formation of AMP and pyrophosphate.

Attenuated vaccine–A virulent organism that has been modified to produce a less virulent form, but nevertheless retains the ability to elicit antibodies against the virulent form.

Attenuation–A mechanism for controlling gene expression in prokaryotes that involves premature termination of transcription.

Attenuator–A nucleotide sequence in the 5′ region of a prokaryotic gene (or in its RNA) that causes premature termination of transcription, possibly by forming a secondary structure.

Authentic protein–A recombinant protein that has all the properties–including any post-translational modifications–of its naturally occurring counterpart.

Autocatalysis–Catalysis in which one of the products of the reaction is a catalyst for the reaction. Usually the catalysis starts slowly and increases as the quantity of the catalyst increases, falling off as the product is used up.

Autocatalytic reaction–*Search in* autocatalysis.

Autoclave–1. An enclosed chamber in which substances are heated under pressure to sterilize utensils, liquids, glassware, etc., using steam. The routine method uses steam pressure of 103.4×10^3 Pa at 121°C for 15 minutes, or longer to allow large volumes to reach the critical temperature.

2. A pressure cooker used to sterilize growth medium and instruments for tissue culture work.

Auto-immune disease–Disorder in which the immune systems of affected individuals produce antibodies against molecules that are normally produced by those individuals (called self antigens).

Auto-immunity–A disorder in the body's defence mechanism in which an immune response is elicited against its own (self) tissues.

Autologous cells–Cells that are taken from an individual, cultured (or stored), and, possibly, genetically manipulated before being infused back into the original donor.

Autolysis–The process of self destruction of a cell, cell organelle, or tissue. It occurs by the action of lysosomic enzymes.

Autonomous(ly) replicating segment (or sequence)–*Search in* ARS.

Autonomous–A term applied to any biological unit that can function on its own, *i.e.*, without the help of another unit, such as a transposable element that encodes an enzyme for its own transposition.

Autopolyploid–A polyploid that has multiple and identical or nearly identical sets of chromosomes (genomes) all derived from the same species. A polyploid species with genomes derived from the same original species.

Autoradiograph–A picture prepared by labelling a substance such as DNA with a radioactive material such as tritiated thymidine and allowing the image produced by decay radiation to develop on a film over a period of time.

Autoradiography–A technique that captures the image formed in a photographic emulsion as a result of the emission of either light or radioactivity from a labelled component that is placed next to unexposed film. The technique is used for detecting the location of an isotope in a tissue, cell or molecule. The sample is placed in contact with a photographic emulsion, usually an X-ray film. The emission of particles from the sample activates the silver halide grains in the emulsion and allows them to reduce to metallic silver when the film is developed. In genetic engineering, autoradiography is most commonly used to detect the hybridization of a radioactive DNA (probe) molecule to denatured DNA in either the Southern transfer or colony hybridization procedures.

Autosome–A chromosome that is not involved in sex determination.

Autotrophic–Self-nourishing organisms capable of utilizing carbon dioxide or carbonates as the sole source of carbon and obtaining energy for life processes from radiant energy or from the oxidation of inorganic elements, or compounds such as iron, sulphur, hydrogen, ammonium and nitrites. *Search in* heterotrophic.

Autotrophy–Autotrophy is the capacity of an organism to use light as the sole energy source in the synthesis of organic material from inorganic elements or compounds. Autotrophic organisms include green photosynthesizing plants and some photosynthetic bacteria. *The opposite word is* heterotrophy.

Auxin (Gr. *auxein*, to increase)–A group of plant growth regulators (natural or synthetic) which stimulate cell division, enlargement, apical dominance, root initiation, and flowering. One naturally produced auxin is indole-acetic acid (IAA).

Auxin-cytokinin ratio–The relative proportion of auxin to cytokinin present in plant-tissue-culture media. Varying the relative amounts of these two hormone groups in tissue culture formulae affects the proportional growth of shoots and roots *in vitro*. As the ratio is increased (increased auxin or decreased cytokinin), roots are more likely to be produced, and as it is decreased root growth declines and shoot initiation and growth are promoted. This relationship was first recognized by C.O. Miller and F. Skoog in the 1950s.

Auxotroph (Gr. *auxein*, to increase + *trophe*, nourishment)–A mutant cell or micro-organism lacking the capacity to form an enzyme or metabolite present in the parental strain, and that consequently will not grow on a minimal medium, but requires the addition of some compound–such as an amino acid or a vitamin–for growth.

Availability–A reflection of the form and location of nutritional elements and their suitability for absorption. In tissue culture media this is related to the abundance of each nutritional element, the osmotic concentration and pH of the medium, the stability and solubility of the item in question, the presence of absorbing agents in the media, and other factors.

Axenic culture–Free of external contaminants and internal symbionts; generally not possible with surface sterilization alone, and incorrectly used to indicate *aseptic culture*.

Axillary bud–A bud found at the axil of a leaf (synonymous with lateral bud).

Axillary bud proliferation–Propagation in culture using protocols and media which promote axillary (lateral shoot) growth. This is a technique for mass production (micro-propagation) of plantlets in culture, achieved primarily through hormonal inhibition of apical dominance and stimulation of lateral branching.

BABS (biosynthetic antibody binding sites)–*Search in* dabs.

BAC (bacterial artificial chromosome)–A cloning vector constructed from bacterial fertility (F) factors; like YAC vectors, they accept large inserts of size 200 to 500 kb. *Search in* cloning vector; YAC.

Bacillus–A rod-shaped bacterium.

Bacillus thuringiensis (Bt)–A bacterium that kills insects; a major component of the microbial pesticide industry.

Back mutation–A second mutation at the same site in a gene as the original mutation. The second mutation restores the wild-type nucleotide sequence.

Backcross–Crossing an organism with one of its parents or with the genetically equivalent organism. The offspring of such a cross are referred to as the backcross generation or backcross progeny. *Search in* testcross.

Bacteria–Plural of bacterium.

Bacterial toxin–A toxin produced by a bacterium, such as Bt toxin by *Bacillus thuringiensis*.

Bacteriocide; bactericide–A chemical or drug that kills bacterial cells.

Bacteriocin–A protein produced by bacteria of one strain and active against those of a closely related strain.

Bacteriophage–A virus that infects bacteria. Also called simply phage. Altered forms are used in DNA cloning work, where they are convenient vectors. The bacteriophages most used are derived from two "wild" phages, called M13 and lambda. Lambda phages are used to clone segments of DNA in the range of around 10-20 kb. They are lytic phages, *i.e.*, they replicate by lysing their host cell and releasing more phages. On a bacteriological plate, this results in a small clear zone–a plaque. Some lambda vectors have also been developed which are expression vectors. The M13 system can grow inside a bacterium, so that it does not destroy the cell it infects but causes it to make new phages continuously. It is a single-stranded DNA phage, and is used for the Sanger di-deoxy DNA sequencing method (*Search in* DNA sequencing). Both of these phages grow on *Escherichia coli* as a host bacterium.

Bacteriostat–A substance that inhibits or slows down growth and reproduction of bacteria.

Bacterium (Gr. *bakterion*, a stick; pl: bacteria)–Common name for the class Schizomycetes: minute (0.5-5μm), unicellular organisms, without a distinct nucleus. Bacteria are prokaryotes, and most of them are identified by means of Gram staining. They are classified on the basis of their oxygen requirement (aerobic vs anaerobic) and shape (spherical = coccus; rodlike = bacillus; spiral = spirillum; comma-shaped = vibrio; corkscrew-shaped = spirochaete; filamentous). Bacteria usually reproduce asexually, by simple cell division, although a few undergo a form of sexual reproduction, termed *conjugation*. A few bacteria can photosynthesize (including green-blue cyano-bacteria), some are saprophytes and others are parasites and can cause diseases. They are major agents of fermentation, putrefaction and decay, and frequently a source of contamination in tissue culture. In plant pathology, strains of bacteria causing disease in specific plant cultivars are called pathovars.

Baculovirus–Baculoviruses are a class of insect virus which have been used to make DNA cloning vectors for gene expression in eukaryotic cells. Baculoviruses have a gene which is expressed at very high levels late in their infection cycle, filling the nucleus of the cell with many-sided bodies full of a protein which is not needed to produce more viruses, but is necessary for the virus's spread in the wild. In a vector cloning system, this gene is replaced by one that the biotechnologist wants expressed. Production of the protein can be up to 50 per cent of the cells' protein content, and several proteins can be made simultaneously, so that multi-sub-unit enzymes can be made by this system. Being an animal expression system, baculoviruses produce proteins that are glycosylated (addition of carbohydrates) like the proteins in animals, making it an attractive option for the production of biopharmaceuticals. In addition, baculoviruses are non-infective and non-pathogenic to vertebrates.

Balanced lethal system–A system for maintaining a recessive lethal allele at each of two loci on the same pair of chromosomes. In a closed population with no crossing-over between the loci, only the double heterozygotes for the lethal mutations survive.

Balanced polymorphism–Two or more types of individuals maintained in the same breeding population.

Bank–*Search in* gene bank.

Bar–A unit used for pressure of fluid. 1 bar = 10^5 Pa. 1 bar is approximately equivalent to 1 atmosphere.

Barr body–A condensed mass of chromatin found in the nuclei of female mammals. It is a late-replicating, inactive X-chromosome. Named after its discoverer, Murray Barr (1908).

Basal–1. Located at the base of a plant or a plant organ.

2. A fundamental formulation of a tissue culture medium.

Basal body–Small granule to which a cilium or flagellum is attached. *The opposite word is* kinetosome.

Base–A cyclic, nitrogen-containing compound that is one of the essential components of nucleic acids. Exists in five main forms (adenine, A; guanine, G; thymine, T; cytosine, C; uracil, U). A and G have a similar structure and are called purines; T, C and U have a similar structure and are called pyrimidines. A base joined to a ribose sugar joined to a phosphate group is a nucleotide–the building block of nucleic acids.

Base analogues–Unnatural purine or pyrimidine bases that differ slightly in structure from the normal bases, but can be incorporated into nucleic acids. They are often mutagenic.

Base collection–In PGR: Defined in the International Undertaking on Plant Genetic Resources (FAO, 1983) as a collection of seed stock or vegetative propagating material (ranging from tissue cultures to whole plants) held for long-term security in order to preserve the genetic variation for scientific purposes and as a basis for plant breeding as multiplication and evaluation. *The opposite word is* active collection.

Base pair (bp)–The two strands that constitute DNA are held together by specific hydrogen bonding between purines and pyrimidines (A pairs with T; and G pairs with C). The size of a nucleic acid molecule is often described in terms of the number of base pairs (symbol: bp) or thousand base pairs (kilobase pairs; symbol: kb; a more convenient unit) it contains.

Base substitution–Replacement of one base by another in a DNA molecule. *Search in* transition; transversion.

Basipetal–*Search in* acropetal.

Basophil–A type of white blood cell (leucocyte), produced by stem cells in the red bone marrow.

Batch culture–A suspension culture in which cells grow in a finite volume of liquid nutrient medium and follow a sigmoid pattern of growth. *The opposite word is* continuous culture; batch fermentation.

Batch fermentation–A process in which cells or micro-organisms are grown for a limited time. At the beginning of the fermentation, an inoculum is introduced into fresh medium, with no addition or removal of medium for the duration of the process.

B cells–An important class of white blood cells that mature in bone marrow and produce antibodies. They are largely responsible for the antibody-mediated or humoral immune response; they give rise to the antibody-producing plasma cells and some other cells of the immune system. *Search in* B lymphocytes.

β–DNA–The normal form of DNA found in biological systems. It exists as a right-handed helix.

Bench-scale process–A small- or laboratory-scale process; commonly used in connection with fermentation.

Beta-DNA–*Search in* β-DNA.

β-galactosidase–An enzyme that catalyses the formation of glucose and galactose from lactose

β-lactamase–1. An ampicillin resistance gene. *Search in* selectable marker.

2. *Search in* β-lactamase

Biennial (*L. biennium*, a period of two years)–In botany, a plant which completes its life cycle within two years and then dies. For most biennial plants, the two growing seasons have to be separated by a period of cold temperature sufficient to induce flowering and fruit formation.

Bifunctional vector–*Search in* shuttle vector.

Binary vector system–A two-plasmid system in *Agrobacterium tumefaciens* for transferring into plant cells a segment of T-DNA that carries cloned genes. One plasmid contains the virulence gene (responsible for transfer of the T-DNA), and another plasmid contains the T-DNA borders, the selectable marker and the DNA to be transferred. *Search in also* cDNA; carrier DNA; plasmid; vector.

Binding–The ability of molecules to stick to each other because of the exact shape and chemical nature of parts of their surfaces. Many biological molecules bind extremely tightly and specifically to other molecules: enzymes to their substrates; antibodies to their antigens; DNA strands to their complementary strands; and so on. Binding can be characterized by a *binding constant* or *association constant* (K_a), or its inverse, the *dissociation constant* (K_d).

Binomial nomenclature–In biology, each species is generally identified by two terms: the first is the genus to which it belongs, and the second is the specific epithet that distinguishes it from others in that genus (*e.g.*, *Quercus suber*, cork oak). The genus name always has an initial capital; the specific epithet is never capitalized, even though it may be derived from a proper name (*e.g.*, keranda nut, *Elaeocarpus bancroftii*). Both terms in the binomial are italicized. Based on the system of classification developed by Carolus Linnaeus.

Binomial expansion–The probability that an event will occur 0, 1, 2,..., n times out of n is given by the successive terms of the expression $(p + q)^n$, where p is the probability of the event occurring, and $q = 1–p$.

Bio–A prefix derived from *bios* and used in scientific words to associate the concept of "living organisms." Usually written with a hyphen before vowels, for emphasis or in neologisms; otherwise usually without a hyphen.

Bio-accumulation–In an organism, concentration of materials which are not components critical for that organism's survival. Usually it refers to the accumulation of metals or other compounds (*e.g.*, DDT, HCH etc.). Many organisms–plants, fungi, protists, bacteria, etc.– accumulate metals when grown in a solution of them, either as part of their defence mechanism against the poisonous effect of those compounds, or as a side-effect of the chemistry of their cell walls. Bio-accumulation is important as part of the microbial mining cycle, removing toxic metals from wastewater, as a purification (bio-remediation) process, etc. *Search in also* biosorption; microbial mining.

Bio-assay–A procedure for the assessment of a substance by measuring its effect in living cells or on organisms. Animals have been used extensively in drug research in bio-assays for the pharmacological activity of drugs. However, bio-assays are now usually developed using bacteria or animal or plant cells, as these are usually much easier to handle than whole animals or plants, are cheaper to make and keep, and avoid the ethical problems associated with testing of animals. Sometimes used to detect minute amounts of substances that influence or are essential to growth.

Bio-augmentation–Increasing the activity of bacteria that decompose pollutants; a technique used in bioremediation.

Biocatalysis; biocatalyst–Use of enzymes to catalyse chemical reactions. *Search in* biotransformation.

Biocontrol–The control of living organisms (especially pests) by biological means. Any process using deliberately introduced living organisms to restrain the growth and development of other, very often pathogenic, organisms, such as the use of spider mites to control cassava mealy bug, or the introduction of myxomatosis into Australia to control rabbits. The term also applies to use of disease-resistant crop cultivars. Biotechnology approaches biocontrol in various ways, such as using fungi, viruses or bacteria which are known to attack an insect or weed pest.

Bioconversion–Conversion of one chemical into another by living organisms, as opposed to their conversion by enzymes (which is biotransformation) or by chemical processes. The usefulness of bioconversion is much the same as that of biotransformation–in particular its extreme specificity and ability to work in moderate conditions. However, bioconversion has several other properties, including the possibility of having several chemical steps. A major commercial application is in the manufacture of steroids. The "basic" steroid molecule, often isolated from plants, is itself a very complicated molecule, and not one that is easy to modify by normal chemical means to produce the very specific molecules needed for drug use. However, a particular type of bioconversion that attacks only specific bits of the molecule can be used. Bioconversion is particularly useful for introducing chemical changes at specific points in large, complex molecules.

Biodegradable–*Search in* biodegradation.

Biodegradation–The breakdown by living organisms of a compound to its chemical constituents. Materials that can be easily biodegraded are colloquially termed *biodegradable.*

Biodiversity–1. The variety of species (species diversity) or other taxa of animals, micro-organisms and plants in a natural community or habitat, or of communities in a particular environment (ecological diversity), or of genetic variation in a species (genetic diversity, q.v.). The maintenance of a high level of biodiversity is important for the stability of ecosystems.

2. The variety of life in all its forms, levels and combinations, encompassing genetic diversity, species diversity and ecosystem diversity. *The opposite word is* agrobiodiversity (Source: FAO, 1999)

Bio-energetics–The study of the flow and the transformations of energy that occur in living organisms with the framework of principles and laws of thermodynamics.

Bio-engineering–The use of artificial tissues, organs and organ components to replace parts of the body that are damaged, lost or malfunctioning.

Bio-enrichment–Adding nutrients or oxygen to increase microbial breakdown of pollutants.

Bio-ethics–The branch of ethics that deals with the life sciences and their potential impact on society. At one extreme, it can be enormously useful in focusing attention on problems that need to be confronted; at the other extreme, it can become a name-calling argument between the "pro-biotechnology" and "anti-biotechnology" schools of thought, which, as it reduces discussion to epithets and clichés, can make better sound bites.

Biofuel–A gaseous, liquid or solid fuel that contains energy derived from a biological source. For example, rapeseed oil or fish liver oil can be used in place of diesel fuel in modified engines. A commercial application is the use of modified rapeseed oil, which–as rapeseed methyl ester (RME)–can be used in modified diesel engines, and is sometimes named bio-diesel. *The opposite word is* biogas.

Biogas–A mixture of methane and carbon dioxide resulting from the anaerobic decomposition of waste such as domestic, industrial and agricultural sewage. a.k.a. *gobar.*

Biogenesis–The principle that a living organism can only arise from other living organisms similar to itself and can never originate from non-living material. *i.e.* life orignates from pre-existing life

Bio-informatics–The use and organization of information of biological interest. In particular, it is concerned with organizing bio-molecular databases, in getting useful information out of such databases, in utilizing powerful computers for analysing such information, and in integrating information from disparate biological sources.

Biolistics (from biological + ballistics)–A technique to insert DNA into cells. The DNA is mixed with small metal particles–usually tungsten or gold–a fraction of a micrometre across. These are then fired into a cell at very high speed. They puncture the cell and carry the DNA into the cell. Biolistics has an advantage over transfection, transduction, etc., because it can apply to any cell, or indeed to parts of a cell. Thus use of biolistics has inserted DNA into animal, plant and fungal cells, and into mitochondria inside cells. a.k.a. microprojectile bombardment.

Biological ageing–*Search in* senescence.

Biological control–*Search in* biocontrol.

Biological containment–Restricting the movement of (genetically engineered) organisms by arranging barriers to prevent them from growing outside the laboratory. Biological containment can take two forms: making the organism unable to survive in the outside environment, or making the outside environment inhospitable to the organism. The latter is rarely suitable for bacteria, which, in principle, could survive almost anywhere. Thus for bacteria and yeasts, the favoured approach is to mutate the genes in the organism so that they require a supply of a specific nutrient that is usually available only in the laboratory. If they get out, they then cannot grow. Making the environment unfriendly to the organism is partly a biological control, partly a physical one. Thus, some of the first genetically engineered rice strains were developed in England (which is too cold for rice to grow) and tried in the field in Arizona (where it is too dry). Biological containment may also involve the use of vector molecules and host organisms which have been genetically disabled such that they can survive only in the peculiar conditions provided by the experimenter and which are unavailable outside the laboratory.

Biological diversity–*Search in* biodiversity (Source: FAO, 1999).

Biomass–1. The cell mass produced by a population of living organisms.

2. The organic mass that can be used either as a source of energy or for its chemical components.

3. All the organic matter that derives from the photosynthetic conversion of solar energy.

Biomass concentration–The amount of biological material in a specific volume.

Biome–1. A major ecological community or complex of communities, extending over a large geographical area and characterized by a dominant type of vegetation.

2. Collection of plant and animal populations in a particular area.

Biometrics–*Search in* biometry.

Biometry–The application of statistical methods to the analysis of biological problems.

Biopesticide–A compound that kills organisms by virtue of specific biological effects rather than as a broader chemical poison. Specific types include bio-insecticides and bio-fungicides. Bio-pesticides differ from biocontrol agents in that bio-pesticides are passive agents, whereas biocontrol agents are active, seeking out the pest to be destroyed. There are some extremely attractive anti-pest materials, such as *Bacillus thuringiensis* (Bt) toxin, which specifically interferes with the absorption of food from the guts of some insects but is harmless to mammals. The rationale behind developing bio-pesticides is that they are more likely to be biodegradable and are targeted at specific elements of the pest's metabolism.

Biopolymer–Any large polymeric molecule (protein, nucleic acid, polysaccharide, lipid) produced by a living organism.

Bioprocess–Any process that uses complete living cells or their components (*e.g.*, enzymes, chloroplasts) to effect desired physical or chemical changes.

Bioreactor–A tank in which cells, cell extracts or enzymes carry out a biological reaction. Often refers to a growth chamber (fermenter, fermentation vessel) for cells or micro-organisms.

Bioremediation–A process that uses living organisms to remove contaminants, pollutants or unwanted substances from soil or water. *The opposite word is* bio-augmentation; bio-enrichment.

Biosensor–A device that uses an immobilized agent (such as an enzyme, antibiotic, organelle or whole cell) to detect or measure a chemical compound. A reaction between the immobilized agent and the molecule being analysed is transduced into an electric signal.

Biosphere–The part of the earth and its atmosphere that is inhabited by living organisms.

Biosynthesis–Synthesis of compounds by living cells, which is the essential feature of anabolism.

Biotechnology–1. The use of biological processes or organisms for the production of materials and services of benefit to humankind. Biotechnology includes the use of techniques for the improvement of the characteristics of economically important plants and animals and for the development of micro-organisms to act on the environment.

2. The scientific manipulation of living organisms, especially at the molecular genetic level, to produce new products, such as hormones, vaccines or monoclonal antibodies.

3. Biotechnology is defined as the application of science and technology to living organisms as well as parts, products and models thereof, to alter living or non-living materials for the production of knowledge, goods and services.

A list based definition of biotechnology comprises the following five categories:

–DNA (the coding): genomics, pharmaco-genetics, gene probes, DNA sequencing/synthesis/amplification, genetic engineering.

–Proteins and molecules (the functional blocks): protein/peptide sequencing/synthesis, lipid/protein engineering, proteomics, hormones and growth factors, cell receptors/signalling/pheromones.

–Cell and tissue culture and engineering: cell/tissue culture, tissue engineering, hybridisation, cellular fusion, vaccine/immune stimulants, embryo manipulation.

–Process biotechnology's: Bio-reactors, fermentation, bioprocessing, bioleaching, biopulping, bio-bleaching, bioremediation, and biofiltration.

–Sub-cellular organisms: gene therapy, viral vectors.

Biotic factor–Other living organisms that are a factor of an organism's environment, and form the biotic environment, affecting the organism in many ways.

Biotic stress–Stress resulting from living organisms which can harm plants, such as viruses, fungi, bacteria, parasitic weeds and harmful insects. *The opposite word is* abiotic stress.

Biotin–A vitamin of the B complex. It is a co-enzyme for various enzymes that catalyse the incorporation of carbon dioxide into various compounds. It is essential for the metabolism of fats. Biotin is attached to pyruvate carboxylase by a long, flexible chain like that of lipoamide in the pyruvate dehydrogenase complex. Adequate amounts are normally produced by the intestinal bacteria in animals. a.k.a. vitamin H (in USA).

Biotin labeling–1. The attachment of biotin to another molecule.

2. The incorporation of a biotin-containing nucleotide into a DNA molecule.

Biotinylated-DNA–A DNA molecule labelled with biotin by incorporation of biotinylated-dUTP into a DNA molecule. It is used as a non-radioactive probe in hybridization experiments, such as Southern transfer. The detection of the labelled DNA is achieved by complexing it with streptavidin (an antibiotic with a high affinity for biotin) to which is attached a colour-generating agent such as horseradish peroxidase that gives a fluorescent green colour upon reaction with various organic reagents.

Biotransformation–The conversion of one chemical or material into another using a biological catalyst: a near synonym is biocatalysis, and hence the catalyst used is called a biocatalyst. Usually the catalyst is an enzyme, or a whole, dead micro-organism that contains an enzyme or several enzymes.

Biotope–A small habitat in a large community.

Biotoxin–A naturally produced toxic compound which shows pronounced biological activity and presumably has some adaptive significance to the organism which produces it.

Bivalent–A pair of synapsed or associated homologous chromosomes (one of maternal origin; the other of paternal origin) that have each undergone duplication. Each duplicated chromosome comprises two chromatids. Thus a bivalent comprises four chromatids.

Blastocyst (also blastocist)–A mammalian embryo (fertilized ovum) in the early stages of development, approximately up to the time of implantation. It consists of a hollow ball of cells.

Blastomere–Any one of the cells formed from the first few cleavages in animal embryology. The embryo usually divides into two, then four, then eight blastomeres, and so on.

Blastula–In animals, an early embryo form that follows the morula stage; typically, a single-layered sheet (blastoderm) or ball of cells (blastocyst).

B-lymphocytes; B-cells–An important class of lymphocytes that mature in bone marrow (in mammals) and the Bursa of Fabricius (in birds), that are largely responsible for the antibody-mediated or humoral immune response; they give rise to the antibody-producing plasma cells and some other cells of the immune system.

Bleach–A fluid, powder or other whitening (bleaching) or cleaning agent, usually with free chlorine ions. Commercial bleach contains calcium hypochlorite or sodium hypochlorite, and is a common disinfectant used for cleaning working surfaces, tools and plant materials in plant tissue culture and grafting.

Bleeding–Used to describe the occasional purplish-black coloration of media due to phenolic products given off by (usually fresh) transfers.

Blot–1. As a verb, this means to transfer DNA, RNA or protein to an immobilizing matrix.

2. As a noun, it usually refers to the autoradiograph produced during the Southern or northern blotting procedures. The variations on this theme depend on the molecules:

–***Southern blot***: the molecules transferred are DNA molecules, and the probe is DNA.

–***Northern blot***: the molecules transferred are RNA, and the probe is DNA.

–***Western blot***: the molecules transferred are protein, and the probe is labelled antibody.

–***Dot blot***: DNA, RNA or protein are dotted directly onto the membrane support, so that they form discrete spots.

–***Colony blot***: the molecules (usually DNA) are from colonies of bacteria or yeast growing on a bacteriological plate.

Blunt end–The end of a DNA duplex molecule in which neither strand extends beyond the other.

Blunt-end cut–To cleave phosphodiester bonds in the backbone of duplex DNA between the corresponding nucleotide pairs on opposite strands. This cleavage process results in both strands finishing at the same residue, *i.e.*, there are no nucleotide extensions on either strand.

Blunt-end ligation–Joining (ligation) of the nucleotides that are at the ends of two blunt-ended DNA duplex molecules.

Boring platform–Sterile bottom half of a Petri dish used for preparing explants with a cork borer.

Bound water–Water held by the cell and not released if freezing occurs in the intercellular space. *The opposite word is* free water.

Bovine somatotrophin (BST)–A bovine growth hormone, this protein is found naturally in cattle, and is the bovine counterpart of human growth hormone, one of the earliest biopharmaceutical products. It has been cloned, using recombinant DNA technology, expressed in large amounts and marketed as an agricultural product to improve the growth rate and protein:fat ratios in farm cattle, and to improve milk yield. Its use is banned in some countries.

Bovine spongiform encephalopathy (BSE)–mad cow disease. *Search in* proteinaceous infectious particle.

Bp–Abbreviation for *base pair.*

Bract–A modified leaf that subtends flowers or inflorescences and may appear to be a petal.

Breed (noun)–(*i*) a sub-specific group of domestic livestock with definable and identifiable external characteristics that enable it to be separated by visual appraisal from other similarly defined groups within the same species or (*ii*) a group of domestic livestock for which geographical and/or cultural separation from phenotypically similar groups has led to acceptance of its separate identity. *The opposite word is* breed at risk; breed not at risk; critical breed; critical-maintained breed; endangered-maintained breed. (Source: FAO, 1999)

Breed at risk–Any breed that may become extinct if the factors causing its decline in numbers are not eliminated or mitigated. (Source: FAO, 1999)

Breeds may be in danger of becoming extinct for a variety of reasons. Risk of extinction may result from, *inter alia*, low population size; direct and indirect impacts of policy at the farm, country or international levels; lack of proper breed organization; or lack of adaptation to market demands. Breeds are categorized as to their risk status on the basis of, *inter alia,* the actual numbers of male and/or female breeding individuals and the percentage of pure-bred females. FAO has established categories of risk status: critical, endangered, critical-maintained, endangered-maintained, and not-at-risk. (Source: FAO, 1999).

Breeding–The process of sexual reproduction and production of offspring.

Breeding value–In quantitative genetics, the part of the deviation of an individual phenotype from the population mean that is due to the additive effects of alleles. In practical terms: if an animal is mated with a random sample of animals from a population, that animal's breeding value for a certain trait is twice the average deviation of its offspring from the population mean for that trait.

Breed not at risk–A breed where the total number of breeding females and males is greater than 1 000 and 20 respectively; or the population size approaches 1 000 and the percentage of pure-bred females is close to 100 per cent, and the overall population size is increasing. *The opposite word is* breed at risk. (Source: FAO, 1999).

Brewing–The process by which beer is made. In the first stage the barley grain is soaked in water and allowed to germinate (malting), during which the natural enzymes of the grain convert the seed starch to maltose, and then to glucose. Grain is then dried, crushed, and added to water at a specific temperature (steeping) and any remaining starch is converted to sugar. The resulting liquid (wort) is the raw material to which yeast is added to convert sugar to alcohol. Hops (female flowers of *Humulus lupulus*) are added during this process to give a characteristic flavour.

Brewer's yeast–Strains of yeast, often *Saccharomyces cerevisiae,* that are used in the production of beer.

Bridge–A filter paper or other substrate used as a wick and support structure for a plant tissue in culture when a liquid medium is used.

Broad-host-range plasmid–A plasmid that can replicate in a number of different bacterial species.

Broad-sense heritability–In quantitative genetics, the proportion of the total phenotypic variation due to genetic variation.

Browning–Discoloration due to phenolic oxidation of freshly cut surfaces of explant tissue. In later stages of culture, such discoloration may indicate a nutritional or pathogenic problem, generally leading to necrosis.

Brucellosis–Disease caused by infection with organisms of the genus *Brucella.*

BSE–Bovine spongiform encephalopathy. *Search in* proteinaceous infectious particle.

BST–*Search in* bovine somatotrophin.

Bt–*Search in Bacillus thuringiensis.*

Bubble column fermenter–A fermentation vessel, or bioreactor, in which the cells or micro-organisms are kept suspended in a tall cylinder by rising air, which is introduced at the base of the vessel.

Bud–A region of meristematic tissue with the potential for developing into leaves, shoots, flowers or combinations; generally protected by modified scale leaves. A terminal (or apical) bud exists at the tip of a stem or branch, while axillary (or lateral) buds develop in the axils of leaves.

Bud scar–A scar left on a shoot when the bud or bud scales drop.

Bud sport–A somatic mutation arising in a bud and producing a genetically different shoot. Bud sports includes changes due to gene mutation, somatic reduction, chromosome deletion or polyploidy.

Budding–1. A method of asexual reproduction in which a new individual is derived from an outgrowth (bud) that becomes detached from the body of the parent.

2. Among fungi, budding is characteristic of the yeast *Saccharomyces cerevisiae.*

3. A form of grafting in which a single vegetative bud is taken from one plant and inserted into stem tissue of another plant so that the two will grow together. The inserted bud develops into a new shoot. *Search in* grafting.

Buffer–A solution that resists change in pH when an acid or alkali is added, or when solutions are diluted.

Buoyant density–The intrinsic density which a molecule, virus or sub-cellular particle has when suspended in an aqueous solution of a salt, such as CsCl, or a sugar, such as sucrose. DNA from different species has a characteristic buoyant density, which reflects the proportion of G=C base pairs. The greater the proportion of G=C, the greater the buoyant density of the DNA.

C–Cytosine residue (pyrimidine base) in either DNA or RNA.

CAAT box (also CAT box)–A conserved sequence found within the promoter region of the protein-encoding genes of many eukaryotic organisms. It has the consensus sequence GGCCAATCT; it occurs around 75 bases prior to the transcription initiation site; and it is one of several sites for recognition and binding of regulatory proteins called transcription factors.

Cabinet–*Search in* growth cabinet.

Calf scours–A watery diarrhoea in calves.

Callus (L. *callum*, thick skin; pl: calluses or calli)–1. A protective tissue, consisting of parenchyma cells, that develops over a cut or damaged plant surface.

2. Mass of unorganized, thin-walled parenchyma cells induced by hormone treatment.

3. Actively dividing non-organized masses of undifferentiated and differentiated cells often developing from injury (wounding) or in tissue culture in the presence of growth regulators.

Callus culture–A technique of tissue culture; it is usually on solidified medium and initiated by inoculation of small explants or sections from established organ or other cultures (the inocula). Callus culture is used as the basis for organogenic (shoot, root) cultures, cell cultures or proliferation of embryoids. Callus cultures can be indefinitely maintained through regular sub-culturing.

Calorie (abbr: cal)–Equivalent to the amount of heat required to raise the temperature of 1 gram of water from 14.5°C to 15.5°C. *The opposite word is* kilocalorie.

Calyx (Gr. *kalyx*, a husk, cup)–All the sepals of a flower considered collectively. The outermost whorl of flower parts.

Cambial zone–Region in stems and roots consisting of the cambium and its recent derivatives.

Cambium (L. *cambium*, one of the alimentary body fluids supposed to nourish the body organs; pl: cambia)–A layer, usually regarded as one or two cells thick, of persistently meristematic tissue between the xylem and phloem tissues, and which gives rise to secondary tissues, thus resulting in an increase in diameter. The two most important cambia are the vascular (fascicular) cambium and the cork cambium.

Cancer–Uncontrolled growth of the cells of a tissue or an organ in a multi-cellular organism. The opposite word is oncogenesis.

Candidate gene–A gene whose function suggests that it may be involved in the genetic variation observed for a particular trait, *e.g.*, the gene for growth hormone is a candidate gene for body weight.

Candidate-gene strategy–An experimental approach in which knowledge of the biochemistry and/or physiology of a trait is used to draw up a list of genes whose protein products could be involved in the trait.

Canine–Pertaining to dogs.

Canola–Any of several cultivars of oilseed rape (more fully: canola oil); the vegetable oil high in mono-unsaturated fatty acid obtained from these cultivars.

Cap–The structure found on the 5′-end of eukaryotic mRNA, and consisting of an inverted, methylated guanosine residue. *Search in* G cap.

Cap site–The site in a gene where translation is initiated, a.k.a. translation initiation site.

Capacitation–The final stage in the maturation process of a spermatozoon, taking place inside the female genital tract as the sperm penetrates the ovum.

Capsid–The protein coat of a virus. The capsid often determines the shape of the virus. *Search in* coat protein.

Carbohydrate–An organic compound based on the general formula Cx (H_2O)y, performing many vital roles in living organisms. Carbohydrates (from 'hydrates of carbon') are the most abundant of the four major classes of biomolecules, which also include proteins, lipids and nucleic acids.The simplest carbohydrates are the sugars (saccharides), including glucose and sucrose. Polysaccharides are carbohydrates of much greater molecular weight and complexity; examples are starch, which serves as energy store in plant seeds and tubers; cellulose and lignin that form the cell walls and woody tissue of plants of plants; glycogen, etc.

Carbowax–*Search in* polyethylene glycol.

Carcinogen–A substance capable of inducing cancer in an organism.

Carcinoma–A malignant tumour derived from epithelial tissue, which forms the skin and the outer cell layers of internal organs.

Carotene (L. *carota*, carrot)–A reddish-orange plastid pigment involved in light reactions in photosynthesis. General formula is C40H56 that are unsaturated hydrocarbons containing 8 isoprene subunits.

Carotenoid–Red to yellow pigments responsible for the characteristic colour of many plant organs or fruits, such as tomatoes, carrots, etc. Oxidation products of carotene are called xanthophylls. Carotenoids serve as light-harvesting molecules in photosynthetic assemblies and also play a role in protecting prokaryotes from the deleterious effects of light.

Carboxypeptidases–Two enzymes (A and B) found in pancreatic juice. Their role is to remove the C-terminal amino-acid from a peptide; the A form removes any amino acid; the B form removes only lysine or arginine. Used when sequencing peptides.

Carpel–Female reproductive organ of flowering plants, consisting of stigma, style and ovary. In some plants, one or more carpels unite to form the pistil.

Carrier–In genetics, typically an individual that has one recessive mutant allele for some defective condition that is "masked" by a dominant normal allele at the same locus, *i.e.*, an individual that is heterozygous for a recessive harmful allele and a dominant normal allele; the phenotype is normal, but the individual passes the defective (recessive) allele to half of its offspring.

Carrier DNA–DNA of undefined sequence content which is added to the transforming (plasmid) DNA used in physical DNA-transfer procedures. This additional DNA increases the efficiency of transformation in electroporation and chemically mediated DNA-delivery systems. The mechanism responsible for this effect is not known. *Search in* binary vector; plasmid; chimeric gene.

Carrier gas–The gas that carries the sample in gas chromatography.

Carrier molecule–1. A molecule that plays a role in transporting electrons through the electron transport chain. Carrier molecules are usually proteins bound to non-protein groups and able to undergoing oxidation and reduction relatively easily, thus allowing electrons to flow.

2. A lipid-soluble molecule that can bind to lipid-insoluble molecules and transport them across membranes. Carrier molecules have specific sites that interact with the molecules they transport. Efficiency

of carrier molecules may be modified by modifying the interacting sites through genetic engineering.

Casein–A group of proteins found in milk.

Casein hydrolysate–Mixture of amino acids and peptides produced by enzymatic or acid hydrolysis of casein. *The opposite word is* organic complex; undefined.

CAT box–*Search in* CAAT box.

Catabolic pathway–A pathway by which an organic molecule is degraded in order to release energy for growth and other cellular processes; degradative pathway.

Catalysis–The process of changing the rate of a chemical reaction by use of a catalyst. Biological catalyst are termed as enzymes.

Catabolism–The metabolic breakdown of large molecules in living organism, with accompanying release of energy. Respiration is an example of this.

Catabolite repression–Glucose-mediated reduction in the rates of transcription of operons that encode enzymes involved in catabolic pathways (such as the *lac* operon).

Catalyst (Gr. *katalyein*, to dissolve)–A substance that promotes a chemical reaction by lowering the activation energy of a chemical reaction, without itself undergoing any permanent chemical change. The process is catalysis. *The opposite word is* catalytic antibody; catalytic RNA.

Catalytic antibody (= abzyme)–An antibody selected for its ability to catalyse a chemical reaction by binding to and stabilizing the transition-state intermediate.

Catalytic RNA (= ribozyme; gene shears)–A natural or synthetic RNA molecule that cuts an RNA substrate.

Cation–A positively charged ion; opposite is *anion*.

Caulogenesis–Stem organogenesis; induction of shoot development from callus.

CD molecules; cluster of differentiation molecules–Any group of antigens that is associated with a specific sub-population of T cells. There are designations for surface molecules on various cells of the immune system, *e.g.*, CD4 is present on the surface of helper T cells.

cDNA; complementary DNA–The double-stranded DNA complement of an mRNA sequence; synthesized *in vitro* from a mature RNA template using reverse transcriptase (to create a single strand of DNA from the RNA template) and DNA polymerase (to create the double-stranded DNA). Preparation of cDNAs is often the first step in cloning DNA sequences of interest. Used as specific and sensitive probes in hybridization studies, because cDNAs usually do not include regulatory or other controlling sequences, and so they can be used to identify (probe) and isolate genes and their associated sequences from genomic DNA. *Search in* binary vector; carrier DNA.

cDNA clone–A double-stranded DNA molecule that is carried in a vector and was synthesized *in vitro* from an mRNA sequence by using reverse transcriptase and DNA polymerase.

cDNA cloning–A method of cloning the coding sequence of a gene, starting with its mRNA transcript. It is normally used to clone a DNA copy of a eukaryotic mRNA. The cDNA copy, being a copy of a mature messenger molecule, will not contain any intron sequences and may be readily expressed in any host organism if attached to a suitable promoter sequence within the cloning vector.

cDNA library–A collection of cDNA clones that were generated *in vitro* from the mRNA sequences isolated from an organism or a specific tissue or cell type or population of an organism. *The opposite word is* library.

CDR (complementarity-determining regions)–These are regions of the variable (V) regions of light and heavy antibody chains that make contact with the antigen. The primary amino acid sequences of these regions are highly variable among antibodies of the same class.

Cell (L. *cella*, small room)–The smallest structural unit of living matter capable of functioning independently; a microscopic mass of protoplasm surrounded by a semi-permeable membrane, usually including one or more nuclei and various non-living products, capable-either alone or by interacting with other cells-of performing all the fundamental functions of life.

Cell culture–The *in vitro* growth of cells derived from multi-cellular organisms. The cells are usually of one type.

Cell cycle–Sequence of stages that a cell passes through between one division and the next. The cell cycle oscillates between mitosis and the interphase, which is divided into G, S, and G_2. In the G phase there is a high rate of biosynthesis and growth; in the S phase there is the doubling of the DNA content as a consequence of chromosome replication; in the G_2 phase the final preparations for cell division (cytokinesis) are made.

Cell differentiation–Continuous loss of physiological and cytological characters of young cells, resulting in getting the characters of adult cells. The unspecialized cells become modified and specialized for the performance of specific functions. Differentiation results from the controlled activation and de-activation of genes.

Cell division–Formation of two or more daughter cells from a single mother cell. The nucleus divides first, followed by the formation of a cell membrane between the daughter nuclei. Division of cytoplasm and nucleus into two or more parts by formation of a cell plate.

Cell-free protein synthesis; cell-free system–*Search in in vitro* translation.

Cell-free transcription–*Search in in vitro* transcription.

Cell-free translation–*Search in in vitro* translation.

Cell fusion–Formation of a single hybrid cell from two cells of different species, cultured *in vitro*. The cells fuse and coalesce, but their nuclei may remain separated. During subsequent cell division, a single spindle is formed so that each daughter cell has a single nucleus containing sets of chromosomes from each parental line. Subsequent divisions often result in the loss of chromosomes and therefore of genes. The cell fusion technique can be used to determine the control of specific genes and their assignment to chromosomes. *The opposite word is* cell hybridization.

Cell generation time–The interval between the beginning of consecutive divisions of a cell. The time that it takes for a population of single-celled organisms to double its cell number. Successive generations of cells or organisms within a population are separated by a time interval called generation time. The cell regeneration time can be determined with the aid of time-lapse microcinematography.

Cell hybridization–The fusion of two or more dissimilar cells leading to the formation of a somatic hybrid. *The opposite word is* cell fusion.

Cell line–A cell lineage that can be maintained in culture. A cell line arises from a primary culture. It implies that cultures from it consist of several lineages of the cells originally present in the primary culture.

Cell-mediated immune response–The activation of T cells of the immune system in response to the presence of a foreign antigen.

Cell membrane–The membrane that separates the cell wall and the cytoplasm, and regulates the flow of material into and out of the cell and it is structuring similar to plasma membrane. *Search in* plasmalemma.

Cell number–The number of cells per unit volume of a culture.

Cell plate–The precursor of the cell wall, formed as cytokinesis starts during cell division. The cell plate develops in the region of the equatorial plate and arises from membranes in the cytoplasm.

Cell sap–Water and dissolved substances, sugar, amino acids, waste substances, etc., in the plant cell vacuole.

Cell selection The process of selecting cells within a group of genetically different cells. Select cells or cell lines are sub-cultured onto fresh medium for continued selection and often are exposed to an increased level of the selection agent. The final objective is to regenerate plants exhibiting the traits selected for at the cellular level.

Cell strain–A strain of cells having specific properties or markers derived from a primary culture of a cell line by selection or cloning. The selected properties must persist during subsequent cultivation. a.k.a. single-cell line.

Cell suspension–Cells in culture in moving or shaking liquid medium, often used to describe suspension cultures of single cells and cell aggregates. *Search in* suspension culture.

Cellular immune response–*Search in* T-cell-mediated (cellular) immune response.

Cellular oncogene (proto-oncogene)–A normal mammalian or avian gene that when mutated or improperly expressed contributes to the development of cancer. *Search in* oncogene.

Cellulase–Enzyme catalysing the breakdown of cellulose.

Cellulose (cell + ose, a suffix indicating a carbohydrate)–A complex carbohydrate composed of long, unbranched chains of beta-glucose molecules, which contribute to the structural framework of plant cell walls. It comprises 40 per cent to 55 per cent by weight of the plant cell wall.

Cellulose nitrate–*Search in* nitrocellulose.

Cellulosome–A multi-protein aggregate that is present in some cellulolytic micro-organisms and contains multiple copies of all the enzymes required to completely break down cellulose. This complex is often found on the outer surface of cellulolytic micro-organisms.

Cell wall–A rigid external coat which surrounds plant cells. It is formed outside the plasmalemma and consists primarily of cellulose.

CentiMorgan (cM)–One percent recombination between two loci. *Search in* map distance; crossing-over unit.

Central dogma–The basic concept that, in nature, genetic information generally can flow only from DNA to RNA to protein. It is now known, however, that information contained in RNA molecules of certain viruses (called retroviruses) can also flow back to DNA.

Central mother cell–A subsurface cell located in a plant apical meristem and characterized by a large vacuole.

Centres of origin–The locations in the world where particular domesticated plants originated. These areas show the highest variation, and are rich in wild alleles.

Centrifugation–Separating molecules by size or density using centrifugal forces generated by a spinning rotor. G-forces of several hundred thousand times gravity are generated in ultracentrifugation. *Search in* density gradient centrifugation.

Centrifuge–A device in which solid or liquid particles of different densities are separated by rotating them in a tube in a horizontal circle. The denser particles tend to move along the length of the tube to a greater radius of rotation, displacing the lighter particles to the other end.

Centriole–An organelle in many animal cells that appears to be involved in the formation of the spindle during mitosis. During cell division, the two centrioles move to opposite sides of the nucleus to form the ends of the spindle.

Centromere–The portion of the chromosome to which the spindle fibres attach during mitotic and meiotic division. It appears as a constriction when chromosomes contract during cell division. After chromosomal duplication, which occurs at the beginning of every mitotic and meiotic division, the two resultant chromatids are joined at the centromere.

Centrosome–A specialized region of a living cell, situated next to the nucleus, where micro-tubules are assembled and broken down during cell division. The centrosome of most animal cells contains a pair of centrioles. During metaphase the centrosome separates into two regions, each containing one of the centrioles.

Cephem-type antibiotic–An antibiotic that shares the basic chemical structure of cephalosporin.

Chain terminator–1. Codons which do not code for an amino acid. They signal ribosomes to terminate protein synthesis. The codons are UAA, UAG and UGA, and have been termed ochre, amber and opal, respectively. Also known as stop codons or termination codons. Often two of these codons are found together at the end of a coding sequence of RNA.

2. In the Sanger method of DNA sequencing, di-deoxynucleoside triphosphates are added as chain terminators in the synthesis of a complementary DNA strand.

Character–A distinctive feature of an organism.

Characterization–Systematic recording of descriptors that are independent of environmental factors.

Charcoal–The black porous residue of partly burnt wood, bones, etc; a form of carbon. When treated to purify it and increase its adsorptive power, it is called activated charcoal in which form it is added to nutrient medium in order to prevent or decrease the effect of browning.

Chelate (noun)–Complex organic molecule that can combine with cations and does not ionize. Chelates can supply micronutrients to plants at slow, steady rates. Usually used to supply iron to plant cells.

Chemically-defined medium–When all of the chemical components of a plant tissue culture medium are fully known and defined. *The opposite word is* undefined; organic complex.

Chemical mutagen–A chemical or product capable of causing genetic mutation in living organisms exposed to it.

Chemiluminescence–The emission of light from a chemical reaction.

Chemostat–A continuous and open culture in which growth rate and cell density are maintained constant by a fixed rate of input of a growth-limiting nutrient. *The opposite word is* phytostat.

Chemotaxis–Motion of a motile cell, organism or part towards or away from an increasing concentration of a particular substance.

Chemotherapy–The treatment of disease, especially infections or cancer, by means of chemicals. In treating cancers, it involves administering chemicals toxic to malignant cells.

Chiasma (Gr. *chiasma*, two lines placed crosswise; pl. chiasmata)–A visible point of junction between two non-sister chromatids of homologous chromosomes during the first meiotic prophase. a.k.a. cross-over. In the diplotene stage of prophase I of meiosis, the four chromatids of a bivalent are associated in pairs, but in such a way that one part of two chromatids is exchanged.

Chimera (or chimaera)–From chimera, a mythological creature with the head of a lion, the body of a goat and the tail of a serpent. An organism whose cells are not all derived from the same zygote.

1. Animal. An individual exhibiting two or more genotypes in patches derived from two or more embryos. An individual derived from two embryos by experimental intervention.

2. Plant. Part of a plant with a genetically different constitution as compared with other parts of the same plant. It may result from different zygotes that grow together, or from artificial fusion (grafting); it may either be periclinal chimera, in which one tissue lies over another as a glove fits a hand; mericlinal chimera, where the outer tissue does not completely cover the inner tissue; and sectoral chimera, in which the tissues lie side by side.

3. A recombinant DNA molecule that contains sequences from different organisms.

Chimeric DNA–A recombinant DNA molecule containing unrelated genes.

Chimeric gene–A semi-synthetic gene, consisting of the coding sequence from one organism, fused to promoter and other sequences derived from a different gene. Most genes used in transformation are chimeric. *Search in* carrier DNA; binary vector; plasmid; transformation; vector.

Chimeric protein–*Search in* fusion protein.

Chimeric selectable marker gene–A gene that is constructed from parts of two or more different genes and allows the host cell to survive under conditions where it would otherwise die.

Chip, DNA–*Search in* DNA chip.

Chi-squared test (χ^2 test)–A significance test used to statistically assess the goodness of fit of observed data to a prediction.

Chitin–A nitrogenous polysaccharide occurring as skeletal material in many invertebrates and fungi.

Chitinase–An enzyme which breaks down chitin.

Chloramphenicol–An antibiotic that interferes with protein synthesis.

Chlorenchyma (Gr. *chloros*, green + *enchyma*, a suffix meaning tissue)–Tissue containing chloroplasts, including leaf mesophyll and other parenchyma cells.

Chlorophyll (Gr. *chloros*, green + *phyllon*, leaf)–One of the two pigments found in chloroplast responsible for the green colour of most plants. It is essential in the absorption of light energy for photosynthesis.

Chloroplast (Gr. *chloros*, green + *plastos*, formed)–Specialized cytoplasmic organelle that contains chlorophyll. Lens-shaped and bounded by a double membrane, chloroplasts contain membranous structures (thylakoids) piled up into stacks, surrounded by a gel-like matrix (stroma). They are the site of solar energy transfer and important reactions of starch or sugar synthesis. Chloroplasts have their own DNA and are inherited cytoplasmically, independent of nuclear genes.

Chloroplastid–*Search in* chloroplast.

Chlorosis (Gr. *chloros*, green + *osis*, diseased statè)–Failure of chlorophyll development, and appearance of yellow colour in plants, because of a nutritional disturbance or because of an infection by a virus, bacteria or fungus that prevents chlorophyll formation.

Chromatid (chromosome + *id*, L. suffix meaning "daughters of")–Each of the two daughter strands comprising a duplicated chromosome. The term remains in use while the two chromatids are still joined at the centromere. As soon as the centromere divides, setting the two chromatids adrift (during anaphase of mitosis; and during anaphase II of meiosis), they are called chromosomes.

Chromatin (Gr. *chroma*, colour)–Substance of which eukaryotic chromosomes are composed. It consists of primarily DNAm with some proteins (mainly histones), and small amounts of RNA. Originally named because of the readiness with which it stains with certain dyes (chromaticity).

Chromatin fibres–A basic organizational unit of eukaryotic chromosomes, consisting of DNA and associated proteins assembled into strands of 30 nm average diameter.

Chromatography (Gr. *chroma* colour + *graphein*, meaning to draw or write)–1. A method for separating and identifying the components of mixtures of molecules having similar chemical and physical properties.

2. The term originally used by Mikhail Tswett (1906) to describe the separation of a mixture of leaf pigments on a calcium carbonate column.

Chromocentre–Body produced by fusion of the heterochromatic regions of the chromosomes in the polytene tissues (*e.g.*, the salivary glands) of certain *Diptera*.

Chromogenic substrate–A compound or substance that contains a colour-forming group.

Chromomeres–Small bodies, described by J. Belling, that are identified by their characteristic size and linear arrangement along a chromosome.

Chromonema (pl: chromonemata)–An optically single thread forming an axial structure within each chromosome.

Chromoplast–Plastid containing pigments, such a chloroplast, or one in which carotenoids predominate.

Chromosomal aberration–Any change in chromosome structure or number. Although it can be a mechanism for enhancing genetic diversity, such alterations are usually fatal or ill-adaptive, especially in animals.

Chromosomal integration site–A chromosomal location where foreign DNA can be integrated, often without impairing any essential function in the host organism.

Chromosomal polymorphism–The occurrence of one to several chromosomes in two or more alternative structural forms within a population; the structurally changed chromosomes are the result of chromosome mutations (*i.e.*, any structural change involving the gain, loss or re-location of chromosome segments).

Chromosome (Gr. *chroma*, colour + *soma*, body)–1. A single DNA molecule, a tightly coiled strand of DNA, condensed into a compact structure *in vivo* by complexing with accessory histones or histone-like proteins.

2. A group of nuclear bodies containing genes which are largely responsible for the differentiation and activity of a eukaryotic cell; one of the bodies into which the nucleus resolves itself at the beginning of mitosis and from which it is derived at the end of mitosis. Chromosomes contain most of the cell's DNA. Chromosomes exist in pairs in eukaryotes–one paternal (from the male parent) and one maternal (from the female parent). Each eukaryotic species has a characteristic number of chromosomes. Bacterial and viral cells contain only a single chromosome, consisting of a single or double strand of DNA or, in some viruses, RNA, without histones.

Chromosome aberration–Abnormal structure or number of chromosomes; includes deficiency, duplication, inversion, translocation, aneuploidy, polyploidy, or any other change from the normal pattern.

Chromosome banding–Staining of chromosomes in such a way that light and dark areas occur along the length of the chromosomes in repeatable patterns. Lateral comparisons identify pairs. Each chromosome can be identified by its banding pattern.

Chromosome mutation–A change in the gross structure of a chromosome, usually causing severely deleterious effects in the organism. They are often due to an error in pairing during the crossing-over stage of meiosis. The main types of chromosome mutation are translocation, duplication, deletion and inversion.

Chromosome jumping–A technique that allows two segments of duplex DNA that are separated by thousands of base pairs (about 200 kb) to be cloned together. After sub-cloning, each segment can be used as a probe to identify cloned DNA sequences that, at the chromosome level, are roughly 200 kb apart. *Search in* positional cloning.

Chromosome theory of inheritance–The theory that chromosomes carry the genetic information and that their behaviour during meiosis provides the physical basis for segregation and independent assortment.

Chromosome walking–A technique that identifies overlapping cloned DNA fragments that form one continuous segment of a chromosome. These fragments can be generated either by random shearing or by partial digestion with a four-base-pair cutter such as *Sau*3A. A series of colony hybridizations is then carried out, starting with some cloned fragment which has already been identified and which is known to be in the region encompassed by the overlapping clones. This identified fragment is used as a probe to pick out clones containing adjacent sequences. These are then used as probes themselves to identify clones carrying sequences adjacent to them and so on. At each round of hybridization one "walks" further along the chromosome from the initial fragment. *Search in* positional cloning.

Chymosin–An enzyme that clots milk; it is used in the manufacture of cheese.

Cilium (pl: cilia; adj: ciliate)–Hairlike locomotor structure on certain cells; a locomotor structure on a ciliate protozoan.

Circadian–Of physiological activity, etc.: occurring or recurring about once a day. *The opposite word is* diurnal.

Circularization–A DNA fragment generated by digestion with a single restriction endonuclease will have complementary 5′ and 3′ extensions (sticky ends). If these ends are annealed and ligated, the DNA fragment will have been converted to a covalently-closed circle, or circularized.

cis configuration–*Search in* coupling.

cis heterozygote–A heterozygote that contains two mutations arranged in a cis configuration (*e.g., a+ b+ / a b*).

cis-acting sequence–A nucleotide sequence that only affects the expression of genes located on the same chromosome.

Cistron–A DNA sequence that codes for a specific polypeptide; a gene. *Search in* DNA; gene.

Claims–The section of a patent that states, in detail, the uses and possible applications of the invention described in the patent.

Class switching–The process during which a plasma cell stops producing antibodies of one class and begins producing antibodies of another class.

Cleave–To break phospho-diester bonds of double-stranded DNA, usually with a type II restriction endonuclease. a.k.a. to cut or digest.

Clonal propagation–Asexual propagation of many new plants (ramets) from an individual (ortet); all have the same genotype.

Clonal selection–The production of a population of plasma cells all producing the same antibody in response to the interaction between a B lymphocyte producing that specific antibody and the antigen bound by that antibody.

Clone (Gr. *klon*, a twig or slip)–1. A group of cells or organisms that are genetically identical as a result of asexual reproduction, breeding of completely inbred organisms, or forming genetically identical organisms by nuclear transplantation.

2. Group of plants genetically identical in which all are derived

from one selected individual by vegetative propagation, without the sexual process.

3. A population of cells that all carry a cloning vehicle with the same insert DNA molecule.

4. Verb: To clone. To insert a DNA segment into a vector or host chromosome. *Search in* cloning.

5. A genetic replica of another organism obtained through a non-sexual (no fertilization) reproduction process.

Cloning by nucleus transfer involves the transfer of a donor cell (from (cultured) cells of embryonic, foetal or adult origin) into the recipient cytoplasm of an enucleated oocyte or zygote, and the subsequent development of embryos and animals. (Based on: FAO, 1999)

Clone bank–*Search in* gene bank.

Cloned strain or line–A strain or line descended directly from a clone.

Cloning–1. The mitotic division of a progenitor cell to give rise to a population of identical daughter cells or clones.

2. Incorporation of a DNA molecule into a chromosomal site or a cloning vector.

3. Animal cloning: the creation of a whole animal by mitotic divisions from a single diploid somatic cell, typically by the process of nuclear transfer. Cloning by nuclear transfer from undifferentiated embryonic cells has been possible for many years, but its widespread application has been hampered by inability to culture embryonic cells from animals other than mice. In 1997, Ian Wilmut and colleagues from Edinburgh showed that it is possible to create a whole animal from a cell taken from differentiated adult tissue, thereby opening up the possibility of widespread animal cloning. *Search in* directional cloning; megabase cloning; molecular cloning; sub-cloning; Dolly.

Cloning site–*Search in* insertion site.

Cloning vector–A small, self-replicating DNA molecule–usually a plasmid or viral DNA chromosome–into which foreign DNA is inserted in the process of cloning genes or other DNA sequences of interest. It can carry inserted DNA and be perpetuated in a host cell. Also called a cloning vehicle, vector, or vehicle.. *Search in* vector.

Cloning vehicle–*Search in* cloning vector.

Closed continuous culture–A continuous culture in which inflow of fresh medium is balanced by outflow of corresponding volumes of spent medium. Cells are separated mechanically from outflowing medium and added back to the culture. *The opposite word is* open continuous culture; batch culture; continuous culture.

Cluster of differentiation–*Search in* CD.

cM–*Search in* centiMorgan; map distance.

Coat protein (= capsid)–The coating protein that encloses the nucleic acid core of a virus.

Coccus (pl: cocci)–A spherical bacterium. Cocci may occur singly, in pairs, in groups of four or more, and in cubical packets.

Coconut milk–Liquid endosperm of the coconut, often used to supply organic nutrients to cultured cells and tissues. *Search in* addendum; organic complex; undefined.

Cocoon–A protective coverage for eggs and/or larvae produced by many invertebrates, such as the silkworm moth.

Co-culture–The joint culture of two or more types of cells, such as a plant cell and a micro-organism, or two types of plant cells. Used in various dual-culture systems or in nurse-culture, q.v.

Coding–The specification of a peptide sequence, by the code contained in DNA molecules.

Coding sequence–That portion of a gene which directly specifies the amino acid sequence of its protein product. Non-coding sequences of genes include control regions, such as promoters, operators and terminators, as well as the intron sequences of certain eukaryotic genes.

Coding strand–The strand of duplex DNA which contains the same base sequence (after substituting U for T) found in the mRNA molecule resulting from transcription of that segment of DNA. a.k.a. sense strand. The mRNA molecule is transcribed from the other strand, known as the template or antisense strand.

Coding strand 5´ ATGAAAGCT TTAGTGGGCGCC CGTAT 3´

Template strand 3´ TACTTTCG AAATCACCCGCGGGCATA 5´

mRNA 5´ AUGAAAGCUUUAGUG GGCGCCCGUAU 3´

Co-dominance–The situation in which both alleles in a heterozygous individual are expressed, so that the phenotype of heterozygotes incorporates the phenotypic effect of each allele. For example, roan coat colour in cattle results from a mixture of red hairs and white hairs, caused by heterozygosity for the red allele and the white allele. Also, protein polymorphisms and micro-satellites show co-dominance: heterozygotes have two bands, whereas homozygotes have only one band.

Co-dominant alleles–Alleles that produce independent effects when in the heterozygous condition.

Codon–A set of three nucleotides in mRNA, functioning as a unit of genetic coding by specifying a particular amino acid during the synthesis of polypeptides in a cell. A codon specifies a transfer RNA carrying a specific amino acid, which is incorporated into a polypeptide chain during protein synthesis. The specificity for translating genetic information from DNA into mRNA, then to protein, is provided by codon-anticodon pairing. *Search in* anticodon; initiation codon; termination codon.

Codon optimization–An experimental strategy in which codons within a cloned gene–ones not generally used by the host cell translation system–are changed by *in vitro* mutagenesis to the preferred codons, without changing the amino acids of the synthesized protein.

Coefficient–A number expressing the amount of some change or effect under certain conditions (*e.g.*, the coefficient of inbreeding).

Co-enzyme–An organic molecule of low molecular weight and usually non-protein, such as a vitamin, that binds to an enzyme and promotes its catalytic activity. *e.g.* NAD^+, $NADP^+$ etc.

Co-evolution–The evolution of complementary adaptations in two species caused by the selection pressure that each exerts on the other. It is common in symbiotic associations, in insect-pollinated plants, etc.

Co-factor–An organic molecule or inorganic ion necessary for the normal catalytic activity of an enzyme.

Co-fermentation–The simultaneous growth of two micro-organisms in one bioreaction.

Co-generation–Production of both electricity and process heat (steam) in an industrial plant.

Cohesion–Holding together; a force holding a solid to a solid or a solid to a liquid, owing to attraction between like molecules.

Cohesive ends–Double-stranded DNA molecules with single-stranded ends which are complementary to each other, enabling the different molecules to join each other. a.k.a. protruding ends; sticky ends; overhang. *The opposite word is* cos ends. *Search in* extension.

Coincidence–The ratio of the observed frequency of double crossings-over to the expected frequency, where the expected frequency is calculated by assuming that the two crossing-over events occur independently of each other.

Co-integrate vector system–A two-plasmid system for transferring cloned genes to plant cells. The cloning vector has a T-DNA segment that contains cloned genes. After introduction into *Agrobacterium tumiefasciens*, the cloning vector DNA undergoes homologous recombination with a resident disarmed Ti plasmid to form a single plasmid carrying the genetic information for transferring the genetically engineered T-DNA region to plant cells.

Co-integrate–A DNA molecule formed by the fusion of two different DNA molecules, usually mediated by a transposable element.

Colchicine (L. *colchicum*, meadow saffron, from colchis, ancient Mingrelia)–An alkaloid obtained from *Colchicum autumnale*, autumn crocus or meadow saffron, which inhibits spindle formation in cells during mitosis, so that chromosomes cannot separate during anaphase, thus inducing multiple sets of chromosomes. Also used to halt mitosis at metaphase–the stage when chromosomes are most visible.

Coleoptile–Protective sheath covering the shoot apex of the embryo in monocotyledenous plants. It protects the plumule as it emerges through the soil.

Coleorhiza (Gr. *koleos*, sheath + *rhiza*, root)–A protective sheath surrounding the radicle of monocotyledenous plants.

Co-linearity–A relationship in which the units in one molecule occur in the same sequence as the units in another molecule which they specify; *e.g.*, the nucleotides in a gene are co-linear with the amino acids in the polypeptide encoded by that gene.

Collection–In PGR: *see* base collection; active collection.

Collenchyma (Gr. *kolla*, glue + *enchyma*, a suffix, derived from parenchyma and denoting a type of cell tissue)–A tissue of living cells, the walls being unevenly thickened with cellulose and hemicellulose, but never lignified; it functions in mechanical support in young, short-lived or non-woody organs and is thus found in midribs and leaf petioles.

Colony hybridization–A technique that uses a nucleic acid probe to identify a bacterial colony with a vector carrying a specific cloned gene or genes.

Colony–1. An aggregate of identical cells (clones) derived from a single progenitor cell.

2. A group of interdependent cells or organisms.

Combinatorial library–During the ligation reaction with cDNAs of light and heavy antibody chains into a bacteriophage lambda vector, many novel combinations consisting of one heavy and one light chain coding region are formed. The library comprises these combinations, each in a separate vector.

Commensalism–The interaction of two or more dissimilar organisms where the association is advantageous to one without affecting the other(s). *The opposite word is* parasitism; symbiosis.

Companion cells–Living cells associated with the sieve cell of phloem tissue in vascular plants.

Comparative gene mapping–The comparison of map locations of genes between species. The results of these comparisons indicate substantial conservation of blocks of genes and even large segments of chromosomes between species. Great use can be made of this conservation of map position. For example, in the case of mammals, it means that if a gene has been mapped in one or both of the intensely-mapped species (humans and mice), then the likely location of that gene in other mammals can be predicted with considerable confidence. Conversely, if a mapped anonymous DNA marker has an effect on a quantitative trait (this being indicative of the marker being linked to a quantitative trait locus (QTL)) in, say, cattle, then knowledge of the comparative map between cattle and humans can identify genes in the homologous region of the human genome that could correspond to the QTL. Such genes are called comparative positional candidate genes.

Comparative positional candidate gene–A gene that is likely to be located in the same region as a DNA marker that has been shown to be linked to a single-locus trait or to a quantitative trait locus (QTL), where the gene's likely location in the genome of the species in question is based on its known location in the map of another species, *i.e.*, is based on the comparative map between the two species.

Competence–Ability of a bacterial cell to take up DNA molecules and become genetically transformed.

Competency–An ephemeral state, induced by treatment with cold cations, during which bacterial cells are capable of taking up foreign DNA.

Competent–A competent cell is capable of developing into a fully functional embryo. The opposite is non-competent.

Complement proteins–Proteins that bind to antibody-antigen complexes and help degrade the complexes by proteolysis.

Complementarity–The relationship between the two strands of a double helix of DNA. Thymine in one strand pairs with adenine in the other strand, and cytosine in one strand pairs with guanine in the other strand.

Complementarity-determining regions–*Search in* CDR.

Complementary DNA–*Search in* cDNA.

Complementary entity–1. One of a pair of nucleotide bases that form hydrogen bonds with each other. Adenine (A) pairs with thymine (T) [or with Uracil (U) in RNA], and guanine (G) pairs with cytosine (C).

2. One of a pair of segments or strands of nucleic acid that will hydridize (join by hydrogen bonding) with each other.

Complementary genes–Two or more interdependent genes, such that (in the case of dominant complementarity) the dominant allele from either gene can only produce an effect on the phenotype of an organism if the dominant allele from the other gene is also present; or (in the case of recessive complementarity) only double homozygous recessive show the effect.

Complementary homopolymeric tailing–The process of adding complementary nucleotide extensions to different DNA molecules, *e.g.*, dG (deoxyguanosine) to the 3′-hydroxyl ends of one DNA molecule and dC (deoxycytidine) to the 3′-hydroxyl ends of another DNA molecule to facilitate, after mixing, the joining of the two DNA molecules by base pairing between the complementary extensions. Also called dG–dC tailing, dA–dT tailing.

Complementary nucleotides–Members of the pairs adenine-thymine, adenine-uracil, and guanine-cytosine that have the ability to hydrogen bond to one another. *Search in* nucleotide.

Complementation–*Search in* genetic complementation.

Complementation test; trans test–Introduction of two mutant chromosomes into the same cell to determine whether the mutants are alleles of the same gene. If the mutations are non-allelic, the genotype will be $m_1 +/+ M_2$, and the phenotype will be wild-type (normal), because each chromosome "covers" for the other. In contrast, if they are allelic, the mutant phenotype will result.

Complete digest–The treatment of a DNA preparation with an endonuclease for sufficient time for all of the potential target sites within that DNA to have been cleaved. *The opposite word is* partial digest.

Composite transposon–A transposable element formed when two identical or nearly identical transposons insert on either side of a non-transposable segment of DNA, such as the bacterial transposon Tn5.

Compound chromosome–A chromosome formed by the union of two separate chromosomes, as in attached-X chromosomes or attached-X-Y chromosomes.

Concatemer–A DNA segment made up of repeated sequences linked end to end.

Concordance–Identity of matched pairs or groups for a given trait, such as sibs expressing the same trait.

Conditional lethal mutation–A mutation that is lethal under one set of environmental conditions–the restrictive conditions–but is viable under another set of environmental conditions–the permissive conditions, *e.g.*, temperature-sensitive mutations.

Conditioning–1. The effects on phenotypic characters of external agents during critical developmental stages.

2. The undefined interaction between tissues and culture medium resulting in the growth of single cells or small aggregates. Conditioning may be accomplished by immersing cells or callus contained within a porous material (such as dialysis tubing) into fresh medium for a period dependent on cell density and a volume related to the amount of fresh medium.

Conidium (pl: conidia)–An asexual spore produced by a specialized hypha in certain fungi.

Conjugation–1. Union of sex cells (gametes) or unicellular organisms during fertilization.

2. The unidirectional transfer of DNA (bacterial plasmid) from one bacterium cell to another and involving cell-to-cell contact. The plasmid usually encodes the majority of the functions necessary for its own transfer.

Conjugative functions–Plasmid-based genes and their products that facilitate the transfer of a plasmid from one bacterium to another.

Consanguinity–Related by descent from a common ancestor.

Consensus sequence–The nucleotide sequence that is present in the majority of genetic signals or elements that perform a specific function.

Conservation of farm animal genetic resources–Refers to all human activities, including strategies, plans, policies and actions, undertaken to ensure that the diversity of farm animal genetic resources is being maintained to contribute to food and agricultural production and productivity; now and in the future. (Source: FAO, 1999).

Constant domains Regions of antibody chains that have the same amino acid sequence in different members of a particular class of antibody molecules.

Constitutive enzyme–An enzyme that is synthesized continually regardless of growth conditions. *The opposite word is* inducible enzyme; repressible enzyme.

Constitutive gene–A gene that is continually expressed in all cells of an organism.

Constitutive promoter–An unregulated promoter that allows for continual transcription of its associated gene. *Search in* promoter.

Constitutive synthesis–Continual production of RNA or protein by an organism.

Constitutive–An organism is said to be constitutive for the production of an enzyme or other protein if that protein is always produced by the cells under all physiological conditions. *Search in* inducible.

Contaminant–Bacterial, fungal or algal micro-organism accidentally introduced into a culture or culture medium. Contaminant may overgrow the plant cells and consequently inhibit their growth. Working under aseptic conditions with a rigorous exclusion of potential contaminants must be practised in plant tissue culture. *Search in* disinfectation; disinfestation.

Contig–A set of overlapping clones that provide a physical map of a portion of a chromosome. *The opposite word is* contiguous map.

Contiguous map; contig map–The alignment of sequence data from large, adjacent regions of the genome to produce a continuous nucleotide sequence across a chromosomal region. *Search in* mapping.

Continuous culture–A suspension culture continuously supplied with nutrients by the inflow of fresh medium. The culture volume is normally constant. *The opposite word is* batch culture; closed continuous culture; continuous fermentation; open continuous culture.

Continuous fermentation–A process in which cells or micro-organisms are maintained in culture in the exponential growth phase by the continuous addition of fresh medium that is exactly balanced by the removal of cell suspension from the bioreactor. *The opposite word is* continuous culture.

Continuous variation Variation not represented by distinct classes. Phenotypes grade into each other, and measurement data are required for analysis. Multiple genes are usually responsible for this type of variation. a.k.a. quantitative variation. *The opposite word is* discontinuous variation.

Control–1. Noun: Unchanged (standard) protocol or treatment for comparison with the experimental treatment. The term is commonly used for untreated organisms.

2. Verb: To direct or regulate cultures with addition of plant growth regulators.

Controlled environment–The environment in which parameters, such as light, temperature, relative humidity and sometimes the partial gas pressure, are fully controlled.

Controlling element–In eukaryotes, transposable elements which control the activity of standard genes. A controlling element may, in the simplest case, inhibit the activity of a gene through becoming integrated in, or close to, that gene. Occasionally, either in germinal or somatic tissue, it may be excised from this site, and due to excision the activity of the gene is more or less restored, while the element may become reintegrated elsewhere in the genome where it may affect the activity of another gene. For example, in maize, a controlling element such as *Ac* or *Ds* is capable of influencing the expression of a nearby gene. *Search in* transposable element.

Conversion–The development of a somatic embryo into a plant. *Search in* regeneration; micropropagation; organogenesis.

Coordinate repression–Correlated regulation of the structural genes in an operon by a molecule that interacts with the operator sequence.

Co-polymers–Mixtures consisting of more than one monomer; for example, polymers of two kinds of organic bases, such as uracil and cytosine (poly-UC) have been combined for studies of the genetic code.

Copy DNA–*Search in* cDNA.

Copy number–The average number of molecules of a plasmid or gene per genome contained in a cell.

Co-repressor–An effector molecule that forms a complex with a repressor and turns off the expression of a gene or set of genes.

Corpus–The corpus is found below the tunica and is a part of the apical meristem. In the corpus, cells divide in all directions, giving them an increase in volume. *Search in* apical meristem.

Correlation–A statistical association between variables.

Cortex–Primary tissue of a stem or root, bounded externally by the epidermis and internally in the stem by the phloem and in the root by the pericycle.

Cos ends–The 12-base, single-strand, complementary extensions of bacteriophage lambda (λ) (DNA).

Co-segregation–When two genetic conditions appear to be inherited together.

Cosmid–A plasmid vector which contains the two cos (cohesive) ends of phage lambda (λ) and one or more selectable markers such as an antibiotic resistance gene. Cosmids exploit certain properties of phage lambda (λ) to enable large, 40-50 kb, DNA fragments to be cloned at high efficiency. Cosmids and cosmid recombinants replicate as plasmids.

Cos sites–*Search in* cos ends.

Cot curve–When duplex DNA is heated, it dissociates into single strands. When the temperature is lowered, complementary strands tend to anneal or re-nature. The extent of re-naturation depends on the product of DNA concentration in moles of nucleotides per litre, and time in seconds. A graph showing the proportion of re-natured DNA against cot is known as a cot curve. The cot at which half the DNA has re-natured is the half-cot, a parameter indicating the degree of complexity of the DNA.

Co-transfection–In baculovirus expression systems, the procedure by which the baculovirus and the transfer vector are simultaneously introduced into insect cells in culture.

Co-transformation–In genetic engineering experiments, it is often necessary to transform with a plasmid for which there is no selectable phenotype and then screen for the presence of that plasmid within the host cell. Co-transformation is a technique in which host cells are incubated with two types of plasmid, one of which is selectable and the other not. Cells which have been transformed with the first plasmid are then selected. If transformation has been carried out at high DNA concentration, then it is probable that these cells will also have been transformed with the second (non-selectable) plasmid. The technique is frequently used in experiments with mammalian cells.

Cotyledons–Leaflike structures at the first node of the seedling stem. In some dicotyledons, they contain stored food for the young plant not yet able to photosynthesize its own food. Often referred to as seed leaves.

Coupling–The phase state in which either two dominant or two recessive alleles of two different genes occur on the same chromosome. Also called cis configuration. *The opposite word is* repulsion.

Covalent bond–A bond in which an electron pair is equally shared by protons in two adjacent atoms.

Covalently closed circle (CCC)–A double-stranded DNA molecule with no free ends. The two strands are interlinked and will remain together even after denaturation. In its native form, a CCC will adopt a supercoiled configuration.

Co-variance–A measure of the statistical association between variables; the extent to which two variables vary together.

CpDNA–The DNA of plant plastids, including chloroplasts.

Critical breed–A breed where the total number of breeding females is less than 100 or the total number of breeding males is less than or equal to five; or the overall population size is close to, but slightly above 100 and decreasing, and the percentage of pure-breed females is below 80 per cent. (*Source*: FAO, 1999)

Critical-maintained breed; endangered–maintained breed–Categories where critical or endangered breeds are being maintained by an active public conservation programme or within a commercial or research facility. (Source: FAO, 1999)

Cross–In genetic studies, the mating of two individuals or populations. Also called mating.

Cross-breeding–Mating between members of different populations (lines, breeds, races or species).

Cross hybridization–The hydrogen bonding of a single-stranded DNA sequence that is partially but not entirely complementary to a single-stranded substrate. Often, this involves hybridizing a DNA probe for a specific DNA sequence to the homologous sequences of different species.

Crossing over–A process in which homologous chromosomes exchange material through the breakage and reunion of two chromatids. It occurs during Meiosis A single crossover represents one reciprocal breakage and reunion event. A double crossover requires two simultaneous reciprocal breakage and reunion events. a.k.a. recombination; recombination event.

Crossover–A recombinant chromosome.

Crossing-over unit–A measure of distance between two loci on genetic maps that is based on the average number of crossing-over events that take place in the interval between those two loci during meiosis. A map interval that is one crossing-over unit in length (a centiMorgan) describes an interval between two loci such that one in every hundred gametes recovered from meiosis is recombinant in that interval, *i.e.*, the allele at the first locus is maternal in origin, while the allele at the other locus in that same gamete is paternal in origin.

Cross pollination–Fertilization of a plant from a plant with a different genetic makeup.

Cross pollination efficiency–Efficiency of pollen from one plant reaching the stigma of another plant.

Crown gall (A.S. *gealla* gall)–A bulbous growth that occurs at the base of certain plants as result of infection, especially by *Agrobacterium tumefaciens;* a bacterial gene carried by the Ti plasmid is transferred by the bacteria into a higher plant cell, where it causes a tumour-like growth. a.k.a. crown gall tumour. *Search in Agrobacterium;* hairy root disease.

Crown–The region at the base of the stem of cereals and forage species from which tillers or branches arise. In woody plants, the root-stem junction. In forestry, the top portions of the tree.

Cryobiological preservation; cryopreservation; freeze preservation.–The preservation of germplasm resources in a dormant state by cryogenic techniques, as currently applied to storage of plant seeds and pollen, micro-organisms, animal sperm, and tissue culture cell lines. *Search in ex situ* conservation; gene bank.

Cryogenic–At very low temperature.

Cryopreservation–*Search in* cryobiological preservation.

Cryoprotectant–Compound preventing cell damage during freezing and thawing processes. Cryoprotectants are agents with high water solubility and low toxicity. Two types of cryoprotectant agent are commonly used: permeating (glycerol and DMSO) and non-permeating (sugars, dextran, ethylene glycol, polyvinyl pyrolidone and hydroxyethyl starch).

Cryptic–1. Structurally heterozygous individuals not identifiable on the basis of abnormal meiotic-chromosome pairing configurations ('cryptic structural hybrids').

2. A form of polymorphism controlled by recessive genes ('cryptic polymorphism').

3. Any mutation which is exposed by a sensitizing mutation and otherwise poorly detected (such mutations probably escape detection because of the plasticity of composition of the corresponding polypeptide).

4. Phenotypically very similar species (cryptic species) which do not hybridize under normal conditions.

5. Cryptic genetic variation refers to the existence of, for example, alleles conferring high performance for a trait, in a breed that has low performance for that trait.

Cultivar (from cultivated + variety) (abbr: cv.)–A category of plants that are, firstly, below the level of a sub-species taxonomically, and, secondly, found only in cultivation. It is an international term denoting certain cultivated plants that are clearly distinguishable from others by stated characteristics and that retain their distinguishing characters when reproduced under specific conditions.

Culture–A population of plant or animal cells or micro-organisms that is grown under controlled conditions.

Culture alteration–A term used to indicate a persistent change in the properties of a culture's behaviour (*e.g.*, altered morphology, chromosome constitution, virus susceptibility, nutritional requirements, proliferative capacity, etc.). The term should always be qualified by a precise description of the change which has occurred in the culture.

Culture medium–Any nutrient system for the cultivation of cells of plants, bacteria or other organisms; usually a complex mixture of organic and inorganic nutrients. *The opposite word is* medium.

Culture room–Room for maintaining cultures and often in a controlled environment of light, temperature and humidity. *The opposite word is* growth cabinet; incubator.

Curing–The elimination of a plasmid from its host cell. Many agents which interfere with DNA replication, *e.g.*, ethidium bromide, can cure plasmids from either bacterial or eukaryotic cells.

Cut–Slang: to make a double-stranded break in DNA, usually with a type II restriction endonuclease. *E.g.*, "The DNA was cut with *Eco*RI and run out on a 1 per cent agarose gel." *The opposite word is* nick; cleave.

Cuticle (L. *cuticula*, diminutive of *cutis*, the skin)–Layer of cutin or wax on the outer surface of leaves and fruits and that reduces water loss.

Cutting (noun)–A detached plant part that under appropriate cultural conditions can regenerate the complete plant without a sexual process.

Cybrid–A cytoplasmic hybrid, originating from the fusion of a cytoplast (the cytoplasm without nucleus) with a whole cell, as in nuclear transfer (although the term is not used in that context). Note that the nucleus and cytoplasm of the fused cell products are from different genetic sources.

Cyclic AMP (cyclic adenosine monophosphate)–A "messenger" that regulates many intracellular reactions by transducing signals from extracellular growth factors to cellular metabolic pathways.

Cyclodextrin–Cyclic polymer of dextrose.

Cystein–An amino acid.

Cytogenetics–Area of biology concerned with chromosomes and their implications in genetics, cellular activity and variability.

Cytokine–In immunology, any of many soluble molecules that cells produce to control reactions between other cells.

Cytokinesis–Cytoplasmic division and other changes exclusive of nuclear division that are a part of mitosis or meiosis. *Search in* cell division.

Cytokinin–Plant growth regulators (hormones) characterized as substances that induce cell division and cell differentiation (*e.g.*, BA, kinetin, and 2-iP). In tissue culture, these substances are associated with enhanced callus and shoot development. The compounds are derivatives of adenine.

Cytology–The study of the structure and function of cells.

Cytolysis–Cell disintegration or dissolution of cells.

Cytoplasm (Gr. *kytos*, a hollow vessel + *plasma*, form)–The living material of the cell, exclusive of the nucleus, consisting of a complex protein matrix or gel. The part of the cell in which essential membranes and cellular organelles (mitochondria, plastids, etc.) reside.

Cytoplasmic genes–DNA-containing bodies in the cell but external to the nucleus.

Cytoplasmic inheritance–Hereditary transmission dependent on the cytoplasm or structures in the cytoplasm rather than the nuclear genes; extrachromosomal inheritance. Thus, plastid characteristics in plants are inherited by a mechanism independent of nuclear genes.

Cytoplasmic male sterility–Genetic defect due to defective functions of mitochondria in the pollen. Fertilization will not occur. Exploited in certain plant breeding strategies, such as F_1-hybrid maize cultivars.

Cytoplasmic organelles–Discrete subcellular structures located in the cytoplasm of cells; these allow division of labour within the cell.

Cytosine–A pyrimidine base found in DNA and RNA.

Cytosol–The fluid portion of the cytoplasm, *i.e.*, the cytoplasm minus its organelles.

Cytotoxic T cell–*Search in* killer T cell.

Cytotype–A maternally inherited cellular condition in *Drosophila* that regulates the activity of transposable P elements.

Dabs (single-domain antibodies)–Antibodies with only one (instead of two) protein chain derived from only one of the domains of the antibody structure. Dabs exploit the finding that, for some antibodies, half of the antibody molecule binds to its target antigen almost as well as the whole molecule. The potential advantages of dabs are that they can be made easily by bacteria or yeasts, and offer a way to clone antibody-like molecules into bacteria, and hence to be able to easily screen millions of antibodies. Related ideas are single-chain antigen binding technology (SCA), biosynthetic antibody binding sites (BABS), minimum recognition units (MRUs), and complementary determining regions (CDRs).

DAD–*Search in* domestic animal diversity.

dA-dT tailing–*Search in* complementary homopolymeric tailing.

Dalton (symbol: Da)–A unit of atomic mass roughly equivalent to the mass of a hydrogen atom. 1.67×10^{-24} g. Named after the famous nineteenth-century chemist, John Dalton (1766-1844)). Used in shorthand expressions of molecular weight, especially as kilo–(kDa) or megadaltons (MDa), which are equal to respectively to 1×10^3 and 1×10^6 daltons.

Darwinian cloning–Selection of a clone from a large number of essentially random starting points, rather than isolating a natural gene or making a carefully designed artificial one. Molecules which are more similar to those needed are selected, mutated to generate new variants, and re-selected. The cycle proceeds until the required molecule is found. The advantage of the system is that the selection is from a vast number of possibilities.

ddNTP–*Search in* di-deoxynucleotide.

Death phase–The final growth phase, during which nutrients have been depleted and cell number decreases. *Search in* growth phases.

Deceleration phase–The phase of declining growth rate, following the linear phase and preceding the stationary phase in most batch-suspension cultures. *Search in* growth phases.

De-differentiation–The process by which cells lose their specialization and proliferate by cell division to form a mass of cells which, in response to appropriate stimuli, may later differentiate again to form either the same cell type or a different one. De-differentiation occurs in response to wounding and in tissue cultures. *Search in* re-differentiation.

Deficiency–Insufficiency or absence of one or more usable forms of enzymatic, nutritional or environmental requirements, so that development, growth or physiological functions are affected.

Defined–1. Fixed conditions of medium, environment and protocol for growth.

2. Precisely known and stated elements of a tissue culture medium. *The opposite word is* undefined.

Degeneracy (of the genetic code)–The specification of one amino acid by more than one codon. It arises from the inevitable redundancy resulting from 64 triplets in a triplet code (4 x 4 x 4 = 64) encoding only 20 amino acids.

Degeneration–1. Changes in cells, tissues or organs due to disease.

2. The reduction in size or complete loss of organs during evolution.

Dehalogenation–The removal of halogen atoms (chlorine, iodine, bromine and fluorine) from molecules, usually during biodegradation.

Dehiscence–The spontaneous and often violent opening of a fruit, seed pod or anther to release and disperse the seeds or pollen.

Dehydrogenase–An enzyme that catalyses the remove of hydrogen atoms in biological reactions.

Dehydrogenation–A chemical reaction in which hydrogen is removed from a compound.

De-ionized water–Water which is free of most inorganic (not completely free, since Na is present in ample quantities) and most organic compounds.

Deletion–A mutation involving the removal of one or more base pairs in DNA sequence. Large deletions are visible as the lack of chromosomal segments.

Deliberate release–Putting something into the outside world; in biotechnology it means putting a genetically modified organism (GMO) into field trials.

Deme–A group of organisms in the same taxon.

De-mineralize To remove the mineral content (salts, ions) from a substance, especially water. Removal methods include distillation and electrodialysis. The process is de-mineralization.

Denaturated DNA–Duplex DNA that has been converted to single strands by breaking the hydrogen bonds of complementary nucleotide pairs. Usually achieved by heating.

Denaturation–Loss of native configuration of a macro-molecule (protein or nucleic acid) by physical or chemical means, usually accompanied by loss of biological activity. Denatured proteins often unfold their polypeptide chains and express changed properties of solubility. The separation of duplex nucleic acid molecules into single strands. Most commonly used by genetic engineers to describe the destruction of hydrogen bonds maintaining the double-stranded nature of all or part of a DNA molecule.

Denature–To induce structural alterations that disrupt the biological activity of a molecule. Often refers to breaking hydrogen bonds between base pairs in double-stranded nucleic acid molecules to produce single-stranded polynucleotides, or altering the secondary and tertiary structure of a protein, destroying its activity.

Denitrification–A chemical process in which nitrates in the soil are reduced to molecular nitrogen, which is released to the atmosphere.

de novo (L. "from the beginning, anew")–Arising, anew, afresh, once more. Also *ex novo*.

Density gradient centrifugation–High-speed centrifugation in which molecules "float" at a point where their density equals that in a gradient of caesium chloride or sucrose. The density gradient may either be formed before centrifugation by mixing two solutions of different density (as in sucrose density gradients) or it can be formed by the process of centrifugation itself (as in CsCl and Cs_2SO_4 density gradients). *Search in* centrifugation.

Deoxyribonucleic acid–*Search in* DNA.

Deoxyribonuclease (DNase)–Any enzyme that hydrolyses DNA.

De-repression–The process of "turning on" the expression of a gene or set of genes whose expression has been repressed (turned off). Displacement of a repressor protein from a promoter region of DNA. When attached to the DNA, the repressor protein prevents RNA polymerase from initiating transcription. The "turning on" of a gene.

Derivative–1. Resulting from or derived from.

2. Term used to identify a variant during meristematic cell division.

Desiccant–Any compound used to remove moisture or water.

Desiccate–To dry, exhaust or deprive of water or moisture. Any chemical used for this purpose is called a desiccant. An apparatus for drying and preventing hygroscopic samples from rehydrating is a desiccator. The process is desiccation.

Dessicator–Apparatus for drying or depriving of moisture.

Desoxyribonucleic acid–Obsolete spelling of deoxyribonucleic acid (DNA).

Desulphurization (USA: desulfurization)–Technology for removing sulphur from oil and coal by use of bacteria. Sulphur residues in fuels end up as sulphur dioxide when the fuel is burned, resulting in acid rain. Bacteria may oxidize sulphites (insoluble) into sulphates (soluble), which can be washed away with the bacteria.

Detergent–Substance which lowers the surface tension of a solution, improving its cleaning properties (*e.g.*, Tween-20TM, a surfactant and wetting agent). *Search in* surfactant; wetting agent.

Determinate growth–Growth determined and limited in time, as in most floral meristems and leaves. The differentiation process is irreversibly established. Determinate growth contrasts with the usual culture growth, which is infinite and indeterminate.

Determination–Process by which undifferentiated cells in an embryo become committed to develop into specific cell types, such as neurons, fibroblasts or muscle cells.

Determined–Describing embryonic tissue at a stage when it can develop only as a certain kind of tissue.

Development–The sum total of events that contribute to the progressive elaboration of an organism. The two major aspects of development are growth and differentiation.

Deviation–1. In statistics: the difference between an actual observation and the mean of all observations.

2. An alteration from the typical form, function or behaviour. Mutation or stress are the common reasons behind deviation.

Dextrin–An intermediate polysaccharide compound resulting from the hydrolysis of starch to maltose by amylase enzyme.

Dextrose–*Search in* glucose.

dG-dC tailing–*Search in* complementary homopolymeric tailing.

Diabetes–A disease associated with the absence or reduced levels of insulin, which is a hormone essential for the transport of glucose to cells.

Diagnostic procedure–A test or assay used to determine the presence of an organism, substance or nucleic acid sequence alteration.

Diakinesis–A stage of meiosis just before metaphase I, in which the separation of homologous chromosomes is almost completed.

Diazotroph–An organism that can fix atmospheric nitrogen.

Dicentric chromosome–A chromosome having two centromeres.

Dichogamy–The condition in which the male and the female reproductive organs of a flower mature at different times, thereby making self-fertilization improbable or impossible.

Dicot–*Search in* dicotyledon.

Dicotyledon (Gr. *dis*, twice + *kotyledon*, a cup-shaped hollow)–A plant with two cotyledons, or seed leaves. One of the two classes of plants in the Angiosperms (the other class is the monocotyledons). Colloquially called a dicot. Examples include many crop plants (potato, pea, beans), ornamentals (rose, ivy) and timber trees (oak, beech, lime).

Di-deoxynucleotide (ddNTP)–A deoxynucleotide that lacks a 3´–hydroxyl group, and is thus unable to form the 3´-5´ phospho-diester bond necessary for chain elongation. Di-deoxynucleotides are used in DNA sequencing and the treatment of viral diseases. Also sometimes referred to as didN. *Search in* nucleotide.

didN–*Search in* di-deoxynucleotide (ddNTP).

Differential centrifugation–A method of separating sub-cellular particles according to their sedimentation coefficients, which are roughly proportional to their size. Cell extracts are subjected to a succession of centrifuge runs at progressively faster rotation speeds. Large particles, such as nuclei or mitochondria, will be precipitated at relatively slow speeds; higher G forces will be required to sediment small particles, such as ribosomes.

Differentially permeable–Referring to a membrane, through which different substances diffuse at different rates; some substances may be unable to diffuse through such a membrane.

Differentiation (L. *differre*, to carry different ways)–A process in which unspecialized cells develop structures and functions characteristic of a particular type of cell.

Development from one cell to many cells, accompanied by a modification of the new cells for the performance of particular functions. In tissue culture, the term is used to describe the formation of different cell types.

Diffusion (L. *diffusus*, spread out)–The movement of molecules from a region of higher concentration to a region of lower concentration.

Digest–To cut DNA molecules with one or more restriction endonucleases. *Search in* cleave.

Dihaploid–An individual which arises from a doubled haploid.

Dihybrid–An individual that is heterozygous for two pairs of alleles; the progeny of a cross between homozygous parents differing at two loci.

Dimer–Association of two molecules.

Dimethyl sulphoxide; dimethyl sulfoxide (C_2H_6OS; m.w. 78.13)–A highly hygroscopic liquid and powerful solvent with little odour or colour. It is an organic co-solvent used in small quantities to dissolve neutral organic substances in tissue culture media preparation. DMSO also has uses as a cryoprotectant.

Dimorphism–The existence of two distinctly different types of individuals within a species. An obvious example is the sexual dimorphism in certain animals.

Diplochromosome–*Search in* endoreduplication.

Diploid (Gr. *diploos*, double + *oides*, like)–1. The status of having two complete sets of chromosomes, most commonly one set of paternal origin and the other of maternal origin.

2. An organism or cell with a double set (2n) of chromosomes (most commonly one of paternal origin, and the other of maternal origin), or referring to an individual containing a double set of chromosomes per cell. Somatic tissues of higher plants and animals are ordinarily diploid in chromosome constitution, in contrast with the haploid gametes.

Diploid cell–A cell which contains two sets of chromosomes. *The opposite word is* haploid cell.

Diplonema (adj: diplotene)–Stage in prophase of meiosis I following the pachytene stage, but preceding diakinesis, in which one pair of sister chromatids begin to separate from the other pair, *i.e.*, the centromeres begin to disjoin.

Diplophase Phase with 2n chromosomes.

Direct embryogenesis–Embryoids form directly in culture, without an intervening callus phase, on the surface of zygotic or somatic embryos or on explant tissues (leaf section, root tip, etc.).

Direct organogenesis–Formation of organs directly on the surface of cultured intact explants. The process does not involve callus formation.

Direct repeat–Two or more stretches of DNA within a single molecule which have the same nucleotide sequence in the same orientation. Direct repeats may be either adjacent to one another or far apart on the same molecule. For example:

TATTA...TATTA

ATAAT...ATAAT

Directed mutagenesis–The process of generation of nucleotide changes in cloned genes by any one of several procedures, including site-specific and random mutagenesis. Also called *in vitro* mutagenesis.

Directional cloning–The technique by which DNA insert and vector molecules are digested with two different restriction enzymes to create non-complementary sticky ends at either end of each restriction fragment, so allowing the insert to be ligated to the vector in a specific orientation and preventing the vector from re-circularizing. *Search in* cloning.

Disarm–To delete from a plasmid or virus genes that are cytotoxic or tumour inducing.

Discontinuous variation–Phenotypic variation involving distinct classes, such as red versus white, tall versus dwarf. *Search in* continuous variation.

Discordant–Members of a pair showing different, rather than similar, characteristics.

Disease (L. *dis*, a prefix signifying the opposite + M.E. *aise*, comfort)–The opposite of ease. Any alteration from the state of metabolism necessary for the normal development and functioning of an organism, usually associated with infection by a pathogen or the malfunction or absence of one or more genes.

Disease resistance–The ability to remain healthy by resisting disease or the disease agent. Disease resistance or tolerance is a subject of intense interest in biotechnology.

Disease-free–A plant or animal certified through specific tests as being free of specified pathogens. Disease-free should be interpreted to mean "free from any known diseases" as "new" diseases may yet be discovered to be present. *The opposite word is* disease indexing.

Disease-indexing–Disease-indexed organisms have been assayed for the presence of known diseases according to standard testing procedures. *The opposite word is* disease-free.

Disinfectation–Full elimination of internal micro-organisms from a culture; disinfectation is rarely obtained. *The opposite word is* disinfestation.

Disinfestation–The elimination or inhibition of the activity of surface-adhering micro-organisms. *The opposite word is* disinfection.

Disjunction–Separation of homologous chromosomes during anaphase I of meiosis; separation of sister chromatids during anaphase of mitosis and anaphase II of meiosis. As soon as the sister chromatids have separated, they are each called a chromosome. *Search in* non-disjunction.

Disomy (adj: disomic)–The presence of a pair of specific chromosome. This is the normal condition, and abnormal occurrences are monosomy, trisomy and nullisomy (with respectively one chromosome of a pair, three or none). There are also abnormal disomic conditions, such as when both chromosomes of the pair were inherited from the same parent.

Dispense–Portion out a nutrient medium into containers, such as test tubes, jars, Erlenmeyer flasks, Petri dishes, etc.

Disaccharide–A carbohydrate consisting of two linked sugar units. *e.g.* Sucrose.

Dissecting microscope–A microscope with a low magnifying power of about 50×, used to examine or excise small plant or animal parts.

Dissection (L. *dissectio,* a dissecting or being dissected)–Separation of a tissue by cutting for analysis or observation.

Dissolve–Pass chemicals into solution.

Distillation (L. *distillatio,* a distilling process)–The process of heating a mixture to separate the more volatile from the less volatile parts, and then cooling and condensing the resulting vapour so as to produce a more nearly pure or refined substance.

Di–sulphide bond–A chemical bond that stabilizes the three-dimensional structure of proteins, and hence the protein's normal function. They form between cysteine residues in the same or different peptide molecules.

Di-sulphide bridge–*Search in* di-sulphide bond.

Ditype–In fungi, a tetrad that contains two kinds of meiotic products (spores), *e.g.*, 2AB and 2ab. *Search in* tetrad (1).

Diurnal–Term describing the occurrence of an event at least once every 24 hours. *The opposite word is* circadian.

Divergent evolution–*Search in* adaptive radiation.

Dizygotic twins–Two-egg twins, *i.e.*, a pair of individuals that shared the same uterus at the same time, but which arose from separate and independent fertilization of two ova.

DMSO–*Search in* dimethyl sulphoxide.

DNA (deoxyribonucleic acid; formerly spelt desoxyribonucleic acid)–The long chain of molecules in most cells that carries the genetic message and controls all cellular functions in most forms of life. The information-carrying genetic material that comprises the genes. DNA is a macro-molecule composed of a long chain of deoxyribonucleotides joined by phospho-diester linkages. Each deoxyribonucleotide contains a phosphate group, the five-carbon sugar 2-deoxribose, and a nitrogen-containing base. The genetic material of most organisms and organelles so far examined is double-stranded DNA; a number of viral genomes consist of single-stranded DNA or single-or double-stranded RNA. In double-stranded DNA, the two strands run in opposite (anti-parallel) directions and are coiled round one another in a double helix. Purine bases on one strand specifically hydrogen bond with pyrimidine bases on the other strand, according to the Watson-Crick rules (A pairs with T; G pairs with C). Hence a constant width for the double helix of 20 Å (2.0 nm) is maintained. In the B-form, DNA adopts a right-handed helical conformation, with each chain making a complete turn every 34 Å (3.4 nm), or once every ten bases. *Search in* mtDNA.

DNA amplification–Multiplication of a piece of DNA in a test-tube into many thousands of millions of copies. The most commonly used process is the polymerase chain reaction (PCR) system, but other systems are being developed, including ligase chain reaction (LCR), nucleic acids sequence-dependent amplification, and the Q-? system.

DNase (deoxyribonuclease)–An enzyme that catalyses the cleavage of DNA. DNase I is a digestive enzyme secreted by the pancreas, that degrades DNA into shorter nucleotide fragments. Many other endonucleases and exonucleases are involved in DNA repair and replication. *The opposite word is* nuclease.

DNA bank–Storage of DNA, which may or may not be the complete genome, but should always be accompanied by inventory information. (*Note:* at the present time, animals cannot be re-established from DNA alone.) (Source: FAO, 1999)

DNA chip–*Search in* DNA micro-array.

DNA cloning–*Search in* gene cloning.

DNA construct–A DNA molecule inserted into a cloning vector, usually a plasmid.

DNA diagnosis–The use of DNA polymorphisms to detect the presence of a specific allele (often associated with a disease or syndrome) or DNA sequence.

DNA fingerprint–The unique pattern of DNA fragments identified originally by Southern hybridization (using a probe that binds to a polymorphic region of DNA) or now by polymerase chain reaction (PCR) (using primers flanking the polymorphic region). *Search in* genetic fingerprinting.

DNA helicase (gyrase)–An enzyme that catalyses the unwinding of the complementary strands of a DNA double helix.

DNA hybridization–The pairing of two DNA molecules, often from different sources, by hydrogen bonding between complementary nucleotides. This technique is frequently used to detect the presence of a specific nucleotide sequence in a DNA sample.

DNA ligase–An enzyme that catalyses a reaction that links two DNA molecules via the formation of a phospho-diester bond between the 3′ hydroxyl and 5′ phosphate of adjacent nucleotides. It plays an important role in DNA repair and replication. DNA ligase is one of the essential tools of recombinant DNA technology, enabling (among other things) the incorporation of foreign DNA into vectors. The ligase enzyme encoded by phage T4 is commonly used in gene-cloning experiments. It requires ATP as a co-factor. T4 is used *in vitro* to join the vector and insert DNAs.

DNA micro-array–A small glass surface to which has been fixed an array of DNA fragments, each with a defined location. A typical DNA chip would contain 10 000 discrete spots (each containing a different DNA fragment) in an area of just a few square centimetres. When a solution of fluorescently labelled DNA fragments is hybridized to the chip, spots to which hybridization occurs are visible as fluorescence. If the spots on the chip are genes (expressed sequence tags, q.v.), hybridization with cDNA from a particular tissue shows which genes are expressed in that tissue. If the spots are short, synthesized oligonucleotides (approximately 25 bases) corresponding to that part of a gene containing a single nucleotide polymorphisms (SNP), with a separate spot for each of the 4 possible bases at that site, hybridization with genomic DNA from an individual plant or animal enables that individual to be genotyped at as many SNP loci as are represented on the chip. The big advantage of DNA chips is the extent to which the process of genotyping can be automated, thereby enabling huge numbers of plants or animals to be genotyped for a huge number of loci.

DNA polymerase–An enzyme that catalyses the synthesis of double-stranded DNA, using single-stranded DNA as a template. *Search in* polymerase.

DNA polymorphism–The existence of two or more alternative forms (alleles) of a chromosomal locus that differ in nucleotide sequence or have variable numbers of repeated nucleotide units. *Search in* allele.

DNA primase–An enzyme that catalyses the synthesis of short strands of RNA that initiate the synthesis of DNA strands.

DNA probe–A labelled (tagged) segment of DNA that is able, after a DNA hybridization reaction, to detect a specific DNA sequence in a mixture of sequences. If the tagged sequence is complementary to any one in the mixture, the two sequences will form a double helix. This will be identified thanks to its label (either by radioactivity or fluorescence).

DNA repair–A variety of mechanisms that repair errors that occur during DNA replication.

DNA repair enzymes–Enzymes that catalyse the repair of DNA.

DNA replication–The process whereby DNA makes exact copies itself, under the action of and control of DNA polymerase.

DNAase–*Search in* DNase.

DNA sequencing–1. Procedures for determining the nucleotide sequence of a DNA fragment. There are two common methods for doing this:

2. The Maxam and Gilbert technique (chemical degradation), that uses different chemicals to break the DNA into fragments at specific bases; or

3. The Sanger technique (called the di-deoxy or chain-terminating method) uses DNA polymerase to make new DNA chains, with di-deoxy nucleotides (chain terminators) to stop the chain randomly as it grows.

In both cases, the DNA fragments are separated according to length by polyacrylamide gel electrophoresis, enabling the sequence to be read directly from the gel.

DNA topo-isomerase–An enzyme that catalyses the introduction or removal of supercoils in DNA.

DNA transformation–*Search in* transfection; transformation.

DNA delivery system–A generic term for any procedure that transports DNA into a recipient cell.

Dolly–The [name of the] first mammal to be created by cloning a cell from an adult animal. In this particular case, the cell came from the mammary tissue of an adult ewe. The creation of Dolly showed that the process of differentiation into adult tissue is not, as previously thought, irreversible. The result was achieved by nuclear transfer. Dolly's birth was announced in 1997. Since then, cattle and mice have also been cloned from adult cells.

Domain–A segment of a protein that has a discrete function or conformation. At the protein level, a domain can be as small as a few amino acid residues or as large as half of the entire protein.

Domestic animal diversity (DAD)–The spectrum of genetic differences within each breed, and across all breeds within each domestic animal species, together with the species differences; all of which are available for the sustainable intensification of food and agriculture production.

(*Source*: FAO, 1999)

Dominance–The type of gene action exhibited by a dominant allele.

Dominant–1. Describing an allele whose effect with respect to a particular trait is the same in heterozygotes as in homozygotes. The opposite is *recessive.*

2. Describing the most conspicuously abundant and characteristic species of a community.

3. Describing an animal that is allowed priority in access to food, mates, etc., by others of its species because of its success in previous aggressive encounters.

Dominant marker selection–Selection of cells via a gene encoding a product that enables only the cells that carry the gene to grow under particular conditions. For example, plant and animal cells that express the introduced Neo^r gene are resistant to the compound G418, while cells that do not carry the Neo^r gene are killed by G418. a.k.a. positive selection; *Search in also* positive selectable marker.

Dominant selectable marker gene–A gene that allows the host cell to survive under conditions where it would otherwise die.

Dominant(–acting) oncogene–A gene that stimulates cell proliferation and contributes to oncogenesis when present in a single copy. *Search in* oncogene.

Donor plant (mother plant)–An explant, graft or cutting used as a source of plant material for micropropagation purposes. a.k.a. ortet. *The opposite word is* explant.

Dormancy (F. *dormir*, from L. *dormire*, to sleep)–An inactive period in the life of an animal or plant during which growth slows or completely ceases. Physiological changes associated with dormancy help the organism survive adverse environmental conditions. Annual plants survive the winter as dormant seeds, while many perennial plants survive as dormant tubers, rhizomes, or bulbs. Hibernation and aestivation in animals help them survive extremes of cold and heat, respectively. *The opposite word is* rest period.

Dosage compensation–A phenomenon whereby inactivation of all but one of the X chromosomes in female mammals results in males and females producing the same quantity of peptide from X-linked genes.

Double crossing-over–Two simultaneous reciprocal breakage and reunion events between the same two chromatids.

Double helix–Describes the coiling of the antiparallel strands of the DNA molecule, resembling a spiral staircase in which the paired bases form the steps and the sugar-phosphate backbones form the rails.

Double fertilization–A process, unique to flowering plants, in which two male nuclei, which have travelled down the pollen tube, separately fuse with different female nuclei in the embryo sac. The first male nucleus fuses with the egg cell to form the zygote; the second male nucleus fuses with the two polar nuclei to form a triploid nucleus that develops into the endosperm.

Double recessive–An organism homozygous for a recessive allele at each of two loci.

Double-stranded complementary DNA (dscDNA)–A double strand DNA molecule created from a cDNA template.

Doubling time–*Search in* generation time.

Downstream–1. In molecular biology, the stretch of nucleotides of DNA that lie in the 3′ direction from the site of initiation of transcription, which is designated as +1 (remembering the convention that the sequence of a DNA molecule is written from the 5′ end to the 3′ end). Downstream nucleotides are marked with plus signs, *e.g.*, +2, +10. Also, to the 3′ side of a particular gene or sequence of nucleotides.

2. In chemical engineering, those phases of a manufacturing process that follow the biotransformation stage. Usually refers to the recovery and purification of the product of a fermentation process. *Search in* downstream processing.

Downstream processing–A general term for all the things which happen in a biotechnological process after the biology, be it fermentation of a micro-organism or growth of a plant. It is particularly relevant to fermentation processes, which produce a large quantity of a dilute mixture of substances, products and micro-organisms. These must be separated, and the product must be concentrated and purified, and converted into a form which is useful. *Search in* downstream.

Drift–*Search in* genetic drift.

Drug–*Search in* therapeutic agent.

Drug delivery–Method by which a therapeutic agent is delivered to its site of action. For traditional therapeutic agents this is another name for *formulation*. However, biotechnology has allowed the development of a range of new therapeutic-agent delivery systems, such as liposomes and other encapsulation techniques, and a range of mechanisms that target a therapeutic agent to a particular cell or tissue. *Search in* therapeutic agent.

Dry weight–The moisture-free weight of tissue obtained by drying at high (oven-drying) or low (freeze-drying) temperatures for an interval sufficient to remove all water and moisture.

dscDNA–*Search in* double-stranded complementary DNA.

dsDNA–Double-stranded DNA.

Dual culture–A culture made of a plant tissue and one organism (such as a nematode) or an obligate parasite/micro-organism (such as a fungus). Dual culture techniques are used for a variety of purposes, including assessing host-parasite interactions and the production of axenic cultures.

Duplex DNA–Double-stranded DNA.

Duplication–The occurrence of a segment more than once in the same chromosome or genome.

EBV–*Search in* estimated breeding value.

EC–*Search in* Enzyme Commission.

Ecdysone–A steroid hormone in insects that stimulates moulting and metamorphosis. It acts on specific genes, stimulating the synthesis of proteins involved in these bodily changes.

Eclosion–1. Emergence of an adult insect from the pupal stage.

2. Beginning of germination of fungal spores.

E. coli–*Search in Escherichia coli.*

Ecological diversity–*Search in* biodiversity.

Ecology–The study of the interactions between organisms and their natural environment, both living and non-living.

Economic trait locus (ETL)–A locus influencing a trait that contributes to income. The plural form (economic trait loci) is also abbreviated as ETL.

Ecosystem–The complex of a living community and its environment, functioning as an ecological unit in nature. *The opposite word is* abiotic factors; biotic factors.

Ecotype–A population or a strain of an organism that is adapted to a particular habitat.

EDTA–*Search in* ethylene-diamine tetra-acetic acid.

Effector cells–Cells of the immune system that are responsible for cell-mediated cytotoxicity.

Effector molecule–A molecule that influences the behaviour of a regulatory molecule, such as a repressor protein, thereby influencing gene expression.

Egg–1. The fertilized ovum (zygote) in egg-laying animals after it emerges from the body.

2. The mature female reproductive cell in animals and plants.

Search also ovum.

EGS–External guide sequence. *Search in* guide sequence.

Elastin–A fibrous protein that is the major constituent of the yellow elastic fibres of connective tissue.

Electro-blotting–The electrophoretic transfer of macromolecules (DNA, RNA or protein) from a gel, in which they have been separated, to a support matrix, such as a nitrocellulose sheet. A transfer usually used in techniques such as Southern and northern blotting.

Electrochemical sensor–Type of biosensor in which a biological process is harnessed to an electrical sensor system, such as an enzyme electrode. Other types couple a biological event to an electrical one via a range of mechanisms, such as those based on oxygen and pH. *Search in* enzyme electrode.

Electromagnetic radiation–Electromagnetic waves, including ultraviolet (UV), X-rays, and gamma radiation (γ–rays). Electromagnetic radiation is used to produce mutant cells or organisms, or, in the case of UV, disinfestation and sterilization, in tissue culture.

Electromagnetic spectrum–The range of wavelengths or frequencies over which electromagnetic radiation extends.

Electron microscope–A microscope that uses an electron beam focused by magnetic "lenses." *Search in* scanning electron microscope; transmission electron microscope.

Electrophoresis–A technique that separates charged molecules–such as DNA, RNA or protein–on the basis of relative migration in an appropriate matrix (such as agarose gel or polyacrylamide gel) subjected to an electric field. *Search in* agarose gel electrophoresis; polyacrylamide gel electrophoresis (PAGE); pulsed-field gel electrophoresis (PFGE).

Electroporation–1. An electrical treatment of cells that induces transient pores, through which DNA can enter the cell.

2. The introduction of DNA or RNA into protoplasts or other cells by the momentary disruption of the cell membrane through exposure to an intense electric field.

ELISA (enzyme-linked immunosorbent assay)–A sensitive technique for accurately determining specific molecules in a mixed sample. The amount of protein or other antigen in a given sample is determined by means of an enzyme-catalysed colour change, avoiding both the hazards and expense of radioactive techniques.

It takes various forms. In the most common form, two antibody preparations are used in ELISA. An antibody (primary) specific to the test protein is adsorbed onto a solid substrate, and a known amount of the sample is added; all molecules of the test protein in the sample are bound by the antibody. A second antibody specific for a second site on the test protein is added; this is conjugated with an enzyme, which catalyses a colour change in the fourth reagent, added finally. The color change is measured photometrically and compared against a standard curve to give the concentration of protein in the sample. ELISA is widely used for diagnostic and other purposes.

Elongation factors–Soluble proteins that are required for polypeptide chain elongation.

Embryo (Gr. *en*, in + *bryein*, to swell)–An immature organism in the early stages of development. In mammals, this occurs in the first months in the uterus. In plants, it is the structure that develops in the megagametophyte, as result of the fertilization of an egg cell, or without fertilization. In aseptic cultures, adventitious embryos show polarization, followed by the growth of a shoot from one end and a root from the other end.

Embryo cloning–The creation of identical copies of an embryo by embryo splitting or by nuclear transfer from undifferentiated embryonic cells.

Embryo culture–The culture of embryos on nutrient media.

Embryogenesis–1. (General) Development of an embryo.

2. (In plants) *In vitro* formation of plants from plant tissues, through a pathway closely resembling normal embryogeny from the zygote; if this development in culture involves somatic cells and not the zygote, it can be indicated by using the term *adventitious embryogenesis* or *somatic embryogenesis*. The generation of embryos has two stages: initiation and maturation. Initiation needs a high level of the group of plant hormones called auxins; maturation needs a lower level. Other chemicals have to be at suitable levels. The procedure involves the explanting of a piece of plant tissue and putting it on a high-auxin medium, where the cells grow into a mass of callus. This is then transferred to a maturation medium, where the callus starts to initialize organs, ultimately growing a root and a shoot.

Embryo multiplication and transfer (EMT)–The cloning of animal embryos and their subsequent transfer to recipients (via artificial inembryonation). The cloned embryos can be clones of an embryo or of an adult.

Embryo sac–A large thin-walled space within the ovule of the seed plant in which the egg and, after fertilization, the embryo develop; the mature female gametophyte in angiosperms. Generally a seven-celled structure. The seven cells are two synergids, one egg cell, three antipodal cells (each with a single haploid nucleus) and one endosperm mother cell with two haploid nuclei.

Embryo sexing–The determination of the sex of an embryo, typically by means of PCR involving amplification from a small sample of embryonic tissue, using primers specific for a locus on the Y chromosome.

Embryo splitting–The splitting of young embryos into several sections, each of which develops into an animal. A form of animal cloning, *i.e.*, of producing animals that are genetically identical. In practice, the number of identicals (identical organisms) that can be produced from a single embryo is less than 10.

Embryo technology–Generic name for any modification of mammalian embryos. It encompasses embryo cloning, embryo splitting, *in vitro* fertilization, and embryo storage.

Embryo transfer (ET)–*Search in* multiple ovulation and embryo transfer.

Embryoid–An embryo-like body developing *in vitro*. It forms a complete, self-contained platelet with no vascular connection with the callus. The term embryoid is no longer commonly used. *Search in* embryo.

Embryonic stem cells–Cells of the early embryo that can give rise to all differentiated cells, including germ line cells.

Empirical–Relating to or based upon practical experience, trial and error, direct observation or observation alone, without benefit of scientific method, knowledge or theory.

EMT–*Search in* embryo multiplication and transfer.

Encapsidation–The process by which a virus' nucleic acid is enclosed in a capsid. *Search in* capsid; coat protein.

Encapsulation–Any method of getting something, usually an enzyme or bacterium, into a small package or capsule while it is still working or alive. It is a method for immobilizing cells for use in a bioreactor.

Encapsulating agents–Anything which forms a shell around an enzyme or bacterium, although the agents used are usually polysaccharides such as alginate or agar. The agents are inert and allow nutrients and oxygen to diffuse into and out of the sphere readily, and are easy to convert from gel (solid) to sol (liquid) or solution form by altering the temperature or the concentration of ions.

Encode–To specify, after decoding by transcription and translation, the sequence of amino acids in a protein.

5′ end–The phosphate group that is attached to the 5′ carbon atom of a sugar (ribose or deoxyribose) of the terminal nucleotide of a nucleic acid molecule.

Endangered breed–A breed where the total number of breeding females is between 100 and 1 000 or the total number of breeding males is less than or equal to 20 and greater than five; or the overall population size is close to, but slightly above 100 and increasing, and the percentage of pure-bred females is above 80 per cent ; or the overall population size is close to, but slightly above, 1 000 and decreasing, and the percentage of pure-bred females is below 80 per cent. (*Source*: FAO, 1999)

Endangered-maintained breed–*Search in* critical-maintained breed.

Endangered species–A plant or animal species in immediate danger of extinction because its population numbers have reached a critical level or its habitats have been drastically reduced.

Endemic–1. Describing a plant or animal species whose distribution is restricted to one or a few specific localities.

2. Describing a disease or a pest that is always present in an area.

End–labelling–The introduction of a radioactive atom at the end of a DNA or RNA molecule. A commonly used method is to use T4 polynucleotide kinase to introduce a ^{32}P atom onto the end of a DNA molecule.

Endocrine gland–Any gland in an animal that manufactures hormones and secretes them directly into the bloodstream to act at distant sites in the body, known as target organs or cells. *The opposite word is* exocrine gland. *Search in* gland.

Endocrine interference–Interference with the normal balance hormones.

Endocytosis–The process by which materials enter a cell without passing through the cell membrane. The membrane folds around material outside the cell, resulting in the formation of a saclike vesicle into which the material is incorporated. This vesicle is then pinched off from the cell surface so that it lies within the cell. *Search in* phagocytosis; pinocytosis.

Endoderm–The internal layer of cells of the gastrula, which will develop into the alimentary canal (gut) and digestive glands of the adult.

Endodermis (Gr. *endon*, within + *derma*, skin)–The layer of living cells, with various characteristically thickened walls and no intercellular spaces, which surrounds the vascular tissue of certain plants and occurs in nearly all roots and certain stems and leaves. The endodermis separates the cortical cells from cells of the pericycle.

Endogamy–The fusion of reproductive cells from closely related parents, *i.e.*, inbreeding. *The opposite word is* exogamy.

Endogenote–The part of the bacterial chromosome that is homologous to a genome fragment (exogenote) transferred from the donor to the recipient cell in the formation of a merozygote.

Endogenous (Gr. *endon*, within, + *genos*, race, kind)–Developed or added from within the cell or organism.

Endomitosis–Duplication of chromosomes without division of the nucleus, resulting in increased chromosome number within a cell. Chromosome strands separate, but the cell does not divide.

Endonuclease–An enzyme that breaks strands of DNA at internal positions; these enzymes are important tools in recombinant DNA technology. *Search in* nuclease.

Endophyte–An organism that lives inside a plant.

Endoplasmic reticulum (Gr. *endon*, within + *plasma*, anything formed or moulded; L. *reticulum*, a small net)–A cytoplasmic net of membranes, adjacent to the nucleus, made visible by the electron microscope. Any system of paired membranes within the cytoplasm. Frequently abbreviated to ER. They are sites of protein synthesis.

Endopolyploidy–The result of nuclear divisions without subsequent cytoplasmic division (cytokinesis); the polyploids so obtained are called endopolyploids. *Search in* polyploidy.

Endoprotease–An enzyme that cleaves the peptide bonds between amino acids within a protein. Cleavage is usually at one or more specific sites.

Endoreduplication–Chromosome reproduction during interphase. 4-chromatid chromosomes (diplo-chromosomes) are seen during this phase.

Endosperm (Gr. *endon*, within + *sperma*, seed)–Nutritive tissue that develops in the embryo sac of most angiosperms. It usually forms after the fertilization of the two fused primary endosperm nuclei of the embryo sac with one of the two male gamete nuclei. Seeds at maturity with endosperms are called aluminous and seeds without it are called exalbuminous. In most diploid plants, *e.g.*, cereals, the endosperm is triploid *(3n)*, but in some (*e.g.*, lily) it is often consumed as the seed matures.

Endosperm mother cell–One of the seven cells of the mature embryo sac, containing the two polar nuclei and, after reception of a sperm cell, gives rise to the primary endosperm cell from which the endosperm develops.

Endotoxin–A component of the cell wall of Gram-negative bacteria that elicits, in mammals, an inflammatory response and fever.

End-product inhibition–The inhibition of an enzyme by a metabolite. The enzyme is sometimes the first enzyme in a biosynthetic pathway, and the metabolite is generally the product of the last step in the pathway. *Search in* feedback inhibition.

Enhancer–1. A substance or object that increases a chemical activity or a physiological process.

2. A DNA sequence that increases the transcription of a eukaryotic gene when they are both on the same DNA molecule. a.k.a. enhancer element; enhancer sequence.

Enhancer element; enhancer sequence–1. A sequence found in eukaryotes and certain eukaryotic viruses which can increase transcription of a gene when located (in either orientation) up to several kilobases from the gene concerned. These sequences usually act as enhancers when on the 5′ side (upstream) of the gene in question. However, some enhancers are active when placed on the 3′ side (downstream) of the gene. In some cases enhancer elements can activate transcription of a gene with no (known) promoter.

2. A substance or object that increases a chemical activity or a physiological process.

3. A major or modifier gene that increases a physiological process.

Enterotoxin–A bacterial protein that, following release into the intestine, causes cramps, diarrhoea and nausea.

Enucleated ovum–Egg cell from which the nucleus has been removed.

Environment–The aggregate of all the external conditions and influences affecting the life and development of an organism.

Enzyme (Gr. *en*, in + *zyme*, yeast or leaven)–A protein produced in living cells, which, even in very low concentration, catalyses specific chemical reactions but is not used up in the reaction. Enzymes are classified into six major groups, according to the type of reaction they catalyse: 1. Oxidoreductase; 2. Transferases; 3. Hydrolases; 4. Lyases; 5. Isomerases; 6. Ligases. The names of most individual enzymes are usually derived from the substrate on which they act, with the suffix–ase. Thus lactase is the enzyme that acts to breakdown lactose; it is classified as a hydrolase. *Search in* Enzyme Commission Number.

Enzyme bioreactor–A reactor in which a chemical conversion reaction is catalysed by an enzyme.

Enzyme Commission (EC) Number–Systematic name and number which identify an enzyme in technical literature. Assigned by the Enzyme Commission, the EC Number consists of four numbers separated by dots: the first classifies the enzyme into one of the six broad groups: 1. Oxidoreductase; 2. Transferases; 3. Hydrolases; 4. Lyases; 5. Isomerases; 6. Ligases. Each group is subdivided into sub-groups, each sub-group into sub-sub-groups, and the last number is specific for the enzyme, *e.g.*, EC 3.1.21.1 is deoxyribonuclease I.

Enzyme electrode–A type of biosensor, in which an enzyme is immobilized onto the surface of an electrode. When the enzyme catalyses its reaction, electrons are transferred from the reactant to the electrode, and so a current is generated. There are two types of enzyme electrodes:

1. ***Ampometric***, where the electrode is kept as near zero voltage as possible. When the enzyme catalyses its reaction, electrons flow into the electrode, and so a current flows; and

2. ***Potentiometric***, when the electrode is held at a voltage which counteracts the voltage determined by the enzyme's tendency to push electrons into its.

Usually enzymes transfer their electrons inefficiently to the electrode, so a mediator compound is coated onto the electrode to help the transfer.

Enzyme-linked immunosorbent assay–*Search in* ELISA.

Enzyme stabilization using antibodies–A method of stabilizing enzymes by binding antibodies to them. The antibodies should not block the active site of the enzyme, as otherwise the protein is stabilized but is inactive as a catalyst. Monoclonal antibodies are usually used as they bind to specific bits of the protein surface. If the enzyme tries to unfold into an inactivate structure, it must not only overcome its own binding energy but also throw off all the bound antibodies; this requires more energy, and so is a correspondingly slower process.

EPD–*Search in* expected progeny difference.

Epicotyl (Gr. *epi*, upon + *kotyledon*, a cup-shaped hollow)–The upper portion of the axis of a plant embryo or seedling, above the cotyledons.

Epidermis (Gr. *epi* upon + *derma*, skin).–1. The outmost layer of cells of the body of an animal. In invertebrates the epidermis is normally only one cell thick and is covered by an impermeable cuticle. In vertebrates the epidermis is the thinner of the two layers of skin.

2. The outermost layer of cells covering a plant. It is overlaid by a cuticle and its functions are principally to protect the plant from injury and to reduce water loss. Some epidermal cells are modified to form guard cells or hairs of various types. In woody plants the functions of the shoot epidermis are taken over by the periderm tissues and in mature roots the epidermis is sloughed off and replaced by the hypodermis.

Epigenesis–Describes the developmental process whereby each successive stage of normal development is built up on the foundations created by the preceding stages of development; an embryo is built up from a zygote, a seedling from an embryo, and so on.

Epigenetic variation–Non-hereditary and reversible variation; often the result of a change in gene expression.

Epigenetic–A term referring to the non-genetic causes of a phenotype.

Epinasty–A process by which the growth of branches or petioles is abnormally pointing downward. This phenomenon is caused by the more rapid growth of the upper side. Epinasty may result from either nutritional deficiencies or irregularities at the plant growth regulator level. Not to be confused with wilting, as epinastic tissues are turgid.

Epiphyte–A plant that grows upon another plant, but is neither parasitic on it nor rooted in the ground.

Episome–A genetic extrachromosomal element (*e.g.*, the fertility factor (F) in *Escherichia coli*) which replicates within a cell independently of the chromosome and is able to integrate into the host chromosome. The step of integration may be governed by a variety of factors and so the term episome has lost favour and been superseded by the wider term *plasmid*. Plasmids and F factors are episomes.

Epistasis–Interaction between genes at different loci, *e.g.*, one gene suppresses the effect of another gene that is situated at a different locus. Suppressed genes are said to be hypostatic. Dominance is associated with members of allelic pairs, whereas epistasis is interaction among products of non-alleles.

Epitope–A specific chemical domain on an antigen that stimulates the production of, and is recognized by, an antibody. Each epitope on a molecule such as protein elicits the synthesis of a different antibody.

Epizootic–A disease affecting a large number of animals simultaneously.

Equational division–Mitotic-type division that is usually the second division in the meiotic sequence; somatic mitosis and the non-reductional division of meiosis. A chromosome division in which the two chromatids of each duplicated chromosome separate longitudinally, prior to being incorporated into two daughter nuclei.

Equatorial plate–The figure formed by the chromosomes in the centre (equatorial plane) of the spindle in mitosis.

Equilibrium–A state of dynamic systems in which there is no net change.

Equilibrium density gradient centrifugation–A procedure used to separate macro-molecules based on their density (mass per unit volume).

Equimolar–Identical molar concentrations. *Search in* molarity; mole.

ER–*Search in* endoplasmic reticulum.

Erlenmeyer flask–A conical flat-bottomed laboratory flask with a narrow neck, designed by E. Erlenmeyer. Widely used for culturing micro-organisms.

ES cells–*Search in* embryonic stem cells.

Escherichia coli–A commensal bacterium inhabiting the colon of many species. *E. coli* is widely used in biology, both as a simple model of cell biochemical function and as a host for molecular cloning experiments. In environmental studies, it is a key indicator of water pollution due to human sewage effluent.

E site–*Search in* exit site.

Essential amino acid–An amino acid that cannot be synthesized by animals and therefore has to be ingested with feed or food.

Essential element–Any of a number of elements required by living organisms to ensure normal growth, development and maintenance.

Essential nutrient–Any substance required by living organisms to ensure normal growth, development and maintenance.

Essential requirement–A nutrient is essential when it is mandatory for growth, development and reproduction. In tissue culture, it comprises inorganic salts, including all of the elements necessary for plant metabolism; organic factors (amino acids, vitamins); usually also endogenous plant growth regulators (auxins, cytokinins and often gibberellins); as well as a carbon source (sucrose or glucose).

EST–*Search in* expressed sequence tag.

Established culture 1. An aseptic viable explant (*Search in* micropropagation stages I and II).

2. A suspension culture subjected to several passages with a constant cell number per unit time.

Estimated breeding value (EBV)–Twice the expected progeny difference. The difference is doubled because breeding value is a reflection of all the genes of an animal, in contrast to progeny difference, which is a reflection of a sample half of an animal's genes. The predicted performance of the offspring of the mating between any two animals is the average of their EBVs (averaged because each parent makes an equal contribution to each offspring).

Estrogen–*Search in* oestrogen.

ET–*Search in* multiple ovulation and embryo transfer.

Ethanol (ethyl alcohol) (C_2H_6O; M.W. 46.07)–Commonly used to disinfest plant tissues, glassware utensils and working surfaces in tissue culture manipulations. The concentration used is 70 per cent (v/v) for disinfecting and 95 per cent (v/v) when flaming tools. Ethanol is also used to dissolve water-insoluble additions (addendums) to culture media.

Ethephon (2-chloroethyl) phosphonic acid ($ClC_2PO_3H_6$; f.w. 144.50)–Through a spontaneous degradation of ethephon, ethylene is produced. Ethephon is a synthetic compound commonly used to treat cultured cells or unripened fruit with ethylene.

Ethidium bromide–A fluorescent dye used to stain DNA and RNA. The dye fluoresces when exposed to UV light.

Ethyl alcohol–*Search in* ethanol.

Ethylene (C_2H_4)–A gaseous plant growth regulator regulating various aspects of vegetative growth, fruit ripening and abscission of plant parts.

Etiolation–An abnormal increase in stem elongation, accompanied by poor or absent leaf development. Physiological etiolation is caused by a lack of chlorophyll, and is typical of plants growing under low light intensity or in complete darkness. It can also be caused by disease.

ETL–*Search in* economic trait locus.

Eucaryote; eucaryotic–*Search in* eukaryote.

Euchromatin–Genetic material that is stained less intensely by certain dyes during interphase, and that comprises many different kinds of genes. *The opposite word is* heterochromatin.

Eugenics–The application of the principles of genetics to the "improvement" of humankind.

Eukaryote–(Gr. *eu*, true + *karyon*, true nucleus) Any organism characterized by having the nucleus enclosed by a membrane. Eukaryotic organisms include animals, plants, fungi and some algae. They also possess membrane-bound functional organelles, such as mitochondria and chloroplasts, in the cytoplasm of their cells. *The opposite word is* prokaryote.

Euploid–An organism or cell having a chromosome number that is an exact multiple of the monoploid *(n)* number. Terms used to identify different levels in an euploid series are diploid, triploid, tetraploid, and so on. *The opposite word is* aneuploid.

Evaluation–Measurement of the characteristics that are important for production and adaptation, either of individual animals or of populations, most commonly in the context of comparative evaluation of the traits of animals or of populations. (*Source*: FAO, 1999)

Evapotranspiration (L. *evaporare, e*, out of, + *vapor*, vapour + F. *transpirer*, to perspire)–The process of water loss in vapour form from a unit surface of land both directly and through leaf surfaces during a specific period of time.

Evolution–The process by which the present diversity of plant and animal life arose from the earliest organisms, a process believed to have been continuing for at least 3 000 million years.

Excinuclease–The endonuclease-containing protein complex that excises a segment of damaged DNA during excision repair.

Excision–1. The natural or *in vitro* enzymatic release (removal) of a DNA segment from a chromosome or cloning vector.

2. Cutting out and preparing a tissue, organ, etc., for culture.

3. Removing adventitious shoots from callus tissue.

Excision repair–DNA repair processes that involve the removal of a damaged or incorrect segment of DNA and its replacement by the synthesis of a new strand using the complementary strand of DNA as template.

Excrete–To transport a compound out of a cell.

Exit site (E site)–The ribosome binding site that contains the free tRNA prior to its release.

ex novo–*Search in de novo.*

Exo III–*Search in* ase III.

Exocrine gland–In animals, a gland that secretes through a duct. *The opposite word is* endocrine gland. *Search in* gland.

Exodeoxyribonuclease III–*Search in* exonuclease III.

Exogamy–The fusion of reproductive cells from distantly related or unrelated organisms, *i.e.*, outbreeding. *The opposite word is* endogamy.

Exogenote–Chromosomal fragment homologous to an endogenote and donated to a merozygote.

Exogenous (Gr. *exe*, out, beyond + *genos*, race, kind)–Produced outside of, originating from, or due to external causes. Opposite of endogenous.

Exogenous DNA–DNA that has been derived from a source organism and has been cloned into a vector and introduced into a host cell. Also referred to as foreign DNA or heterologous DNA.

Exon–A segment of a eukaryotic gene that is transcribed as part of the primary transcript and is retained, after processing, with other exons to form a functional mRNA molecule. *Search in* DNA; intron; split gene; splicing.

Exon amplification–A procedure that is used to amplify exons.

Exonuclease–An enzyme that digests DNA or RNA, beginning at the ends of strands. It requires a free end in order to degrade a DNA or RNA molecule. 5′ exonucleases require a free 5′ end and degrade the molecule in the 5′→3′ direction. 3′ exonucleases require a free 3′ end and degrade the molecule in the opposite direction. *The opposite word is* endonuclease.

Exonuclease III–An *E. coli* enzyme that removes nucleotides from the 3′-hydroxyl ends of double stranded DNA.

Exopolysaccharide–A high-molecular-weight polymer that is composed of sugar residues and is secreted by a micro-organism into the surrounding environment.

Exotoxin–A toxin released by a bacterium into the medium in which it grows.

Expected progeny difference (EPD)–The predicted performance of the future offspring of an animal for a particular trait, calculated from measurement(s) of the animal's own performance and/or the performance of one or more of its relatives, for the trait in question and/or for one or more correlated traits. Typically, the prediction is expressed as a deviation from a well-defined base population, assuming the animal in question is mated to a sample of animals whose average genetic merit equals that of the base population. The predicted performance of the offspring of the mating between any two animals is the sum of their EPDs.

Explant–A plant part aseptically excised and prepared for culture in a nutrient medium. *The opposite word is* donor plant.

Explantation–The removal of cells, tissues or organs of animals and plants for observation of their growth and development in appropriate culture media.

Explant donor–The source plant or mother plant from which is taken the explant used to initiate a culture. *The opposite word is* donor plant.

Exponential phase–The growth stage where cells undergo their maximum rate of cell division. The exponential phase follows the lag phase, and precedes the linear growth phase. *Search in* growth phases.

Export–To transport a compound out of a cell.

Express–To transcribe and translate a gene's message into a peptide product.

Expressed sequence tag (EST)–Short cDNA sequence. So-called because it represents part of the sequence (*i.e.*, the "tag" of a sequence) of an expressed gene.

Expression library–A population of different DNA molecules encoding peptides, that has been cloned into one kind of expression vector. *Search in* library.

Expression system–Combination of host and vector which provides a genetic context for making a cloned gene function, *i.e.*, produce peptide, in the host cell.

Expression vector–A cloning vector that has been constructed in such a way that, after insertion of a DNA molecule, its coding sequence is properly transcribed and the RNA is translated. The cloned gene is put under the control of a promoter sequence for the initiation of transcription, and often also has a transcription termination sequence at its end. Such promoters are termed *high level*; examples include P1 (the leftward promoter of phage l) and the promoter of the yeast PGK (phosphoglycerate kinase) gene.

Expressivity–Degree of expression of a trait controlled by a gene. A particular gene may show different degrees of expression in different individuals. *Search in* variable expressivity.

***ex situ* conservation**–A conservation method which entails the actual removal of germplasm resources (seeds, pollen, sperm, individual organisms) from the original habitat or natural environment. *The opposite word is* gene bank; cryobiological preservation; *in situ* conservation.

***ex situ* conservation of farm animal genetic diversity**–All conservation of genetic material *in vivo* but out of the environment in which it developed, and *in vitro* including, *inter alia* the cryoconservation of semen, oocytes, embryos, cells or tissues. Note that *ex situ* conservation and *ex situ* preservation are considered here to be synonymous. (*Source*: FAO, 1999)

***ex situ* preservation**–*Search in ex situ* conservation.

Extension–Single-stranded DNA region consisting of one or more nucleotides at the end of a strand of duplex DNA. a.k.a. protruding end; sticky end; overhang; cohesive end. *Search in* extension.

3′-extension–A short single-stranded nucleotide sequence on the 3′-hydroxyl end of a double-stranded DNA molecule. a.k.a. 3′ protruding end; 3′ sticky end; 3′ overhang.

5′-extension–A short single-stranded nucleotide sequence on the 5′-hydroxyl end of a double-stranded DNA molecule. a.k.a. 5′ protruding end; 5′ sticky end; 5′ overhang.

Extinct breed–A breed where it is no longer possible to recreate the breed population. Extinction is absolute when there are no breeding males (semen), breeding females (oocytes), nor embryos remaining. (*Source*: FAO, 1999)

External guide sequence (EGS)–*Search in* guide sequence.

Extinction–The irreversible condition of a species or other group of organisms of having no living representatives in the wild, which follows the death of the last surviving individual of that species or group. Extinction may occur on a local or global level; it can result from various human activities, including the destruction of habitats or the overexploitation of species that are hunted or harvested as a resource.

Extrachromosomal–Things that are not part of the chromosomes; DNA units in the cytoplasm that control cytoplasmic inheritance.

Extrachromosomal inheritance–*Search in* cytoplasmic inheritance.

Extrachromosomes–Self-replicative genetic elements separate from main chromosome(s) of a cell. This definition usually excludes viruses, but the division is somewhat arbitrary. In bacteria, plasmids are the principal extrachromosomes; they encode functions which are not essential to the growth and division of the host cell. In eukaryotes, extrachromosomes may be either essential or dispensable. They may inhabit (*i*) the nucleus, *e.g.* extrachromosomal rDNA molecules, yeast 2μm plasmid; (*ii*) the cytosol, *e.g.*, dsRNA molecules in fungi; or (*iii*) the cytoplasmic organelles, *e.g.*, mitochondrial DNA, chloroplast DNA. Eukaryotic extrachromosomal elements may be recognized genetically by their failure to show segregation at meiosis.

Exude–Slowly discharge; leak liquid material (exudate such as tannins or oxidized polyphenols) through pores or cuts, or by diffusion into the medium. In some woody plant species, exudation is associated with a lethal browning of explants.

ex vitro (L. "from glass")–Organisms removed from tissue culture and transplanted; generally plants to soil or potting mixture.

***ex vivo* gene therapy**–The delivery of a gene or genes to the isolated cells of an individual. After culturing, the transformed cells are introduced back into the individual by transfusion, infusion or injection, to alleviate a genetic disorder.

F factor–A bacterial episome that confers the ability to function as a genetic donor in conjugation; the fertility factor in bacteria.

F_1–The first filial generation in a cross between any two parents; the first generation of descent from a given mating.

F_2–The second filial generation, produced by crossing two members of the F_1, or by self-pollinating the F_1. The "grandchildren" of a given mating.

FACS–*Search in* fluorescence activated cell sorting.

Factorial mating–A mating scheme in which each male parent is mated with each female parent. Made possible in animals by means of *in vitro* embryo production. Such a mating scheme substantially reduces the rate of inbreeding in a selection programme.

FAD (Flavin Adenine Dinucleotide)–A co-enzyme important in various biochemical reactions. It comprises a phosphorylated vitamin B_2 (riboflavin) molecule linked to the nucleotide adenine monophosphate (AMP). It functions as a hydrogen acceptor in dehydrogenation reactions, being reduced to $FADH_2$. This in turn is oxidized to FAD by the electron transport chain, thereby generating ATP (two molecules of ATP per molecule of $FADH_2$).

False fruit–*Search in* pseudocarp.

False-negative–A negative assay result that should have been positive.

False-positive–A positive assay result that should have been negative.

Farm animal genetic resources–Those animal species that are used, or may be used, for the production of food and agriculture, and the populations within each of them. These populations within each species can be classified as wild and feral populations, landraces and primary populations, standardized breeds, selected lines, and any conserved genetic material. (Source: FAO, 1999)

Farm animal genetic resources utilization–*Search in* utilization of farm animal genetic resources.

Fascicle–*Search in* vascular bundle.

Fed-batch fermentation–Culture of cells or micro-organisms where nutrients are added periodically to the bioreactor.

Feedback inhibition–The process by which the accumulated end product of a biochemical pathway stops synthesis of that product. A late metabolite of a synthetic pathway regulates synthesis at an earlier step of the pathway. *The opposite word is* end-product inhibition.

Fermentation–The anaerobic breakdown by micro-organisms of complex organic substances, especially carbohydrates like glucose. The process is energy-yielding. Fermentation is often misused to describe large-scale aerobic cell culture in specialized vessels (fermenters, bioreactors) for secondary product synthesis.

Fermentation substrates–Materials used as food for growing micro-organism. The fermentation substrates and the trace materials needed, together with chemicals added to make the fermentation easier, form the culture medium.

Fermenter–*Search in* bioreactor.

Fertile–Of an organism: Capable of breeding and reproduction.

Fertilization–(L. *fertilis*, capable of producing fruit) The union of two gametes from opposite sexes to form a zygote; it involves the fusion of nuclei of gametes (karyogamy) and the fusion of cytoplasm (plasmogamy). Typically, each gamete contains a haploid set of chromosomes. Hence, after fusion of the nuclei, the resulting nucleus contains a diploid set of chromosomes.

Several categories are distinguished:

1. ***Self–fertilization (selfing)***–fusion of male and female gametes from the same euploid organism.

2. ***Cross–fertilization (crossing)***–fusion of male and female gametes from different euploid individuals.

3. **Double fertilization**–in angiosperms, the fusion of one male gamete with the ovum at about the same time as the second male gamete fuses with the female polar nuclei (or secondary nucleus) to form the endosperm.

Fertilizer–Any substance that is added to soil in order to increase its productivity. Fertilizers can be of natural origin, such as composts, or they can be inorganic (artifical fertilizer) chemical, particularly nitrates and phosphates.

Fetus–*Search in* foetus.

Feulgen's test–A histochemical test in which the distribution of DNA in the chromosomes of dividing cell nuclei can be observed. A tissue section is first treated with dilute hydrochloric acid to remove the purine bases of the DNA, thus exposing the aldehyde groups of the sugar deoxyribose. The section is then immersed in Schiff's reagent, which combines with the aldehyde groups to form a magenta-colored compound.

F factor–A bacterial episome that confers the ability to function as a genetic donor in conjugation; the fertility factor in bacteria.

Fibres–Elongated cells with tapering, pointed ends; the cells interlock to form a strong, rigid tissue; pits in the walls are usually very narrow and not very numerous.

Fibril–Microscopic to sub-microscopic cellulose thread that is part of the cellulose matrix of plant cell walls.

Fibrous root–Root system in which both primary and lateral roots have approximately equal diameters. Opposite is *tap root.*

Field gene bank–*Search in* gene bank (2).

Filial generation–*Search in* F_1; F_2.

Filter bioreactor; mesh bioreactor–Cells are grown on an open mesh of an inert material, which allows the culture medium to flow past it but retains the cells. This is similar in idea to membrane and hollow fibre reactors, but can be much easier to set up, being similar to conventional tower bioreactors, but with the meshwork replacing the central reactor space.

Filter sterilization–Process of sterilizing a liquid by passage through a filter with pores so small that they prevent the passage of micro-organisms and microbial spores.

Filtration–1. Separation of solids from liquids by using a porous material that only allows passage of the liquid or of solids of a smaller than the pore size of the filter. The material passing the filter forms the filtrate.

2. Removal of cell aggregates to obtain a filtrate of single cells that can be utilized as plating inocula.

Fission–(L. *fissilis*, easily split) Asexual reproduction involving the division of a single-celled individual into two new single-celled individuals of equal size.

Fitness–The number of offspring left by an individual, often compared with the average of the population or with some other standard, such as the number of offspring left by a particular genotype.

Fixation–An event that occurs when all the alleles at a locus except one are eliminated from a population. The remaining allele, now occurring with a frequency of 100 per cent, is said to have been fixed.

FLAG–*Search in* affinity tag.

Flagellum–(pl: flagella; adj: flagellate) A whiplike organelle of locomotion in certain cells; locomotor structures in flagellate protozoa.

Flaming–A technique for sterilizing instruments. The instrument is dipped in alcohol (usually 95 per cent (v/v) ethanol) and then the alcohol on the instrument is ignited, thus heat-sterilizing the tool surface.

Flanking region–The DNA sequences extending either side of a specific sequence.

Floccule–Micro-organism aggregate or colloidal particle floating in or on a liquid. Contaminated liquid media are usually cloudy, illustrating this flocculation phenomenon.

Flocculant–Agent that causes small particles to aggregate (flocculate).

Fluorescence-*in situ* hybridization (FISH)–Hybridization of cloned DNA to intact chromosomes, where the cloned DNA has been labelled with a fluorescent dye. This is the major method of physical mapping of cloned DNA fragments on chromosomes.

Fluorescence-activated cell sorting (FACS)–The use of laser beams to detect differences in fluorescence between different types of cells in a mixture, and the subsequent deflection of cells into separate bins corresponding to each type of cell in the mixture. One of the popular uses of this technology is in sperm sexing.

Fluorescent probe–Probe whose response is based on the fluorescence intensity of individual cells or cell components.

Flow cytometry–A technique used to sort cells or other biological materials by means of flow through apertures of defined size or by laser sorting.

Flower–The structure in angiosperms (flowering plants) that bears the organs for sexual reproduction.

Flush end–*Search in* blunt end.

Flush-end cut–*Search in* blunt-end cut.

Foetus–Pre-natal stage of a viviparous animal, between the embryonic stage and parturition in which the main features are clearly recognizable.

Food processing enzyme–Enzyme used to control food texture, flavour, appearance and, to a certain extent, nutritional value. Amylases break down complex polysaccharides to simplex sugars; proteases tenderize meat proteins. Biotechnology can assist the development of new food enzymes by finding or engineering enzymes which fit better with the other processes which the food must undergo, like cooking or canning.

Fog–Fine particles of liquid suspended in the air, such as of water in a fog chamber used for acclimatizing recent *ex vitro* transplants. *The opposite word is* mist propagation.

Fold-back–The structure formed when a double-stranded DNA molecule containing an inverted repeat sequence is denatured and then allowed to re-anneal at low DNA concentrations. The repeated sequence permits the formation of a double-stranded region within each of the separated strands of the original molecule.

Folded genome–The condensed intracellular state of the DNA in the nucleoid of a bacterium. The DNA is segregated into domains, and each domain is independently negatively supercoiled.

Follicle–Any enclosing cluster of cells that protects and nourishes a cell or structure within. Thus a follicle in the ovary contains a developing egg cell, while a hair follicle envelops the root of hair.

Follicle stimulating hormone (FSH)–A hormone, secreted by the anterior pituitary gland in mammals, that stimulates, in female mammals, ripening of specialized structures in the ovary (Graafian follicles) that produce ova and, in males, the formation of sperm in the testis. It is a major constituent of fertility drugs.

Foot-candle–An obsolete photometric measure of light intensity. Now superseded by the lux (symbol: lx) (1 fc » 10.7 lx). *Search in* lux; photon.

Forced cloning–The insertion of foreign DNA into a cloning vector in a predetermined orientation.

Foreign DNA–A DNA molecule that is incorporated into either a cloning vector or a chromosomal site. *Search in* exogenous DNA.

Fortify–To add strengthening components or beneficial ingredients to a nutrient medium.

Formulation–1. For traditional therapeutic agents, this refers to the method by which a therapeutic agent is delivered to its site of action.

2. For tissue culture, *Search in* medium; medium formulation.

Fouling–The coating or plugging (by materials or micro-organisms) of equipment, thus preventing it from functioning properly.

Founder animal–In transgenesis research, an organism that carries a transgene in its germ line and can be used in matings to establish a pure-breeding transgenic line, or one that acts as a breeding stock for transgenic animals.

Founder principle–The possibility that a new, small, isolated population may be genetically different from the "parent" population, because the founding individuals (being a small, random sample from the large, "parent" population) could be quite different from typical members of the "parent" population.

Four-base cutter; four-base-pair-cutter; four-cutter–A type II restriction endonuclease that binds (and subsequently cleaves) DNA at sites that contain a sequence of four nucleotide pairs that is uniquely recognised by that enzyme. Because any sequence of four bases occurs more frequently by chance than any sequence of six bases, four-base cutters cleave more frequently than do six-base cutters. Thus, four-base cutters create smaller fragments than six-base cutters.

Fractionation–Separation.

Fragment–*Search in* restriction fragment.

Frameshift mutation–A mutation that changes the reading frame of an mRNA, either by inserting or deleting nucleotides.

Free-living conditions–Natural or greenhouse conditions where the plantlets are transferred from *in vitro* conditions to soil mixtures. In such instances, plantlets must manufacture their own food supply for survival. *Search in* acclimatization; hardening off.

Free water–Water released by a cell when freezing occurs in intercellular spaces. *The opposite word is* bound water.

Freeze-drying–The process of drying a tissue or an organ in a frozen state under vacuum. Tissues are freeze-dried to measure their dry weight or to preserve them for future analysis. Freeze-drying is the standard way of preserving micro-organisms for long periods of time. *The opposite word is* lyophilization.

Freeze preservation–*Search in* cryobiological preservation.

Frequency distribution–A graph showing either the relative or absolute incidence of classes in a population. The classes may be defined by either a discrete or a continuous variable; in the latter case, each class represents a different interval on the scale of measurement.

Fresh weight–The weight, including the water content, of a plant or plant part at the time of harvest. *The opposite word is* wet weight.

Friable–A term commonly used to describe a crumbling or fragmenting callus. A friable callus is easily dissected and readily dispersed into single cells or clumps of cells in solution.

FSH–*Search in* follicle stimulating hormone.

Functional gene cloning–*Search in* candidate gene strategy.

Fungicide–An agent, such as a chemical, that kills fungi and that are used to control pathogenic fungi.

Furfural; furfuraldehyde–Used industrially as a solvent and as a raw material for synthetic resin.

Fusion biopharmaceuticals–Biopharmaceutical proteins formed as a result of fusion proteins. Their advantages are:

–synergistic activities in one molecule; thus when the molecule binds to a cell, it does two things at once;

–the adverse effect or poor stability of one molecule may be offset by the properties of the other; and

–one molecule acts as a targeting mechanism to bring the other to the site where it is meant to act.

Search in fusion protein: immunotoxin.

Fusion gene–A hybrid gene created by joining portions of two different genes (to produce a new protein) or by joining a gene to a different promoter (to alter or regulate gene transcription).

Fusion protein–A polypeptide made from a recombinant gene that contains portions of two or more different genes. The different genes are joined so that their coding sequences are in the same reading frame: the genetic apparatus reads the gene fusion as a single gene, and so produces a fusion protein. Also known as hybrid protein or chimeric protein. These proteins are used:

–to add an affinity tag to a protein;

–to produce a peptide as part of a larger protein, which is then cut up after it has been made by cloning;

–to produce a protein with the combined characteristics of two natural proteins; or

–to produce a protein where two different activities are physically linked.

G cap–The 5′-terminal methylated guanine nucleoside that is present on many eukaryotic mRNAs; it is joined, after transcription, to the mRNA. *Search in* cap.

Gall–A tumorous growth in plants.

Gamete–A mature reproductive cell which is capable of fusing with a cell of similar origin but of opposite sex to form a zygote from which a new organism can develop.

Gametes have a haploid chromosome content. In animals, a gamete is a sperm or egg; in plants, it is pollen, spermatic nucleous, or ovum.

Gamete and embryo storage–Storage of ova, sperm or fertilized embryos outside their original source. Almost invariably this means cryopreservation in liquid nitrogen.

Gametic (phase) disequilibrium–In relation to any two loci, the occurrence of haplotypes (gametes) with a frequency greater than or less than the product of the frequency of the two relevant alleles. *Search in also* gametic (phase) equilibrium.

Gametic (phase) equilibrium–In relation to any two loci, the occurrence of haplotypes (gametes) with a frequency equal to the product of the frequency of the two relevant alleles, *e.g.*, loci A and B are in linkage equilibrium if the frequency of the haplotype (gamete) A_iB_i equals the product of the frequencies of alleles A_i and B_i. a.k.a. linkage equilibrium, but this is a misleading term, because the concept applies equally to linked and unlinked loci.

Gametoclone–A plant regenerated from a tissue culture originating from gametic tissue.

Gametogenesis–The process of the formation of gametes.

Gametophyte–That phase of the plant life cycle that bears the gamete producing organs; the cells have *n* chromosomes. In angiosperms, the pollen grain is the male gametophyte and the embryo sac is the female gametophyte.

Gametophytic incompatibility–A phenomenon controlled by the complex S locus, in which a pollen grain cannot fertilize an ovule produced by a plant that carries the same S allele as the pollen grain. Thus S_1 pollen cannot fertilize an ovule of a plant possessing the S_1 allele.

Gap–1. Period of time, during the cell cycle, between M and S phases.

2. A missing section on one of the strands of double-stranded DNA. The DNA will therefore have a single-stranded region.

Gapped DNA–A duplex DNA molecule with one or more internal single-stranded regions.

Gas transfer–The rate at which gases are transferred from gas into solution. It is an important parameter in fermentation systems because it controls the rate at which the organism can metabolize. Gas transfer can be done by several methods, including use of small gas bubbles, that diffuse faster than larger ones, thanks to their larger surface area per unit of volume; or spreading the liquid out, for example in a thin sheet, or in a thin permeable tube, as in hollow fibre bioreactor.

Gastrula–An early animal embryo consisting of two layers of cells; an embryological stage following the blastula. In which the embryo usually consists of two germ layers (ectoderm + endoderm) with a cavity surrounded by endoderm.

G cap–*Search in* cap.

GC island–A segment of DNA that is rich in G=C base pairs and often precedes a transcribed gene in the genomes of vertebrate organisms.

GDP–Guanosine diphosphate, *Search in* guanosine.

Gel–A lyophilic colloid that has coagulated to a rigid or jelly-like solid or colloidal mass. It is used for the electrophoretic separation of nucleic acids or proteins, and for encapsulation. *Search in* encapsulating agent; gel electrophoresis.

Gel electrophoresis–An analytical method for separating molecules according to their size. Samples are put at one end of a slab of polymer gel; an electric field across the gel pulls the molecules through it; the smaller molecules pass more easily and so move towards the other end faster; the various sizes of molecules end up at different positions according their size. Gels are made from different materials, but common combinations are:

Molecule:	DNA	RNA	Protein
Gel matrix:	agarose	agarose or poly-acrylamide	Polya-crylamide Or starch

Various chemicals can be included in the gel to help separation, such as the detergent sodium dodecyl sulphate (SDS) in protein gels to unfold proteins, or urea in DNA sequencing gels, which unfolds DNA. *Search in* iso-electric focusing gels; polyacrylamide gel electrophoresis (PAGE); pulsed field gel electrophoresis (PFGE).

Gelatin–A glutinous, proteinaceous gelling and solidifying agent. Gelatin is produced by boiling animal connective tissues, which partially hydrolyses the collagen. Gelatin is used to gel or solidify nutrient solutions for tissue culture. *The opposite word is* agar.

Gelatinization–Steam cooking of milled grain, a process that increases the surface area of the starch and converts the original mash to a material with a gel-like consistency.

GelriteTM–The brand name of a *Pseudomonas*-derived refined polysaccharide used as a gelling agent and agar substitute.

GEM–Genetically engineered microorganism.

Gene (Gr. *gen*, race, offspring)–Conceptually, the unit of heredity transmitted from generation to generation during sexual or asexual reproduction. More generally, the term is used in relation to the transmission and inheritance of particular identifiable traits. Since the molecular revolution, it is now known that a gene is a segment of nucleic acid that encodes peptide or RNA.

Search in polygenes; dominant gene; fusion gene; gene amplification; gene expression; gene flow; gene pool; gene splicing; gene translocation; recessive gene; regulatory gene.

Gene addition–The addition of a functional copy of a gene to the genome of an organism.

Gene amplification–The selective production of multiple gene copies without a proportional increase in others.

Gene bank–1. The physical location where collections of genetic material in the form of seeds, tissues or reproductive cells of plants or animals are stored.

2. Field gene bank: A facility established for the *ex situ* storage and maintenance, using horticultural techniques, of individual plants. Used for species whose seeds are recalcitrant, or for clonally propagated species of agricultural importance, such as apple varieties).

3. A collection of cloned DNA fragments from a single genome. Ideally the bank should contain cloned representatives of all the DNA sequences in the genome.

4. A population of micro-organisms, each of which carries a DNA molecule that was inserted into a cloning vector. Ideally, all of the cloned DNA molecules together represent the entire genome of another organism. Also called gene library, clone bank, bank, library. This term is sometimes also used to denote all of the vector molecules, each carrying a piece of the chromosomal DNA of an organism, prior to the insertion of these molecules into a population of host cells.

Gene cloning–The process of synthesizing multiple copies of a particular DNA sequence using a bacteria cell or another organism as a host. The gene of interest is inserted into a self-replicating DNA molecule (DNA vector, often a plasmid) and the resulting recombinant DNA molecule is amplified in an appropriate host cell. Used in genetic engineering. molecular cloning and cloning. *Search in* DNA; host.

Gene conservation; genetic resources conservation–The conservation of species, populations, individuals or parts of individuals, by *in situ* or *ex situ* methods, to provide a diversity of genetic materials for present and future generations.

Gene conversion–A process, often associated with recombination, during which one allele is replicated at the expense of another, leading to non-Mendelian segregation ratios. In whole tetrads, for example, the ratio may be 6:2 or 5:3 instead of the expected 4:4.

Gene expression–The process by which a gene produces RNA and protein, and hence exerts its effects on the phenotype of an organism.

Gene flow–The spread of genes from one breeding population to another (usually) related populations by migration, possibly leading to changes in allele frequency.

Gene frequency–*Search in* allele frequency.

Gene imprinting–The differential expression of a single gene according to its parental origin.

Gene insertion–The incorporation of one or more copies of a gene into a chromosome.

Gene interaction–Modification of gene action by a non-allelic gene.

Gene library–Collection of cloned DNA fragments that ideally includes all the genetic information of an organism. If the original source of the DNA is the genomic DNA from an organism, then it is called a genomic library; if the DNA is from cDNA made by enzymatic copying of RNA, then the library includes representative fragments from all genes that were being expressed at the time the RNA was sampled, and would be called a cDNA library. a.k.a. gene bank; library; clone bank; bank; library.

Gene linkage–*Search in* linkage.

Gene modification–Chemical change to a gene's DNA sequence. *Search in* DNA.

Gene pool–1. The total genetic information in all the genes in a breeding population at a given time.

2. In PGR: Use is made of the concept of 1°, 2° and 3° (primary, secondary and tertiary) gene pools. In general, members of a 1° gene pool are inter-fertile; those of the 2° gene pool can cross with the 1° gene pool under special circumstances; with the 3° gene pool, extreme techniques are required to achieve crossing.

Search in germplasm.

Gene probe–A single-stranded DNA or RNA fragment used in genetic engineering to search for a particular gene or other DNA sequence. The probe has a base sequence complementary to the target sequence and will thus attach to it by basepairing. By labelling the probe, it can be identified after subsequent separation and purification.

Genera–*Search in* genus.

Generally Regarded as Safe (GRAS)–Designation given to foods, drugs, and other materials that have been used for a considerable period of time and have a history of not causing illness to humans, even though extensive toxicity testing has not been conducted. More recently, certain host organisms for recombinant DNA experimentation have been given this status.

Generate–To propagate or (mass) proliferate. The process is generation or regeneration.

Generation time–*Search in* cell generation time.

Generative–*Search in* somatic cell; somatic embryo.

Generative nucleous–One of the two male gametes in the pollen tube of angiosperms.

Gene recombination–The appearance of gene combinations in the progeny that differ from the combinations present in the parents.

Gene replacement–The incorporation of a transgene into a chromosome at its normal location by homologous recombination, thus replacing the copy of the gene originally present at the locus.

Gene sequencing–The process of elucidating the nucleotide sequence of a gene. *Search in* DNA sequencing.

Gene shears–*Search in* ribozyme.

Gene splicing–A stage in the processing of mRNA, occurring only in eukaryote cells, in which intervening sequences (introns) are removed from the primary RNA transcript (hnRNA), and the coding exons are joined together to form the mature mRNA molecule.

Gene therapy–The treatment of inherited diseases by introducing into the cells of affected individuals the wild-type copies of the defective gene causing the disorder. If reproductive cells are modified, the procedure is called germ-line or heritable gene therapy. If cells other than reproductive cells are modified, the procedure is called somatic-cell or non-inheritable gene therapy.

Genetically engineered organism (GEO)–sometimes used for genetically modified organism, q.v.

Genetically modified organism (GMO)–An organism that has been modified by the application of recombinant DNA technology.

Genetic assimilation–Eventual extinction of a natural species as massive pollen flow occurs from another related species and the natural species becomes more like the related species. *The opposite word is* gene flow.

Genetic code–1. The set of 64 nucleotide triplets (codons) that specify the 20 amino acids and termination codons (UAA, UAG, UGA).

2. The relationships between the nucleotide base-pair triplets of an mRNA molecule and the 20 amino acids that are the building blocks of proteins.

Search in base pair; nucleic acid; nucleotide.

Genetic complementation–When two DNA molecules that are in the same cell together produce a function that neither DNA molecule can supply on its own.

Genetic disease–A disease that has its origin in changes to the genetic material. Usually refers to diseases that are inherited in a Mendelian fashion, although non-inherited forms may also result from DNA mutation.

Genetic distance–A measure of the genetic similarity between any pair of populations. Such distance may be based on phenotypic traits, allele frequencies or DNA sequences. For example, genetic distance between two populations having the same allele frequencies at a particular locus, and based solely on that locus, is zero. The distance for one locus is maximum when the two populations are fixed for different alleles. When allele frequencies are estimated for many loci, the genetic distance is obtained by averaging over these loci. (*Source*: FAO, 1999)

Genetic distancing–The collection of the data on phenotypic traits, marker allele frequencies or DNA sequences for two or more populations, and estimation of the genetic distances between each pair of populations. From these distances, the best representation of the relationship among all the populations may be obtained. (*Source*: FAO, 1999)

Genetic diversity–The heritable variation within and among populations which is created, enhanced or maintained by evolutionary forces. *Search in* biodiversity; mutation; genetic drift; gene re-shuffling processes.

Genetic drift–Change in allele frequency from one generation to another within a population, due to the sampling of finite numbers of genes that is inevitable in all real (finite) populations. The smaller the population, the greater is the genetic drift. Sooner or later (depending on the size of the population), genetic drift results in loss of alleles from a population, and hence leads to a loss of genetic variation. Because of this, the minimization of genetic drift is an important consideration for conservation of genetic resources. *Search in* genetic diversity.

Genetic engineering–Changes in the genetic constitution of cells (apart from selective breeding) resulting from the introduction or elimination of specific genes through modern molecular biology techniques. This technology is based on the use of a vector for transferring useful genetic information from a donor organism into a cell or organism that does not possess it. *Search in* gene cloning.

A broader definition of genetic engineering includes selective breeding and other means of artificial selection.

Genetic equilibrium–Condition in a group of interbreeding organisms in which the allele frequencies remain constant over time.

Genetic fingerprinting–A technique in which an individual's DNA is analysed to reveal the pattern of repetition of particular nucleotide sequences throughout the genome. *Search in* DNA fingerprint.

Genetic heterogeneity–The situation in which different mutant genes produce the same phenotype.

Genetic immunization–Delivery to a host organism of a cloned gene that encodes an antigen. After the cloned gene is expressed, it elicits an antibody response that protects the organism from infection by a virus, bacterium or other disease-causing organism.

Genetic information–Information contained in a nucleotide base sequence in chromosomal DNA or RNA.

Genetic linkage–*Search in* linkage map; linkage.

Genetic map–The linear array of genes on a chromosome, based on recombination frequencies (linkage map) or physical location (physical or chromosomal map). *Search in* mapping.

Genetic mapping–Determining the linear order of genes and/or DNA markers along a chromosome. a.k.a. mapping.

Genetic marker–A DNA sequence used to "mark" or track a particular location (locus) on a particular chromosome. *The opposite word is* marker gene.

Genetic polymorphism–*Search in* polymorphism.

Genetic resources conservation–*Search in* gene conservation.

Genetics–The science that deals with study of heredity and variation.

Genetic selection–The process of selecting genes, cells, clones, etc., within populations or between populations or species. Genetic selection usually results in differential success rates of different genotypes, reflecting many variables, including selection pressure and genetic variability in populations.

Genetic transformation–The transfer of extracellular DNA among and between species by using bacterial or viral vectors.

Genetic variation–Differences between individuals attributable to differences in genotypes.

Gene tracking–Following the inheritance of a particular gene from generation to generation.

Gene translocation–The movement of a gene from one chromosomal location to another.

Genome–1. The entire complement of genetic material (genes + non-coding sequences) present in each cell of an organism, or in a virus or organelle.

2. A complete set of chromosomes (hence of genes) inherited as a (haploid) unit from one parent.

Genomic DNA library; genomic library–A collection of clones containing the genomic DNA sequences of an organism. Typically, these molecules are propagated in bacteria or phage. The library is an important tool used in the process of isolating genes. *Search in* library.

Genotype (from gene + type)–1. The genetic constitution (gene makeup) of an organism.

2. The pair of alleles at a particular locus, *e.g.*, *Aa* or *aa*.

3. The sum total of all pairs of alleles at all loci that contribute to the expression of a quantitative trait.

–The opposite word is phenotype.

Genus (pl: genera)–A somewhat arbitrary group of closely related species, where perceived relationship is typically based on physical resemblance.

GEO (Genetically engineered organism)–*Search in* GMO.

Geotropism (Gr. *ge*, earth + *tropos*, turning)–A growth curvature induced by gravity.

Germ–1. In botany, a common name for a plant embryo.

2. Colloquial: A disease-causing micro-organism.

Germ cell–Any cell in the series of cells (the germ line) that eventually produces gametes. In mammals, germ cells are found in the germinal epithelium of the ovaries and testes.

Germ cell gene therapy–The repair or replacement of a defective gene within the gamete-forming tissues, resulting in a heritable change in an organism's genetic constitution.

Germicide–Any chemical agent used to control or kill any pathogenic and non-pathogenic micro-organisms.

Germinal epithelium–1. A layer of epithelial cells on the surface of the ovary that are continuous with the mesothelium.

2. The layer of epithelial cells lining the seminiferous tubules of the testis, which gives rise to spermatogonia. *Search in* spermatogenesis.

Germination (L. *germinare*, to sprout)–1. The initial stages in the growth of a seed to form a seedling. The embryonic shoot (plumule) and embryonic root (radicle) emerge and grow upwards and downwards respectively. Food reserves for germination come from endosperm tissue within the seed and/or from the seed leaves (cotyledons).

2. The growth of spores (fungal; algal) and pollen grains.

Germ layers; primary germ layers–The layers of cells in an animal embryo at the gastrula stage, from which the various organs of the animal's body will be derived.

Germ line cells–Cells that produce gametes.

Germ line gene therapy–The delivery of a gene or genes to a fertilized egg or an early embryonic cell. The transferred gene(s) is present in all or some of the nuclei of the cells of the mature individual, including possibly the reproductive cells, and alters the phenotype of the individual that develops.

Germ line–A lineage of "generative" cells (= germ track) ancestral to the gametes (sperm and egg cells proper) which, during the development of an organism (animal or plant), are set aside as potential gamete-forming tissues. These ancestral cells, together with the gametes, are called germ cells, as opposed to somatic cells. Location, nature and time of formation of potential gamete-forming tissues are species specific, and may vary greatly from one species to another.

Germplasm–1. The genetic material that forms the physical basis of hereditary and which is transmitted from one generation to the next by means of the germ cells.

2. An individual or clone representing a type, species or culture, that may be held in a repository for agronomic, historic or other reasons.

Gestation–The period in animals bearing live young (especially mammals) from fertilization of the egg to birth of the young (parturition).

GH–*Search in* growth hormone.

Gibberellins–Plant growth regulators involved in elongation, enhancement of flower, fruit and leaf size, germination, vernalization and other processes. They also break dormancy and affect cell expansion in presence of auxin.

Gland–A group of cells or a single cell in animals or plants that is specialized to secrete a specific substance. In animals, there are two types of glands: endocrine glands discharge their products directly into the blood vessels; exocrine glands secrete through a duct or network of ducts into a body cavity or onto the body surface. In plants, glands are specialized to secrete certain substances produced by the plant. The secretions may be retained within a single cell, or secreted to the outside.

Glaucus–A surface with a waxy, white coating. In most cases, this waxy covering can be rubbed off.

Globulins–Common proteins in blood, eggs and milk, and as a reserve protein in seeds. Globulins are insoluble in water, but soluble in salt solutions. Alpha, beta and gamma globulins can be distinguished in blood serum. Gamma globulins are important in developing immunity to diseases.

GLP (Good Laboratory Practice); GMP (Good Manufacturing Practice)–They are codes of practice designed to reduce to a minimum the chance of accidents which could affect a research project or a manufactured product. The GLP and GMP prescriptions are quite voluminous, but boil down to a few key points. The essential point of both GLP and GMP is that everything is recorded, and that only established procedures are used and by people who have been trained to use them.

Glucocorticoid–A steroid hormone that regulates gene expression in higher animals.

Glucose invertase; glucose isomerase–Enzymes that catalyse the interconversion of the two sugars, glucose and fructose. As fructose is chemically more stable than glucose, a mixture of glucose and fructose with the enzyme will end up almost entirely as fructose. This is valuable for the food industry, as fructose is substantially sweeter than glucose, and so more sweetness per gram is achieved using fructose. The usual use for glucose isomerase is to take glucose made by hydrolysis of corn starch and turn it into a mixture of mostly fructose with some glucose. The corn starch is broken down using amylases. The result is called high-fructose corn syrup (HFCS). Invertase takes sucrose and turns it into glucose and fructose.

Gluten–A mixture of two proteins, gliadin and glutenin, occurring in the endosperm of wheat grain. Their amino acid composition varies, but glutamic acid (33 per cent) and proline (12 per cent) predominate. The composition of wheat glutens determines the strength of the flour and whether or not it is suitable for biscuit or bread making. Sensitivity of the lining of the intestine to gluten occurs in coeliac disease, a condition that must be treated by a gluten-free diet.

Glycolysis (Gr. *glycos*, sugar (sweet) + *lysis*, dissolution)–Sequence of reactions that converts glucose into pyruvate, with the formation of ATP and NADH.

Glycosylation–The covalent addition of sugar or sugar-related molecules to proteins or polynucleotides.

GM food–Food that comprises, in whole or in part, material that has been modified by the application of recombinant DNA technology.

GMO–*Search in* genetically modified organism.

GMP (good manufacturing practice)–*Search in* GLP/GMP.

GMP (guanosine monophosphate)–*Search in* guanosine.

Gobar–*Search in* biogas.

Golgi apparatus–An assembly of vesicles and folded membranes within the cytoplasm of plant and animal cells that stores and transports secretory products (such as enzymes and hormones) and plays a role in formation of a cell wall (when this is present). It is named after its discoverer, the Italian cytologist Camillo Golgi (1843-1926).

Gonad–Any of the usually paired organs in animals that produce reproductive cells (gametes). The most important gonads are the male testis, which produces spermatozoa, and the female ovary, which produces ova (egg cells). The gonads also produce hormones that control secondary sexual characteristics.

Good laboratory practice–*Search in* GLP/GMP.

Good manufacturing practice–*Search in* GLP/GMP.

G-proteins–Proteins with an important role in relaying signals in mammalian cells. The proteins occur on the inner surface of the plasma membrane and transmit signals from outside the membrane, via transmembrane receptors, to adenylate cyclase, which catalyses the formation of the second messenger, cyclic AMP, inside the cell. G-proteins derive their name from their ability to bind to guanine nucleotides, namely GTP and GDP; the GTP-protein complex is able to activate adenylate cyclase, whereas the GDP-protein complex is unable. G-proteins are activated when the signalling molecule (typically a hormone) binds to the transmembrane receptor.

Graft (to)–To place a detached branch (scion) in close cambial contact with a rooted stem (rootstock) in such a manner that scion and rootstock unite to form a single plant.

Graft inoculation test–A test based on the use of a suspected viral carrier which is grafted to an indicator plant. If symptoms appear in the indicator plant, the viral assay is positive.

Graft union–The point at which a scion from one plant is joined to a stock from another plant.

Gram molecular weight–*Search in* mole.

Gram-negative organism–Any prokaryotic organism that does not retain the first stain (crystal violet) used in Gram's staining technique. It does retain the second stain (safranin O) and therefore has a pink colour when viewed under a light microscope. Retention of the stain is due to the structure of the cell wall.

Gram-positive organism–Any prokaryotic organism that retains the first stain used in the Gram technique, which gives a purple-black colour when viewed under a light microscope. Retention of the stain is due to the structure of the cell wall.

Gram staining–A technique to distinguish between two major bacterial groups, based on stain retention by their cell walls. Bacteria are heat-fixed, then stained with crystal violet, followed by iodine solution, and then rinsed with alcohol or acetone. Gram-positive bacteria are stained bright purple, while Gram-negative bacteria are decolourized.

Grana (L. *granum*, a seed)–Structures within chloroplasts, seen as green granules with the light microscope and as a series of parallel lamellae with the electron microscope. Disc- or sac-like structures found in chloroplasts composed of stacked membranes and containing the chlorophyll and carotenoid pigments directly involved in photosynthesis (singular: granum).

GRAS–*Search in* generally regarded as safe.

Gratuitous inducer–A substance that can induce transcription of a gene or genes, but is not a substrate for the induced enzyme(s).

Gravitropism–*Search in* geotropism.

Green Revolution–Name given by William Goud to the dramatic increase in crop productivity during the third quarter of the 20th century, as result of integrated advances in genetics and plant breeding, agronomy, and pest and disease control.

Gro-luxTM–A wide-spectrum fluorescent lamp suitable for plant growth purposes.

Growth–An increase in cell size or cell number, or both, resulting in an increase in dry weight.

Growth cabinet–A cupboard used for incubating tubes or culture vessels under controlled environmental conditions. The degree of control over temperature, light and humidity is a function of the quality of the cabinet. *The opposite word is* culture room; incubator.

Growth curve–*Search in* growth phases.

Growth factor–Any of various chemicals, particularly polypeptides, that have a variety of important roles in the stimulation of new cell growth and cell maintenance. They bind to the cell surface on receptors. Specific growth factors can cause new cell proliferation.

Growth hormone (GH); somatotrophin; somatotropin–A hormone, secreted by the mammalian pituitary gland, that stimulates protein synthesis and growth of the long bones in the legs and arms. It also promotes the breakdown and use of fats as an energy source, rather than glucose. Production of growth hormone is greatest during early life. Its secretion is controlled by the opposing actions of two hormones from the hypothalamus: somatocrinin (growth-hormone-releasing hormone), which promotes its release; and somatostatin (growth-hormone-inhibiting hormone), that inhibits the release of growth hormone from the anterior pituitary gland.

Growth inhibitor–Any substance inhibiting the growth of an organism. The inhibitory effect can range from mild inhibition (growth retardation) to severe inhibition or death (toxic reaction). Two plant growth regulators that may act as inhibitors are ethylene and abscisic acid. The concentration of the inhibitor, the length of exposure to it, and the relative susceptibility of the organisms exposed to the inhibitor, are all important factors which determine the extent of the inhibitory effect.

Growth phases; growth phase curve–The characteristic periods in the growth of a bacterial culture, as indicated by the shape of a graph of viable cell number versus time, namely: lag phase; logarithmic (or exponential) phase ; stationary phase; death phase.

Growth rate–Increase in mass per unit of time. *Search in* growth.

Growth regulator–A synthetic or natural compound that at low concentrations elicits and controls growth responses in a manner similar to hormones.

Growth retardant–A chemical that selectively interferes with normal hormonal promotion of growth and other physiological processes, but without appreciable toxic effects.

Growth ring–Any of the rings that can be seen in a cross-section of a woody stem, such as a tree trunk. It represents the xylem formed in one year as a result of fluctuating activity of the vascular cambium.

Growth substance–Any organic substance, other than a nutrient, that is synthesized by plants and regulates growth and development. They are usually made in a particular region, such as the shoot tip, and transported to other regions, where they take effect.

GTP–Guanosine triphosphate. *Search in* guanosine.

Guanine–One of the four nucleic acid bases. A purine derivative that is one of the major component bases of nucleotides and the nucleic acids, DNA and RNA.

Guanosine–A nucleoside consisting of one guanine molecule linked to a D-ribose sugar molecule. The derived nucleotides, guanosine mono-, di-, and triphosphate (GMP, GDP, and GTP, respectively), are important in various metabolic reactions.

Guard-cell–Specialized epidermal cell that occurs as a pair around a stoma and controls opening and closing of the stoma through changes in turgor. *Search in* stoma.

Guide RNA–An RNA molecule that contain sequences that function as a template during RNA editing.

Guide sequence–An RNA molecule (or a part of it) which hybridizes with eukaryotic mRNA and aids in the splicing of intron sequences. Guide sequences may be either external (EGS) or internal (IGS) to the RNA being processed and may hybridize with either intron or exon sequences close to the splice junction. *Search in* split gene; exon.

Gymnosperm–Any plant whose ovules and the seeds into which they develop are borne unprotected, rather than enclosed in ovaries, as are those of the flowering plants (the term gymnosperm means naked seed).

Gynandromorph–An individual in which one part of the body is female and another part is male; a sex mosaic.

Gynogenesis–Female parthenogenesis; after fertilization of the ovum, the male nucleus is eliminated and the haploid individual (described as gynogenetic) so produced possesses the maternal genome only. *The opposite word is* androgenesis; parthenogenesis; anther culture.

Gyrase–*Search in* DNA helicase.

Habituation–The phenomena that, after a number of sub-cultures, cells can grow, without the addition of specific factors, such as no longer needing exogenous growth regulators in the tissue culture medium. Such cells are autonomous.

Haemoglobin–Conjugated protein compound containing iron, located in erythrocytes of vertebrates; important in the transportation of oxygen to the cells of the body.

Haemolymph–The mixture of blood and other fluids in the body cavity of an invertebrate.

Hair–A single or multicellular, absorptive (root hair) or secretory (glandular hair) and sometimes only a superficial outgrowth (covering hair) of the epidermal cells. The term trichome is often used but includes outgrowths from tissue below the epidermal layer. Distinguishing between hairs and trichomes can be difficult. Trichomes are usually connected to the vascular system, whereas hairs lack a vascular connection.

Hairy root culture–A fairly recent development in plant culture, consisting of highly branched roots of a plant. A plant tissue is treated with a culture of the bacterium *Agrobacterium rhizogenes*, which transfers part of its own plasmid DNA to the cells of an infected plant. This alters the plant's metabolism, including alterations in hormone levels, which in turn cause the explant to grow highly branched roots from the sites of infection. The roots branch much more frequently than the usual root system of that plant, and are also covered with a mass of tiny root hairs. Their most significant feature is that they produce secondary metabolites at levels similar to those made in the original plant. Thus they can be used as replacement plants for making such compounds as food flavours or fragrances.

Hairy root disease–A disease of broad-leaved plants, where a proliferation of root-like tissue is formed from the stem. Hairy root disease is a tumorous state similar to crown-gall, and is induced by the bacterium *Agrobacterium rhizogenes* containing an Ri plasmid. *Search in* *Agrobacterium*, crown gall.

Halophyte–A plant that can tolerate a high concentration of salt in the growing medium.

Halothane–A volatile anaesthetic.

Hanging droplet technique–*Search in* microdroplet array.

Haploid (Gr. *haploos*, single + *oides*, like)–A cell or organism containing only one representative from each of the pairs of homologous chromosomes found in the normal diploid cell. Having a single complete set of chromosomes, or referring to an individual or generation containing such a single set of chromosomes per cell. Usually a gamete. *Search in* monoploid.

Haploid cell–A cell containing only one set, or half the usual (diploid) number, of chromosomes.

Haplotype–A group of alleles, each from a different locus in the same region of a chromosome, that exist in the same double helix.

Haplozygous–*Search in* hemizygous.

Haptoglobin–A serum protein, alpha globulin, that interacts with haemoglobin during recycling of the iron molecule of haemoglobin.

Hardening off–Adapting plants to outdoor conditions by gradually withholding water, lowering the temperature, increasing light intensity, or reducing the nutrient supply. The hardening-off process conditions plants for survival when transplanted outdoors. The term is also used for gradual acclimatization to *in vivo* conditions of plants grown *in vitro*, *e.g.*, gradual decrease in humidity. *The opposite word is* acclimatization; free-living conditions.

Hardy-Weinberg equilibrium–The frequencies of genotypes at a locus resulting from random mating at that locus; for two alleles, A_1 and A_2, with respective frequencies p and q, the Hardy-Weinberg equilibrium frequencies are p^2 A_1A_1, $2pq$ A_1A_2, q^2 A_2A_2. Despite the simplifying assumptions required to predict these frequencies, most loci in most populations are in Hardy-Weinberg equilibrium. Thus the Hardy-Weinberg law, which predicts these frequencies, is one of the great unifying themes of biology.

Harvesting–1. The process involved in gathering in ripened crops.

2. The collection of cells from cell cultures or of organs from donors for the purpose of transplantation.

Heat pump–An apparatus that extracts heat from a fluid or gas that is marginally above ambient temperature. Heat pumps are commonly used to heat (or cool) greenhouses and laboratories.

Heat therapy–*Search in* thermotherapy.

Helix–Any structure with a spiral shape. The Watson and Crick model of DNA is in the form of a double helix.

Helminths–Parasitic worms, especially internal parasites of man and animals.

Helper cells–T cells that respond to an antigen displayed by a macrophage by stimulating B and T lymphocytes to develop into antibody-producing plasma cells and killer T cells, respectively.

Helper plasmid–A plasmid that provides a function or functions to another plasmid in the same cell.

Helper T cells–*Search in* helper cells.

Helper T lymphocytes–*Search in* helper cells.

Helper virus–A virus that provides a function or functions to another virus in the same cell.

Hemicellulase–An enzyme that degrades hemicellulose to galactose; hemicellulase is available as a commercial product.

Hemicellulose (Gr. *hemi*, half + cellulose)–Any cellulose-like carbohydrate, but with differing chemical composition. Together with pectin and lignin, hemicelluloses form the cell wall matrix.

Hemizygous; haplozygous–The condition in which only one allele of a pair is present, as in sex linkage or as a result of deletion. Genes present only once in the genotype and not in pairs; as in haploids, in differential segments of sex chromosomes, or in diploids as a result of aneuploidy or loss of chromosome segments.

Hemoglobin–*Search in* haemoglobin.

Hemolymph–*Search in* haemolymph.

Hemophilia–*Search in* haemophilia.

HEPA filter (high efficiency particulate air filter)–A filter capable of screening out particles larger than 0.3 µm. HEPA filters are used in laminar air flow cabinets (hoods) for sterile transfer work. *Search in* prefilter.

Herbicide–Any substance that is toxic to plants; usually applied to agrochemicals intended to kill specific unwanted plants, such as weeds which is chemically related to hormone.

Herbicide resistance–The ability of a plant to withstand herbicide. Herbicide resistance has been one of the early targets of plant genetic engineering. If a herbicide is sprayed onto a field planted with such resistant crops, then all the plants except the crop would be killed, thus providing an effective method of weed control without having to develop herbicides specific to each weed type. There is substantial concern in some quarters about the widespread use of this technology, which is essentially giving the plant kingdom the ability to evade man's most effective herbicides. The concern are that, firstly, such engineering will lead to increased use of the herbicides, at a time when it is generally accepted that the use of chemicals should be kept as low as possible, and that, secondly, there is the possibility that resistant crop plants will escape to become weeds, or that their resistant genes could be transferred to other species, including weeds.

Heredity–Resemblance among individuals related by descent; transmission of traits from parents to offspring.

Heritability–***In the narrow sense***: 1. the proportion of phenotypic superiority of parents that is seen in their offspring; 2. the proportion of the total phenotypic variation due to variation in breeding values.

In the broad sense: the proportion of the total phenotypic variation due to genetic variation. The degree to which a given trait is controlled by inheritance.

Search in broad-sense heritability; narrow-sense heritability.

Hermaphrodite–1. Animal that has both male and female reproductive organs, or a mixture of male and female attributes.

2. A plant whose flowers contain both stamen and carpels.

Heteroalleles–Mutations that are functionally allelic but structurally non-allelic; mutations at different sites in a gene.

Heterochromatin–Regions of chromosomes that stain darkly even during interphase; thought to be for the most part genetically inactive. *The opposite word is* euchromatin.

Heteroduplex–A double-stranded DNA molecule or DNA-RNA hybrid, where each strand is of a different origin, and consequently containing one or more mismatched (non-complementary) base pairs. A DNA duplex is prepared by the hybridization of single-stranded DNA molecules derived from two different sources. Where the two DNAs have identical or very similar sequences, a double-stranded molecule will be established, whereas where the two DNAs differ in sequence, single-stranded regions will remain. A heteroduplex will be revealed as single-strand "bushes" when DNA is observed electron microscopically. A map of homologous and non-homologous regions of the two molecules may thereby be constructed. This process is known as heteroduplex mapping.

Heterogametic sex–Producing unlike gametes with regard to the sex chromosomes. In mammals, the XY male is heterogametic, and the XX female is homogametic.

Heterogeneity–*Search in* genetic heterogeneity.

Heterogeneous nuclear RNA (hnRNA)–Large RNA molecules, which are unedited mRNA transcripts, or pre-mRNAs found in the nucleous of a eukaryotic cell. *Search in* RNA.

Heterokaryon–Cell with two or more different nuclei in a common cytoplasm as a result of cell fusion. Opposite: homokaryon.

Heterologous–From a different source, as in heterologous DNA.

Heterologous probe–A DNA probe that is derived from one species and used to screen for a similar DNA sequence from another species.

Heterologous protein–*Search in* recombinant protein.

Heteroplasmy–A cellular condition in which two genetically different types of organelles are present. *The opposite word is* homoplasmy.

Heteroploid–Term given to a cell culture when the cells comprising the culture possess nuclei containing chromosome numbers other than the diploid number.

Heteropyknosis (adj: heteropyknotic)–Property of certain chromosomes, or of their parts, to remain more dense during the cell cycle and to stain more intensely than other chromosomes or parts.

Heterosis (Gr. *heteros*, different + *osis*, suffix for "a state of")–*Search in* hybrid vigour.

Heterozygote (adj: heterozygous) (Gr. *heteros*, different + *zygon*, yoke)–An individual that has different alleles at the same locus in its two homologous chromosomes.

Hfr–High-frequency recombination strain of *Escherichia coli;* in such strains, the F episome is integrated into the bacterial chromosome.

HGH–Human growth hormone.

High efficiency particulate air filter–*Search in* HEPA.

Histocompatibility–The degree to which tissue from one organism will be tolerated by the immune system of another organism.

Histocompatibility–complex; histocompatibility system The collection of genes coding for peptides present on the surface of nucleated cells; these peptides are responsible for the differences between genetically non-identical individuals that cause rejection of tissue grafts between such individuals. These peptides were originally called histocompatibility antigens. They are now called histoglobulins, reflecting their structural similarity to immunoglobulins.

Histoglobulin–*Search in* histocompatibility complex.

Histology (Gr. *histos*, cloth, tissue + *logos*, discourse)–Science that deals with the microscopic structure of animal and plant tissues.

Histone–Group of water-soluble proteins rich in basic amino acids, closely associated with DNA in plant and animal chromatin. Histones are involved in the coiling of DNA in chromosomes and in the regulation of gene activity. Histones from reversible complexes with DNA, called nucleohistones.

HIV (Human immunodeficiency virus)–The retrovirus that causes AIDS in humans.

HLA (Human-leukocyte-antigen system)–*Search in* major histocompatibility antigens.

hnRNA–*Search in* heterogeneous nuclear RNA.

Holoenzyme–*Search in* apoenzyme.

Holometabolous–An insect that undergoes complete metamorphosis to the adult from a morphologically distinct larval stage.

Hollow fibre–A tube of a porous material, having an internal diameter of a fraction of a millimetre, and so its ratio of surface area to internal volume is very large. This has had two types of application. Firstly, hollow fibres can be used as filters. Because they have a huge surface area, they take much longer to clog up than normal filters. Secondly, they are used in the hollow fibre bioreactor, in which cells are kept inside the hollow, porous fibres, and the culture medium is circulated outside the reactor. The fibres let nutrients in and products out (as they are in solution), but do not allow the passage of cells. Hollow fibre bioreactors are very effective for maintaining mammalian cells in culture because they have a very large surface area for the cells to grow on without needing a large reactor to hold them, and because the nutrient reaching the cells can be kept fresh. The reactor also provides an easy way of removing the product that the cells are making: such as monoclonal antibodies. Hollow fibre reactors are less use when the cells themselves have to grow, because it is hard to get at the inside of the fibre to remove surplus cells.

Homeobox–A DNA sequence found in several genes that are involved in the specification of organs in different body parts in animals; they are characteristic of genes that influence segmentation in animals. The homeobox corresponds to an amino acid sequence in the polypeptide encoded by these genes; this sequence is called the homeodomain.

Homeodomain–*Search in* homeobox.

Homeotic mutation–A mutation that causes a body part to develop in an inappropriate position in an organism, such as the mutation in *Drosophila* that causes legs to develop on the head in place of antennae.

Homoalleles–Mutations that are both functionally and structurally allelic; mutations at the same site in the same gene.

Homodimer–A protein with two identical polypeptide chains.

Homogametic sex–Producing similar gametes with regard to the sex chromosomes. In mammals, the XY male is heterogametic, and the XX female is homogametic. *Search in* heterogametic sex.

Homogenotization–A genetic technique used to replace one copy of a gene, or other DNA sequence within a genome, with an altered copy of that sequence. The DNA is first cloned and then altered in some way, *e.g.*, a transposon is inserted into a gene. The mutated gene copy can be used to replace the original gene by recombination *in vivo*. The incorporation of the mutated gene is usually selected, for example, by virtue of its containing a transposon-encoded antibiotic resistance. *Search in* replacement.

Homokaryon–Cell with two or more identical nuclei as a result of fusion. Opposite: heterokaryon.

Homologous–From the same source, or having the same evolutionary function or structure.

Homologous chromosomes–Chromosomes that occur in pairs and are generally similar in size and shape: one comes from the male parent and the other from the female. Such chromosomes contain the same linear array of genes.

Homologous recombination–The exchange of DNA fragments between two DNA molecules or chromatids of paired chromosomes (during crossing over) at the site of identical nucleotide sequences.

Homology–The degree of identity between individuals, or characters. The degree of identity between the nucleotide sequences of two nucleic acid molecules or the amino acid sequences of two protein molecules. Although sequence determination is the ultimate test of homology, useful estimates can be provided by either DNA-DNA or DNA-RNA hybridization.

Homomultimer–*Search in* homopolymer.

Homoplasmy–A cellular condition in which all copies of an organelle are genetically identical. *Search in* heteroplasmy.

Homopolymer–A nucleic acid strand that is composed of one kind of nucleotidc, *e.g.*, GGGGGGG. *Search in* polymer.

Homopolymeric tailing–*Search in* tailing.

Homozygote; (adj: homozygous) (Gr. *homos*, one and the same + *zygon*, yoke)–An individual that has two copies of the same allele at a particular locus in its two homologous chromosomes. *Search in* allele; genotype; heterozygote.

Hormone (Gr. *hormaein*, to excite)–A specific organic product, produced in one part of a plant or animal body, and transported to another part where, at low concentrations, it promotes, inhibits or quantitatively modifies a biological process.

Host-specific toxin–A metabolite produced by a pathogen which has a host specificity equivalent to that of the pathogen. Such toxins are utilized for *in vitro* selection experiments to screen for tolerance or resistance to the pathogen.

Host–An organism that contains another organism, or a cloning vector.

Humoral immune response–The production of antibody by B cells of the immune system in response to the presence of a foreign antigen. *Search in* antibody-mediated immune response.

Hup⁺–*Search in* hydrogen-uptake positive.

Hybrid (L. *hybrida*)–1. The offspring of two parents that are genetically different. A cross between two genetically unlike individuals.

2. A heteroduplex DNA or DNA-RNA molecule.

Hybrid arrested translation–A method used to identify the proteins encoded by a cloned DNA sequence. A crude cellular mRNA preparation, composed of many individual types of mRNA, is hybridized with cloned DNA. Only mRNA molecules homologous to the cloned DNA will anneal to it. The rest of the mRNA molecules are put into an *in vitro* translation system and the protein products are compared with the proteins obtained by use of the whole mRNA preparation. *The opposite word is* hybrid released translation.

Hybrid cell–The mononucleate cell which results from the fusion of two different cells, leading to the formation of a synkaryon.

Hybrid dysgenesis–A syndrome of abnormal germ-line traits, including mutation, chromosome breakage, and sterility, which results from activity of transposable elements.

Hybrid released translation–A method used to detect the proteins encoded by cloned DNA. The cloned DNA is bound to a nitrocellulose filter and a crude preparation of mRNA is hybridized to the filter-bound DNA. Only mRNA sequences homologous to the cloned DNA will be retained on the filter. These mRNA molecules can then be removed by high temperature or by using formamide. The purified mRNA is then placed in an *in vitro* translation system and the proteins encoded by the message can be analysed by electrophoresis through a polyacrylamide gel. *Search in* hybrid arrested translation.

Hybrid seed–1. Seed produced by crossing genetically dissimilar parents.

2. In plant breeding, used colloquially for seed produced by specific crosses of carefully selected pure lines, such that the F_1 crop displays hybrid vigour. As the F_1 crop is heterozygous, it does not breed true and so new seed must be purchased each season.

Hybrid selection–The process of choosing individuals possessing desired characteristics from among a hybrid population.

Hybrid vigour; heterosis–The extent to which the performance of a hybrid in one or more traits is better than the average performance of the two parental populations.

Hybridization–1. Interbreeding of species, races, varieties and so on, among plants or animals; a process of forming a hybrid by cross pollination of plants or by mating animals of different types.

2. The production of offspring of genetically different parents, normally from sexual reproduction, but also asexually by the fusion of protoplasts or by transformation.

3. The pairing of two polynucleotide strands, often from different sources, by hydrogen bonding between complementary nucleotides. *Search in* northern hybridization; Southern hybridization.

Hybridoma–A hybrid cell, derived from a B (antibody producing) lymphocyte fused to a tumour cell, which grows indefinitely in tissue culture and is selected for the secretion of the specific antibody produced by that B cell.

Hydrate–A compound formed by the incorporation of water.

Hydrogen bond–A relatively weak bond formed between a hydrogen atom (which is covalently bound to a nitrogen or oxygen atom) and a nitrogen or oxygen atom with an unshared electron pair. Weak interactions between electro-negative atoms and hydrogen atoms (electro-positive) that are linked to other electro-negative atoms.

Hydrogen-uptake positive (Hup$^+$)–A term describing a micro-organism that is capable of assimilating (or taking up) hydrogen gas.

Hydrolysis–A reaction in which a molecule of water is added at the site of cleavage of a molecule to two products.

Hydrophobic interactions–Association of non-polar groups with each other when present in aqueous solutions because of their insolubility in water.

Hydroponics–The growing of plants in aerated water containing all the essential mineral nutrients, with no soil. Also called soilless gardening or cultivation.

3′-hydroxyl end–The hydroxyl group that is attached to the 3′ carbon atom of the sugar (ribose or deoxyribose) of the terminal nucleotide of a nucleic acid molecule.

Hyperploid–A genetic condition in which a chromosome or a segment of a chromosome is over-represented in the genotype. Opposite: hypoploid.

Hypersensitive sites–Regions in the DNA that are highly susceptible to digestion with endonucleases.

Hypertonic–A solution with an osmotic potential greater than that of living cells, leading to water loss from, shrinkage or plasmolysis of cells in a hypertonic situation. Opposite: hypotonic.

Hypervariable region–The parts of both the heavy and light chains of an antibody molecule that enable it to bind to a specific site on an antigen.

Hypervariable segment–A region of a protein that varies considerably between strains or individuals.

Hypochlorite–Generic term for aqueous solutions of sodium hypochlorite, potassium hypochlorite or calcium hypochlorite, which are oxidizing agents and used for disinfecting surfaces and surface-sterilizing tissues, and for bleaching.

Hypocotyl (Gr. *hypo*, under + *kotyledon*, a cup-shaped hollow)–Portion of an embryo or seedling below the cotyledons, which is a transitional area between stem and root.

Hypomorph–A mutation that reduces but does not completely abolish gene expression.

Hypoplastic–Reduction in plant growth or development (dwarfing, stunting) resulting from an abnormal condition associated with a disease or nutritional stress.

Hypoploid–A genetic condition in which a chromosome or segment of a chromosome is underrepresented in the genotype. Opposite: hyperploid.

Hypostatic; hypostasis–*Search in* epistasis.

Hypothalamic peptides–Peptides generated in the vertebrate forebrain and concerned with regulating the body's physiological state.

Hypothesis (Gr. *hypothesis*, foundation)–A tentative theory or supposition provisionally adopted to explain certain facts and to guide in the investigation of other facts. Once proven by rigorous scientific investigation, it becomes a theory or a law.

Hypotonic–Osmotic potential less than that of living cells. Cells placed in a hypotonic solution display swelling and turgidity.

ICSI–*Search in* intracytoplasmic sperm injection.

Identical twins–*Search in* monozygotic twins.

Ideogram–*Search in* karyogram.

Illuminate–To supply or brighten with light. Illumination is an absolute requirement for tissue cultures. Fluorescent lights are commonly employed. The light intensity is dependent on the light source and the requirements of the culture.

Imaginal disc–A mass of cells in the larvae of *Drosophila* and other holometabolous insects that gives rise to particular adult organs, such as antennae, eyes or wings.

Imbibition (L. *imbibere*, to drink)–1. The absorption of liquids or vapours into the ultramicroscopic spaces or pores found in materials.

2. The initial water uptake by seeds starting germination.

Immediate early gene–A viral gene that is expressed promptly after infection.

Immobilized cells–Cells entrapped in matrices such as alginate, polyacrylamide and agarose designed for use in membrane and filter bioreactors.

Immortalization–The genetic transformation of a cell type into a cell line which can proliferate indefinitely.

Immortalizing oncogene–A gene that upon transfection enables a primary cell to grow indefinitely in culture.

Immune response–The processes, including the synthesis of antibodies, that are used by vertebrates to respond to the presence of a foreign antigen. *Search in* primary immune response; secondary immune response.

Immunity–The state of relative insusceptibility of an animal or plant to infection by disease-producing organisms or to the harmful effects of their poisons.

Immunization–The production of immunity in an individual by artificial means. Active immunization involves the introduction, either orally or by infection, of specially treated bacteria, viruses or their toxins so as to stimulate the production of antibodies.

Immuno-affinity chromatography–A purification technique in which an antibody is bound to a matrix and is subsequently used to bind and separate a protein from a complex mixture. *The opposite word is* affinity chromatography.

Immunoglobulin–One of a group of proteins (globulins) in the body that act as antibodies. They are produced by specialized cells (B lymphocytes) and are present in blood serum and other body fluids.

Immunosuppression–The suppression of immune response. Immunosuppression is necessary following organ transplants in order to prevent the host rejecting the grafted organ.

Immunoassay; immunodiagnostics–An assay system which detects proteins by using an antibody specific to that protein. A positive result is seen as a precipitate of an antibody-protein complex. The antibody can be linked to a radioactive atom or to an enzyme which catalyses an easily monitored reaction such as a change in colour.

Immunochemical control–Use of immune agents to combat infections.

Immunodiagnostics–*Search in* immunoassay.

Immunogen–*Search in* antigen.

Immunogenicity–The ability to elicit an immune response.

Immunosensor–A biosensor having an antibody as biological part.

Immunosuppressor–A substance, an agent or a condition that prevents or diminishes the immune response.

Immunotherapy–The use of an antibody or a fusion protein containing the antigen binding site of an antibody to cure a disease or enhance the well-being of a patient.

Immunotoxin–Protein drugs consisting of an antibody joined to a toxin molecule. Immunotoxins can be made by linking toxin and antibody molecules chemically, or by fusing the genes for the toxin and the antibody. The antibody portion of the molecule facilitates binding to a target molecule or cell, and the toxin inactivates or kills the target molecule or cell.

Impeller–An agitator that is used for mixing the contents of a bioreactor.

Inactivated agent–A virus, bacterium or other organism that has been treated to prevent it from causing a disease.

Inbred line–The product of inbreeding, *i.e.*, the mating of individuals that have ancestors in common; in plants and laboratory animals, it refers to populations resulting from at least 6 generations of selfing or 20 generations of brother-sister mating, that are for all practical purposes, completely homozygous. In farm animals, the term is sometimes used to describe populations that have resulted from several generations of the mating of close relatives, without having reached complete homozygosity.

Inbreeding–Matings between individuals that have one or more ancestors in common, the extreme condition being self-fertilization is most intense type of inbreeding, which occurs in many plants and some primitive animals.

Inbreeding-depression–Reduction in vigour, yield, etc., of a population that is commonly seen as the level of inbreeding increases. The traits that show greatest inbreeding depression are those that are most closely associated with viability and reproductive ability.

Inclusion body–Protein that is overproduced in a recombinant bacterium and forms a crystalline array inside the bacterial cell.

Incompatibility–1. Selectively-restricted mating competence, which limits fertilization, such as lack of proper functions by otherwise normal pollen grains or certain pistils, a condition that may be caused by a variety of factors.

2. A physiological interaction resulting in graft rejection or failure.

3. A function of a related group of plasmids. Plasmids which are closely related share similar replication functions and this leads to the exclusion of one or the other plasmid if they are present in the same cell; thus such plasmids are incompatible. Plasmids are placed in incompatibility groups by this simple reaction and, in general, plasmids belonging to one incompatibility group are very closely related.

Incompatibility group–A classification scheme indicating which plasmids can co-exist within a single cell. Plasmids must belong to different incompatibility groups to co-exist within the same cell. Plasmids that belong to the same incompatibility group are unstable when placed in the same cell. A plasmid cloning vector should always belong to an incompatibility group different from that of the host bacterium's endogenous plasmids.

Incomplete digest–*Search in* partial digest.

Incomplete dominance–A type of gene action in which heterozygotes have a phenotype that is distinctly different from, and intermediate to, the homozygous phenotypes. *Search in* heterozygote; phenotype.

Incomplete penetrance–When some individuals in a population have a specific genotype that causes an abnormality but are not affected.

Incubation (L. *incubare*, to lie on)–1. The hatching of eggs by means of heat, either natural or artificial.

2. Period between infection and appearance of symptoms induced by parasitic organisms.

3. The culture of cells and organisms.

Incubator–An apparatus in which environmental conditions (light, photoperiod, temperature, humidity, etc.) are fully controlled, and used for hatching eggs, multiplying micro-organisms, culturing plants, etc. *The opposite word is* culture room; growth cabinet.

Indehiscent–Describing a fruit or fruiting body that does not open to release its seeds or spores when ripe.

Independent assortment–The random distribution of alleles (from different loci) to the gametes that occurs when genes are located in different chromosomes or far apart on large chromosomes. The distribution of alleles at one locus is independent of other the distribution of alleles at another locus. *The opposite word is* linkage.

Indeterminate growth–1. Unlimited growth potential for a definite or indefinite period. Some apical meristems can produce unrestricted numbers of lateral organs.

2. In legumes, used to describe plant architecture.

Indirect embryogenesis–Embryo formation on callus tissues derived from zygotic or somatic embryos, seedling plant or other tissues in culture. *The opposite word is* direct embryogenesis.

Indirect organogenesis–Organ formation on callus tissues derived from explants.

Inducer–A low-molecular-weight compound or a physical agent that is bound by a repressor so as to produce a complex that can no longer bind to the operator; thus, the presence of the inducer turns on the expression of the gene(s) controlled by the operator.

Inducible–A gene or gene-product is said to be inducible if its transcription or synthesis is increased by exposure of the cells to an effector. Effectors are usually small molecules whose effects are specific to particular operons or groups of genes. *See* constitutive.

Inducible enzyme–An enzyme that is synthesized only in the presence of the substrate that acts as an inducer.

Inducible gene–A gene that is expressed only in the presence of a specific metabolite, the inducer.

Induction (L. *inducere*, to lead in)–Act or process of causing to occur; process whereby a cell or tissue influences neighbouring cells or tissues. Turning on transcription of a specific gene or operon. Getting an organism to make a protein by exposing it to some stimulus.

Induction media–1. Media used to induce the formation of organs or other structures.

2. Media causing variation or mutation in the tissues exposed to it.

Inembryonation–*Search in* artificial inembryonation.

I/E region–*Search in* integration-excision region.

Inert–A support structure that makes no chemical contribution, and whose only function is support. Physiologically it is a neutral or immobile unit.

Infection–The invasion of any living organisms by disease-causing micro-organisms, which proceed to establish themselves, multiply, and produce various clinical signs in their host.

Infectious agent–Typically, a proliferating virus, bacterium or parasite that causes a disease in plants or animals.

Infiltrate–To force the passage of liquid into tissue pores or space, such as by applying a vacuum, then releasing it, during the disinfectation procedure.

Inflorescence (L. *inflorescere*, to begin to bloom)–1. The way the flowers are arranged on the stalk of a plant.

2. The flowers of a plant collectively.

3. The flowering or blooming process.

IGS (Internal guide sequence)–*Search in* guide sequence.

Inhibitor–Any substance or object that retards a chemical reaction; a major or modifier gene that interferes with a reaction or with the expression of another gene.

Inheritance–The transmission of particular characteristics and/or genes from generation to generation.

Initial–Cells in a meristem that remain permanently meristematic and form tissues of particular structure and function.

Initiation–1. Early steps or stages of a tissue culture process (culture growth, organogenesis, embryogenesis)

2. Early stages of biosynthesis.

Initiation codon; initiator–The codon AUG which specifies the first amino acid methionine of a polypeptide chain. In bacteria, the initiation codon is either AUG, which is translated as *n*-formyl methionine or, rarely, GUG (valine). In eukaryotes, the initiation codon is always AUG and is translated as methionine. The term is also used to describe the corresponding sequence in DNA, namely ATG. *Search in* start codon; initiator codon.

Initiation factors–Soluble proteins required for the initiation of translation. At least three structurally and functionally different initiation factors have been identified in bacterial protein synthesis.

Initiator–*Search in* initiation codon.

Inoculate–Deliberately introduce something into. The process is inoculation. Not the same as *contamination*.

1. In bacteriology, tissue culture, etc., placing inoculum into (or onto) medium to initiate a culture.

2. In immunology, to immunize.

Inoculation cabinet–Small room or cabinet for inoculation (of tissue or micro-organism cultures) operations, often with a current of sterile air to carry contaminants away from the work area.

Inoculum (pl: inocula)–1. A small piece of tissue cut from callus, or an explant from a tissue or organ, or a small amount of cell material from a suspension culture, transferred into fresh medium for continued growth of the culture. *Search in also* minimum inoculum size.

2. Microbial spores or parts (such as mycelium).

3. Vaccine.

Inorganic compound–A chemical compound that generally is not derived from living processes; compounds that do not contain carbon.

Inositol (hexahydroxycyclohexane; $C_6H_6(OH)_6$)–1. A cyclic acid that is constituent of certain cell phosphoglycerides.

2. A water-soluble nutrient frequently referred to as a "vitamin" in plant tissue culture. Also acts as a growth factor in some animals and micro-organisms.

Inositol lipid–A membrane-anchored phospholipid that transduces hormonal signals by stimulating the release of any of several chemical messengers. *Search in* phospholipid.

Insecticide–A substance that kills insects.

Insert–A DNA molecule that is incorporated into a cloning vector.

Insertion element–Generic term for DNA sequences found in bacteria capable of genome insertion. Postulated to be responsible for site-specific phage and plasmid integration.

Insertion mutations–Changes in the base sequence of a DNA molecule resulting from the random integration of DNA from another source. *Search in* DNA; mutation.

Insertion sequence–*Search in* IS element.

Insertion site; cloning site–A unique restriction site in a vector DNA molecule into which foreign DNA can be inserted. The term is also used to describe the position of integration of a transposon or insertion site element.

in situ (Latin for "in place")–Meaning in the natural place or in the original place.

1. Experimental treatments performed on cells or tissue rather than on extracts from them.

2. Assays or manipulations performed with intact tissues.

***in situ* colony; *in situ* plaque hybridization**–A procedure for screening colonies or plaques growing on plates or membranes for the presence of specific DNA sequences by the hybridization of nucleic acid probes to the DNA molecules present in these colonies or plaques.

***in situ* conservation**–A conservation method that attempts to preserve the integrity of genetic resources by conserving them within the evolutionary dynamic ecosystems of their original habitat or natural environment. *The opposite word is ex situ* conservation.

***in situ* conservation of farm animal genetic diversity**–All measures to maintain live animal breeding populations, including those involved in active breeding programmes in the agro-ecosystem where they either developed or are now normally found, together with husbandry activities that are undertaken to ensure the continued contribution of these resources to sustainable food and agricultural production, now and in the future. (*Source*: FAO, 1999)

***in situ* plaque hybridization**–*Search in in situ* colony.

Instability–A random-type variation or a lack of steadiness. Due to genetic instability, cell lines lose certain characteristics or functions in culture.

Insulin–A polypeptide hormone secreted by the Langerhans islets of the pancreas, and that regulates the level of sugar in the blood.

Integrating vector–A vector that is designed to integrate cloned DNA into the host cell chromosomal DNA.

Integration–The recombination process which inserts a small DNA molecule (usually by homologous recombination) into a larger one. If the molecules are circular, integration involves only a single crossing-over; if linear, then two crossings-over are required. A well known example is the integration of phage λ (lambda) DNA into the *E. coli* genome.

Integration–excision (I/E) region–The portion of bacteriophage lambda (λ) DNA that enables bacteriophage lambda (λ) DNA to be inserted into a specific site in the *E. coli* chromosome and excised from this site.

Integument–One of the layers that enclosed the ovule, and is the precursor of the seed coat.

Intensifying screen–A plastic sheet impregnated with a rare-earth compound, such as calcium tungstate, which absorbs ß radiation and emits light. When placed on one side of a piece of X-ray film with a radioactive sample on the other side, the intensifying screen will capture some of the ß emissions which pass through the film, blackening the X-ray film and so greatly enhancing the sensitivity of the detection. An intensifying screen is used in Southern and northern blotting procedures.

Interaction–In statistics, an effect that cannot be explained by the additive action of contributing factors; a departure from strict additivity.

Intercalary–(L. *intercalare*, to insert) Meristematic tissue or growth not restricted to the apex of an organ, *i.e.*, growth at nodes.

Intercalary growth–A pattern of stem elongation typical of grasses. Elongation proceeds from the lower internodes to the upper internodes through the differentiation of meristematic tissue at the base of each internode.

Intercalating agent–A chemical capable of inserting between adjacent base pairs in a DNA molecule.

Intercellular space–Pore space between cells, especially typical of leaf tissues.

Interfascicular cambium–Cambium that arises between vascular bundles.

Interference–Crossing over at one point that alters the chance of another crossing-over nearby; detected by studying the pattern of crossings-over with three or more linked genes. Interference is positive or negative depending on whether the chance of another crossing-over nearby is reduced or increased, respectively.

Interferon–A family of small proteins that stimulate viral resistance in cells.

Intergeneric–A cross between two different genera.

Intergenic regions–DNA sequences located between genes; they comprise a large percentage of a genome and have no known function.

Interleukin–A group of proteins that transmit signals between immune cells and are necessary for mounting normal immune responses.

Internal guide sequence (IGS)–*Search in* guide sequence.

Intervening sequence–*Search in* intron.

Internode (L. *inter*, between + *nodus*, a knot)–The region of a stem between two successive nodes.

Interphase–The stage in the cell cycle when the cell is not dividing; the stage follows telophase of one division and extends to the beginning of prophase in the next division. DNA replication occurs during this stage.

Intersex–An organism displaying a mixture of male and female attributes.

Interspecific–Between two different species, *e.g.*, an interspecific cross is a cross between two species.

Intervening sequence–*Search in* intron.

Intracellular (L. *intra*, within + cell)–Occurring within a cell.

Intracytoplasmic sperm injection (ICSI)–The injection, using micromanipulation, of a single sperm into the cytoplasm of a mature oocyte.

Intrageneric–Within a genus, such as a hybrid resulting from a cross between species within one genus.

Intragenic complementation–Complementation that occurs between two mutant alleles of a gene; common only when the product of the gene functions as a homomultimer.

Intraspecific–Within a species or its populations, including subspecies, such as an intraspecific cross, or variation.

Introgression–The introduction of new gene(s) into a population by crossing between two populations, followed by repeated backcrossing to that population while retaining the new gene(s).

Intron; intervening sequence–A segment of DNA sequence of a eukaryotic gene, not represented in the mature (final) mRNA transcript, because it is spliced out of the primary transcript before it can be translated; a process known as intron splicing. Some genes of higher eukaryotes contain a large number of introns, which make up the bulk of the DNA sequence of the gene.

Introns are also found in genes whose RNA transcripts are not translated, namely eukaryotic rRNA and tRNA genes. In these cases the intron sequence does not appear in the functional RNA molecule. *The opposite word is* exon.

Invariant–Constant, unchanging; usually referring to the portion of a molecule that is the same across species.

Invasiveness–Ability of a plant to spread beyond its introduction site and become established in new locations, where it may have a deleterious effect on organisms already existing there.

Inversion–A chromosome re-arrangement or chromosomal change that reverses the order of a linear array of genes in it.

Inverted repeat–Two regions of a nucleic acid molecule which have the same nucleotide sequence but in an inverted orientation, such as

5′ GCACTTG......CAAGTGC 3′

3′ CGTGAAC......GTTCACG 5′

Because they contain exactly the same message when read in either direction, inverted repeats are said to be palindromes.

in vitro (L. for "in glass")–Living in test tubes, outside the organism or in an artificial environment, typically in glass vessels in which cultured cells, tissues, organs or whole plants may reside.

***in vitro* embryo production (IVEP)**–The combination of ovum pickup, *in vitro* maturation of ova, and *in vitro* fertilization. A potential means of overcoming the variability between donors in number of ova collected in embryo-transfer programmes.

***in vitro* fertilization (IVF)**–A widely used technique in human and animal science, whereby the egg is fertilized with sperm outside the body. Usually, the fertilized egg is cultured outside the body for a few days (to confirm that fertilization has occurred) before re-implantation into a female.

***in vitro* maturation (IVM)**–Culture of immature ova in the laboratory, usually until they are ready for *in vitro* fertilization.

***in vitro* mutagenesis**–*Search in* directed mutagenesis.

***in vitro* transcription; cell–free transcription**–The specific and accurate synthesis of RNA in the test tube using purified DNA preparations as a template. So-called "coupled systems" may be obtained from *E. coli* which carry out both mRNA synthesis and its translation into protein. For eukaryotes, separate cell-free systems have to be set up to demonstrate the activity of the three functionally distinct RNA polymerase complexes.

***in vitro* translation; cell-free translation**–The synthesis of proteins in the test-tube from purified mRNA molecules using cell extracts containing ribosomal sub-units, the necessary protein factors, tRNA molecules and aminoacyl tRNA synthetases, ATP, GTP, amino acids and an enzyme system for re-generating the nucleoside triphosphates. Prokaryotic translation systems are usually prepared from *E. coli* or the thermophilic bacterium, *Bacillus stearothermophilus*. Eukaryotic systems usually employ rabbit reticulocyte lysates or wheat germ.

in vivo (L. for "in living")–The natural conditions in which organisms reside. Refers to biological processes that take place within a living organism or cell under normal conditions.

***in vivo* gene therapy**–The delivery of a gene or genes to a tissue or organ of an individual to alleviate a genetic disorder.

Ion–A charged particle.

Ion channel–An integral protein within a cell membrane, through which selective ion transport occurs.

Ionic bonds–Attractions between oppositely charged chemical groups.

Ionizing radiation–The portion of the electromagnetic spectrum that results in the production of positive and negative charges (ion pairs) in molecules. X-rays and gamma rays are examples of ionizing radiation.

IPTG (Isopropyl-b-d-thiogalacto-pyranoside)–An inducer of the *lac* (lactose) operon. In recombinant DNA technology, IPTG is often used to induce cloned genes that are under the control of the *lac* repressor-*lac* promoter system.

Irradiation (L. *in*, into + *radius*, ray)–1. Exposure to any form of radiation.

2. Treatment with a rays, such as ultraviolet rays, etc.

3. Apparent enlargement of objects, due to difference in illumination.

4. In food technology: exposure of food to controlled low levels of ionizing radiation to sterilize or disinfest it, without inducing radioactivity in the target. If in sealed containers, the product is sterile and can be stored under ambient conditions. Used for continuous flow disinfestation of grain at unloading facilities.

IS element (insertion sequence)–A short (800-1400 nucleotide pairs) DNA sequence found in bacteria that is capable of transposing to a new genomic location; other DNA sequences that are bounded by IS elements may also be transposed.

Isoalleles–Different forms of a gene that produce the same phenotype or very similar phenotypes.

Isochromosome–A chromosome with two identical arms and identical genes. The arms are mirror images of each other.

Isodiametric–Term commonly used to describe cells with equal diameters.

Iso-electric focusing gels–Variant of DNA gel electrophoresis, which separates macromolecules on the basis of their iso-electric point rather than their size.

Iso-enzyme–*Search in* isozyme.

Isoform–A member of a family of closely related proteins that have some amino acid sequences in common and some different.

Isogamy–Sexual reproduction involving the production fusion of gametes that are similar in size and structure.

Isogenic stocks–Strains of organisms that are genetically identical; completely homozygous.

Isolating mechanism–Any of the biological properties of organisms that prevent interbreeding (and therefore exchange of genetic material) between members of different species that inhabit the same geographical area.

Isolation medium–An optimum medium suitable for explant survival, growth and development.

Isoleucine–*Search in* amino acid.

Isomerase–Any of a class of enzymes that catalyse the re-arrangement of the atoms within a molecule, thereby converting one isomer into another.

Iso-osmotic–*Search in* isotonic.

Isotonic; Iso–osmotic–Solutions having the same osmotic potential; the same molar concentration. For protoplasts to survive, the medium they are suspended in must be isotonic with them.

Isotope–One of two or more forms of an element that have the same number of protons (atomic number) but differing numbers of neutrons (mass numbers). Radioactive isotopes are commonly used to make DNA probes and metabolic tracers.

Isozyme–A variant of a particular enzyme. In general, all the isozymes of a particular enzyme have the same function and sometimes the same activity, but they differ in amino-acid sequence. With the help of isozyme analysis, based on electrophoresis techniques that can separate the different variants, related species or cultivars can be distinguished.

IVEP–*Search in in vitro* embryo production.

IVF–*Search in in vitro* fertilization.

IVM–*Search in in vitro*

J–1. Joining segment

2. Joule

Jiffy potTM–Pots made from wood pulp and peat, commonly used for transplanting tissue-culture-derived plants into soil medium.

JIVET–*Search in* juvenile *in vitro* embryo technology.

JIVT–*Search in* juvenile *in vitro* embryo technology.

J/m^2; J m^{-2} (joules per square metre)–A unit of light measurement.

Joining segment (J)–A small DNA segment that links genes in order to yield a functional gene encoding an immunoglobulin.

Joule (SI symbol: J)–The amount of energy needed to apply a force of 1 newton over a distance of 1 metre.

Jumping genes–*Search in* transposable elements.

Jumping library–*Search in* chromosome jumping.

Juvenility–Early phase of development in which an organism is juvenile and incapable of sexual reproduction.

Juvenile hormone–A hormone secreted by insects from a pair of endocrine glands close to the brain. It inhibits metamorphosis and maintains the larval features.

Juvenile *in vitro* embryo technology (JIVT; JIVET)–A technology involving collection of immature ova from young animals, *in vitro* maturation of those ova, *in vitro* fertilization, and then transfer of the resultant embryos into recipient females. A method of achieving rapid generation turnover.

Kanamycin–An antibiotic of the aminoglycoside family that poisons translation by binding to the ribosomes.

Kanr Kanamycin-resistance gene–*Search in* selectable marker.

Kappa chain–One of two classes of antibody light chains. *Search in* lambda chain.

Karyogamy–The fusion of nuclei or nuclear material that occurs during sexual reproduction. *Search in* fertilization.

Karyogram–A diagram representing the characteristic features of the chromosomes of a species.

Karyokinesis–The division of a cell nucleus. *Search in* meiosis; mitosis, which may not be followed by cytokinesis.

Karyotype–The chromosome constitution of a cell, an individual, or of a related group of individuals, as defined both by the number and the morphology of the chromosomes, usually in mitotic metaphase; chromosomes arranged in order of length and according to position of centromere; also, the abbreviated formula for the chromosome constitution, such as 47, XX + 21 for human trisomy-21 (Down's syndrome).

kb Kilobase pairs–*Search in* base pair; kilobase.

k_{cat}–The catalytic rate constant that characterizes an enzyme-catalysed reaction. The larger the k_{cat} value, the faster the conversion of substrate into product.

k_{cat} /K_m–The catalytic efficiency of an enzyme-catalysed reaction. The greater the value of k_{cat}/K_m, the more rapidly and efficiently the substrate is converted into product.

Killer T cells–T cells that carry T-cell receptors and kill cells displaying the recognized antigens.

kilobase (kb)–A length unit equal to 1 000 base pairs of a double-stranded nucleic acid molecule. One kilobase of double-stranded DNA has a mass of about 660 kilodalton.

kilobase pairs (kb)–Used interchangeably with kilobase.

kilocalorie (kcal)–It is equal to 1000 cal. *Search in* calorie.

kilodalton (kDa)–A unit of atomic mass equal to 1 000 daltons. *Search in* dalton.

kilojoule (kJ)–It is equal to 1 000 J; 1 kcal = 4.184 kJ. *Search in* joule.

Kinase–An enzyme that can transfer a phosphate from a high energy phosphate such as ATP, to an organic molecule.

Kinetics–Dynamic processes involving motion.

Kinetin (Gr. *kinetikos*, causing motion)–One of the cytokinins, a group of growth regulators that characteristically promote cell division in plants.

Kinetochores–Two parallel bushlike filaments situated in the centromere of a chromosome.

Kinetosome–Granular cytoplasmic structure which forms the base of a cilium or flagellum. *The opposite word is* basal body.

Kinin–A substance or growth hormone which inducing cell division. In plant systems, the prefix cyto–has been added (cytokinin) to distinguish it from kinin in animal systems.

Klenow fragment–A product of proteolytic digestion of the DNA polymerase I from *E. coli*; it has both polymerase and 3′-exonuclease activities but not 5′-exonuclease activity.

K_m–A dissociation constant that characterizes the binding of an enzyme to a substrate. The smaller the value of K_m, the tighter the binding of the enzyme to the substrate. Also called the Michaelis constant.

Knockout–An animal resulting from an embryonic stem cell in which a normal functional gene has been replaced by a non-functional form of the gene. This technique is used extensively in mice–much can be learned about the function of a gene by studying the phenotype of animals that lack the peptide product of that gene.

Label–A compound or atom that is either attached to or incorporated into a macromolecule and is used to detect the presence of a compound, substance, or macromolecule in a sample. a.k.a. tag.

Labelling–The process of replacing a stable atom in a compound with a radioactive isotope of the same element to enable it to be detected by autoradiography or other techniques. Increasingly, radioactive labelling is being replaced by fluorescent labelling. The method is used to trace the path of the labelled compound through a biological or chemical system.

***lac* repressor–*lac* promoter system**–*Search in* IPTG.

Lactose–Milk sugar, a disaccharide with one unit each of glucose and galactose.

Lag phase–1. A state of apparent inactivity preceding a response; called also a latent phase.

2. The initial growth phase, during which cell number remains relatively constant, prior to rapid growth.

3. The first of five growth phases of most batch-propagated cell suspension cultures, being the phase in which inoculated cells in fresh medium adapt to the new environment and prepare to divide. *Search in* growth phases.

Lagging strand–The strand of DNA that is synthesized discontinuously during replication (because DNA synthesis can proceed only in the 5′ to 3′ direction).

Lambda chain–One of two classes of antibody light chains. *The opposite word is* kappa chain.

Lamella (L. diminutive of *lamina*, a thin blade; pl: lamellae)–A double-membrane structure, plate or vesicle that is formed by two membranes lying parallel to each other.

Lamina (L. *lamina*, a thin plate)–Blade or expanded part of a leaf.

Laminar air-flow cabinet; laminar air-flow hood–Cabinet for inoculation of cultures. The working area is kept sterile by a continuous, non-turbulent flow of sterilized air through a HEPA filter. *Search in* HEPA filter; prefilter.

Laminarin–A storage polysaccharide of the brown algae.

Lampbrush chromosomes–Large diplotene chromosomes present in oocyte nuclei, and particularly conspicuous in amphibians. These chromosomes have extended regions called loops, which are active sites of transcription.

Landrace–An early, cultivated form of a crop species, evolved from a wild population.

Latent agent–Something, usually a virus, that is present in a host organism without producing any symptoms.

Latent bud–An inactive bud not held back by rest or dormant period, but which may start growth if stimulated.

Latent phase–*Search in* lag phase.

Lateral bud–A bud produced at the base of a leaf petiole. *Search in* axillary bud.

Lateral meristem–A meristem giving rise to secondary plant tissues, such as the vascular and cork cambia. The term is sometimes used to refer to an axillary meristem.

Lawn–In biotechnology: a uniform and uninterrupted layer of bacterial growth, in which individual colonies cannot be observed.

Layering–Technique for vegetative propagation, in which new plants produce adventitious roots before being severed from the parent plant.

LCR–*Search in* ligase chain reaction.

LD_5 (lethal dose$_{5\ percent}$)–The amount of a chemical required to kill 50 per cent of the test population. The higher the LD_{50}, the lower the presumed toxicity of the chemical.

Leader peptide–*Search in* signal sequence.

Leader sequence–A sequence of nucleotides at the 5′ end of an mRNA that is not translated into protein.

Leading strand–The strand of DNA that is synthesized continuously during replication.

Leaf blade–The usually flattened portion of the leaf.

Leaf bud cutting–A cutting that includes a short section of stem with attached leaf.

Leaf margin–The edge of a leaf.

Leaf primordium (L. *primordium*, a beginning)–A lateral outgrowth from the apical meristem, which will become a leaf when fully developed and expanded.

Leaf roll–Virus diseases characterized by symptomatic curling of the host's leaves.

Leaf scar–Mark left on a stem after leaf abscission.

Leaflet–Expanded leaflike part of a compound leaf.

Lectin–Any of a group of proteins, derived from plants, that can bind to specific oligosaccharides on the surface of cells, causing cells to clump together.

Leptonema (adj: leptotene)–Stage in meiosis immediately preceding synapsis, in which the chromosomes appear as single, fine, threadlike structures (but they are really double because DNA replication has already taken place).

Lethal allele; lethal gene–A mutant form of a gene that eventually results in the death of an organism if expressed in the phenotype.

Lethal gene–*Search in* lethal allele.

Lethal mutation–*Search in* lethal allele.

Library (gene library)–A collection of cells, usually bacteria or yeast, that have been transformed with recombinant vectors carrying DNA inserts from a single species. *Search in* cDNA library; expression library; genomic library; bank; gene bank.

Life cycle–The complete sequence of events undergone by organisms of a particular species, from the fusion of gametes in one generation to the same stage in the following generation.

Ligand–A molecule that can bind to another molecule in or on cells.

Ligase–*Search in* DNA ligase.

Ligase chain reaction (LCR)–A technique for determining the presence or absence of a specific nucleotide pair within a target gene.

Ligate (to)–The process of joining two or more DNA fragments.

Ligation–The joining of two linear nucleic acid molecules by the formation of phospho-diester bonds. In cloning experiments, a restriction fragment is often ligated to a linearized vector molecule using T_4 DNA ligase.

Lignification (L. *lignum*, wood + *facere*, to make)–Impregnation of a cell wall with lignin.

Lignin (L. *lignum*, wood)–A group of high-molecular-weight amorphous materials, comprising polymers of phenylpropanoid compounds, giving strength to certain tissues. Wood is composed of lignified xylem cells (about 15 to 30 per cent by weight).

Lignocellulose–The combination of lignin, hemicellulose and cellulose that forms the structural framework of plant cell walls.

Lineage–Several individuals originating from a common descent, such as the production of a cell line from a single cell plated *in vitro*.

Linear phase–When cell numbers constantly increases. The linear phase is located between the exponential growth and the deceleration phases. *Search in* growth phases.

Linearized vector–*Search in* ligation.

LINEs (long interspersed nuclear elements)–Families of long (average length = 6 500 bp), moderately repetitive (about 10,000 copies).

LINEs are cDNA copies of functional genes present in the same genome; also known as processed pseudo-genes.

Linkage–The tendency of non-allelic genes to be inherited together more than would be expected if they were assorting independently. Linkage exists between two loci when they are located sufficiently close on the same chromosome that some gametes are produced without crossing-over occurring between the two loci.

Linkage disequilibrium–*Search in* gametic (phase) disequilibrium.

Linkage equilibrium–*Search in* gametic phase equilibrium.

Linkage map–A linear or circular diagram that shows the relative positions of genes on a chromosome as determined by recombination fraction q.v.. *Search in* genetic map.

Linked genes; linked markers–Genes or markers that show linkage q.v..

Linker–A synthetic double-stranded oligonucleotide that carries the sequence for one or more restriction endonuclease sites.

Lipases–An enzymes which break down lipids into their component fatty acid and "head group" moieties. The lipases used in biotechnology are almost invariably digestive lipases, meant to break down the fats in food. They can be used to break down complex fats into their components, which are then used to make other materials.

Lipid (Gr. *lipos*, fat)–Any of a group of fats or fat-like compounds insoluble in water and soluble in fat solvents.

Lipofection–Delivery into eukaryotic cells of DNA, RNA or other compounds that have been encapsulated in an artificial phospholipid vesicle.

Lipopolysaccharide (LPS)–A compound containing lipid bound to a polysaccharide; often a component of microbial cells walls.

Liposome–A microscopic artificial membrane vesicle consisting of a spherical phospholipid bilayer. Liposomes can be incorporated into living cells and used to transport relatively toxic drugs into diseased cells, where they can exert their maximum effect. DNA molecules may be entrapped in, or bound to the surface of, the vesicles, and subsequent fusion of the liposome with the cell membrane will deliver the DNA into the cell. Liposomes have been used to develop an efficient transfection procedure for *Streptomyces* bacteria.

Liquefaction–Enzymatic digestion (often by amylase) of gelatinized starch to form lower molecular weight polysaccharides.

Liquid media–Media without a solidifying agent.

Liquid membranes–Thin films made up of liquids (as opposed to solids) which are stable in another liquid (usually water). Thus the liquid must not dissolve in the water, but nevertheless must be prevented from collapsing into a lot of small droplets.

Liquid nitrogen–Nitrogen gas condensed to a liquid with a boiling point of about–196°C. Very commonly the medium in which containers of genetic material are stored. *Search in* cryopreservation.

Litmus–A pH indicator paper (range 4.5-8.3) impregnated with an extracted lichen pigment. It turns red in acidic and blue in alkaline solutions. However, the use of litmus paper as an indicator is not a precise method of pH measurement.

Live vaccine–A living, non-virulent form of a micro-organism or virus that is used to elicit an antibody response that will protect the inoculated organism against infection by a virulent form of the micro-organism or virus. Also a living, non-virulent micro-organism or virus that express a foreign antigenic protein and is used to inoculate humans or animals. The latter organisms are also called a live recombinant vaccines.

locus (pl: loci)–A site on a chromosome.

Lod score–The logarithm of the odds of linkage between two loci. Calculated from pedigree data, as the log (to base 10) of the ratio of the probability of the observed pedigree assuming linkage with a specified recombination fraction ?, to the probability of the observed pedigree assuming no linkage, *i.e.*, recombination fraction = 0.5. A lod score (also called a z value) is thus calculated as $z = \log_{10}$ {Prob(data | ?)/Prob(data | 0.5)}. A lod score of +3.00 (which is odds of 1000:1) or greater is regarded as acceptable evidence for linkage;–2.00 (which is the log of 1:100) or less indicates that no linkage exists.

Log phase–*Search in* logarithmic phase.

Logarithmic phase (log(arithmic) or exponential growth phase)–The steepest slope of the growth curve; the phase of vigorous growth, during which cell number doubles every 20-30 minutes. *Search in* growth phases.

Long-day plant–Plant requiring short nights before flowering is initiated.

Long interspersed nuclear elements–*Search in* LINEs.

Long template–A DNA strand that is synthesized during the polymerase chain reaction and has a primer sequence at one end but is extended beyond the site that is complementary to the second primer at the other end.

Long terminal repeat (LTR)–A string of bases that occurs at each end of the "genome" of a retrovirus that has become integrated into the host genome. Involved in the integration process.

Loop bioreactors–Fermenters in which the fermenting material is cycled between a bulk tank and a smaller tank or loop of pipes. The circulation helps to mix the materials and to ensure that gas injected into the fermenter is well distributed in the liquid. The reactors are also very useful for photosynthetic fermentations, where they allow the photosynthesizing organisms to be passed along a large number of small pipes, where the light can get to them easily, rather than inside a single volume, where only the organisms near the edges get much light.

LPS–*Search in* lipopolysaccharide.

LTR–*Search in* long terminal repeat.

Lux (SI symbol: lx)–The unit of measurement for illuminance (*i.e.*, the amount of illumination) impinging upon a surface. 1 lx is the illuminance impinging upon a surface of 1 m^2, each point of which is at a distance of 1 m away from a uniform point source of light of 1 cd (candela). It supersedes the foot-candle. *Search in* foot-candle; photon.

Luxury consumption–Nutrient absorption by an organism in excess of that required for optimum growth and productivity.

Luteinizing hormone–A pituitary hormone which causes growth of the yellow body of the ovary and also stimulates activity of the interstitial cells of the testis.

Lymphocyte–A general class of white blood cells that are important components of the immune system of vertebrates.

Lymphokine–Generic name for proteins that are released by lymphocytes to act on other cells involved in the immune response. The term includes interleukins and interferons.

Lymphoma–Cancer originating in the lymph nodes, spleen and other lympho-reticular sites.

Lyase–Any of a class of enzymes that catalyse either the cleavage of a double bond and the addition of new groups to a substrate, or the formation of a double bond.

Lyophilize–Rapid freezing followed by a dehydration under high vacuum. The process is lyophilization. *The opposite word is* freeze-drying.

Lysis (Gr. *lysis*, a losing)–The destruction or breakage of cells of biological matter usually restricted to break down of living cells which is either by viruses or by chemical or physical treatment.

Lysogen–A bacterial cell whose chromosome contains integrated viral DNA.

Lysogenic–A type or phase of the virus life cycle during which the virus integrates into the host chromosome of the infected cell, remaining essentially dormant for some period of time. *Search in* lysogen.

Lysogenic bacteria–Bacteria harbouring temperate (non-virulent, lysogenic) bacteriophages.

Lysogeny–A condition in which a bacteriophage genome (pro-phage) survives within a host bacterium, either as part of the host chromosome or as part of an extrachromosomal element, and does not initiate lytic functions.

Lytic–A phase of the virus life cycle during which the virus replicates within the host cell, releasing a new generation of viruses when the infected cell lyses.

Lytic cycle–The steps in viral production that usually lead to cell lysis.

Lysosome–A membrane-bound sac within the cytoplasm of animal cells that contains enzymes responsible for the digestion of material in food vacuoles, the dissolution of foreign particles entering the cell and, on the death of the cell, the breaking down of all cell structures. The digestive system of the cell.

M13–A single-stranded DNA bacteriophage used as a vector for DNA sequencing.

M13 strand–The single-stranded DNA molecule that is present in the infective form of bacteriophage M13.

mAB–*Search in* monoclonal antibody.

Macerate–To disintegrate tissues to obtain a cell dissociation. Cutting, soaking or enzymatic actions are commonly used.

Macromolecule–Molecule of large molecular weight, such as proteins, nucleic acids and polysaccharides.

Macronutrient (Gr. *makros*, large + L. *nutrire*, to nourish)–For growth media: an essential element normally required in concentrations >0.5 millimole/l. *e.g.* carbon, hydrogen, oxygen, nitrogen, phosphorus, potassium, sulphur etc.

Macrophages–Large, white blood cells that ingest foreign substances and display on their surfaces antigens produced from the foreign substances, to be recognized by other cells of the immune system.

Macropropagation–Production of plant clones from growing parts.

Major histocompatibility antigen–A cell-surface macromolecule that allows the immune system to distinguish foreign or "non-self" from "self". A better term is histoglobulin (*Search in* histocompatibility antigen). These are the antigens that must be matched between donors and recipients during organ and tissue transplants to prevent rejection.

Major histocompatibility complex–The large cluster of genes that encode the major histocompatibility antigens in mammals.

Malignant–Having the properties of cancerous growth.

Malt extract–A mixture of organic compounds from malt, used as a culture medium adjunct. *Search in* organic complex; undefined.

Malting–A process of generating starch-degrading enzymes in grain by allowing it to germinate in a humid atmosphere. *Search in* brewing.

Mammary glands–The milk-producing organs of female mammals, which provide food for the young.

Mammary tumours–Tumours of the milk glands.

Management of farm animal genetic resources–The sum total of technical, policy and logistical operations involved in understanding (characterization), using and developing (utilization), maintaining (conservation), accessing, and sharing the benefits of animal genetic resources. (Source: FAO, 1999)

Mannitol ($C_6H_{14}O_6$; f.w. 182.17)–A sugar alcohol widely distributed in plants. Mannitol is commonly used as a nutrient and osmoticum in suspension medium for plant protoplasts.

Mannose ($C_6H_{12}O_6$; f.w. 180.16)–A hexose component of many polysaccharides and mannitol. Mannose is occasionally used as a carbohydrate source in plant tissue culture media.

Map–1. *Verb*: To determine the relative positions of loci on a DNA molecule. Linkage mapping is done by estimating the recombination fraction between loci, from the genotypes of offspring of particular matings. The further apart two loci are on a chromosome, the greater will be the frequency of recombination between them up to a maximum of 50 per cent, the situation observed when they are sufficiently far apart on a chromosome that recombinant gametes are as frequent as non-recombinant gametes, or when they are on different chromosomes. Physical mapping is usually performed by the use of *in situ* hybridisation of cloned DNA fragments to metaphase chromosomes.

2. *Noun*: A diagram showing the relative positions of, and distances between, loci.

Map distance–The standard measure of distance between loci, expressed in centiMorgans (cM). Estimated from recombination fraction via a mapping function. For small recombination fractions, map distance equals the percentage of recombination (recombination frequency) between two genes. 1 per cent recombination = 1 cM. Sometimes called a map unit.

Mapping–Determining the location of a locus (gene or genetic marker) on a chromosome. *Search in* continuous map; linkage map; physical map.

Mapping function–A mathematical expression relating observed recombination fraction to map distance expressed in centiMorgans. Two common mapping functions are those developed by Haldane (1919; *J. Genet.* **8**: 299-309) and Kosambi (1944; *Ann. Eugen.*, **12**: 172-175). In both functions, the relationship between recombination fraction and map distance is approximately linear for recombination fractions less than 10 per cent ; as recombination fraction increases above 10 per cent (up to its maximum of 50 per cent), map distance is increasingly greater than recombination fraction.

Map unit–*Search in* map distance; crossing-over unit.

Marker–An identifiable DNA sequence that facilitates the study of inheritance of a trait or a gene. Such markers are used in mapping the order of genes along chromosomes and in following the inheritance of particular genes: genes closely linked to the marker will generally be inherited with it. Markers must be readily identifiable in the phenotype, for instance by controlling an easily observable feature (such as eye colour) or by being readily detectable by molecular means, *e.g.*, microsatellite markers. *See* gene tracking.

Marker-assisted introgression–The use of DNA markers to increase the speed and efficiency of introgression of a new gene or genes into a population. The markers will be closely linked to the gene(s) in question.

Marker-assisted selection (MAS)–The use of DNA markers to increase the response to selection in a population. The markers will be closely linked to one or more quantitative trait loci.

Marker gene–A gene of known function and known location on the chromosome. *The opposite word is* genetic marker.

Marker peptide–A portion of fusion protein that facilitates its identification or purification.

MAS–*Search in* marker-assisted selection.

Mass selection–As practised in plant and animal breeding, the choosing of individuals for reproduction from the entire population on the basis of individual phenotypes.

Maternal effect–An effect attributable to some aspect of performance of the mother of the individual being evaluated.

Maternal inheritance–Inheritance controlled by extrachromosomal (cytoplasmic) factors that are transmitted through the egg.

Matric potential–A water potential component, always of negative value, resulting from capillary, imbibitional and adsorptive forces. *Search in* pressure potential.

Maturation–The formation of gametes or spores.

MCS–*Search in* polylinker.

MDA–Multiple drop array. *Search in* microdroplet array.

Mean In statistics, the arithmetic average; the sum of all measurements or values in a sample divided by the sample size.

Media–*Search in* culture medium; medium.

Median–In a set of measurements, the central value above and below which there are an equal number of measurements.

Medium (pl: media)–1. In plant tissue culture, a term for the liquid or solidified formulation upon which plant cells, tissues or organs develop. *Search in* culture medium.

2. In general terms, it could also means a substrate for plant growth, such as nutrient solution, soil, sand, etc., *e.g.*, potting medium.

Medium formulation–In tissue culture, the particular formula for the culture medium. It commonly contains macro-elements and micro-elements (high and low salt), some vitamins (B vitamins, inositol), plant growth regulators (auxin, cytokinin and sometimes gibberellin), a carbohydrate source (usually sucrose or glucose) and often other substances, such as amino acids or complex growth factors. Media may be liquid or solidified with agar; the pH is adjusted (ca. 5-6) and the solution is sterilized (usually by filtration or autoclaving). Some formulations are very specific in the kind of explant or plant species that can be maintained; some are very general.

Megabase (abbr: Mb)–A length of DNA consisting of 10^6 base pairs (if double-stranded) or 10^6 bases (if single-stranded). 1 Mb = 10^3 kb = 10^6 bp.

Megabase cloning–The cloning of very large DNA fragments. *Search in* cloning.

Megadalton (MDa)–One megadalton is equal to 10^6 daltons. *Search in* dalton.

Megaspore; macrospore–A haploid (n) spore developing into a female gametophyte in heterosporous plants.

Meiosis (Gr. *meioun*, to make smaller)–The special cell division process by which the chromosome number of a reproductive cell becomes reduced to half (n) the diploid (2n) or somatic number. Two consecutive divisions occur. In the first division, homologous chromosomes became paired and may exchange genetic material (via crossing over) before moving away from each other into separate daughter nuclei (reduction division). These new nuclei divide by mitosis to produce four haploid nuclei. Meiosis results in the formation of gametes in animals or of spores in plants. It is an important source of variability through recombination.

Meiotic analysis–A technique used to analyse chromosome-pairing relationships.

Meiotic drive–Any mechanism that causes alleles to be recovered unequally in the gametes of a heterozygote

Meiotic product(s)–*Search in* gametes.

Melanin–Pigment, as typically produced by specialised epidermal cells called melanocytes which determining skin colour, hair and eyes colour etc.

Melting temperature (abbr: T_m)–The temperature at which a double-stranded DNA or RNA molecule denatures into separate single strands. The T_m is characteristic of each DNA species and gives an indication of its base composition. DNAs rich in G:C base pairs are more resistant to thermal denaturation than A:T rich DNA since three hydrogen bonds are formed between G and C, but only two between A and T.

Membrane bioreactors–Bioreactors where cells grow on or behind a permeable membrane, which lets the nutrients for the cell through but retains the cells themselves. A variations on this theme is the hollow-fibre reactor.

Memory cells–Long-lived B and T cells that mediate rapid secondary immune responses to a previously encountered antigen.

Mendelian population–A natural, interbreeding unit of sexually reproducing plants or animals sharing a common gene pool.

Mendelism–The theory of heredity that forms the basis of classical genetics, proposed by Gregor Mendel in 1866 and formulated in two laws (*Search in* Mendel's Laws).

Mendel's Laws–Two laws summarizing Gregor Mendel's theory of inheritance. The Law of Segregation states that each hereditary characteristic is controlled by two 'factors' (now called alleles), which segregate and pass into separate germ cells. The Law of Independent Assortment states that pairs of 'factors' segregate independently of each other when germ cells are formed. *Search in* independent assortment; linkage.

Mericlinal–Refers to a chimera with tissue of one genotype partly surrounded by that of another genotype.

Mericloning–A propagation method using shoot tips in culture to proliferate multiple buds, which can then be separated, rooted and planted out.

Meristele–The vascular cylinder tissue in the stem. *Search in* stele.

Meristem (Gr. *meristos,* divisible)–Undifferentiated but determined tissue, the cells of which are capable of active cell division and differentiation into specialized and permanent tissue such as shoots and roots.

Meristem culture–A tissue culture containing meristematic dome tissue without adjacent leaf primordia or stem tissue. The term may also imply the culture of meristemoidal regions of plants, or meristematic growth in culture.

Meristem tip–An explant comprising the meristem (meristematic dome) and usually one pair of leaf primordia. Also refers to explants originating from apical meristem tip or lateral or axillary meristem tip. Do not confuse the meristem tip with the term "shoot tip," which is much larger and usually has more immature leaves and stem tissue.

Meristem tip culture–Cultures derived from meristem tip explants. The use of meristem tip culture is for virus elimination or axillary shoot proliferation purposes, but less commonly for callus production.

Meristemoid–A localized group of cells in callus tissue, characterized by an accumulation of starch, RNA and protein, and giving rise to adventitious shoots or roots.

Merozygote–Partial zygote produced by a process of partial genetic exchange, such as transformation in bacteria.

Mesh bioreactor–*Search in* filter bioreactor.

Mesoderm–The middle germ layer that forms in the early animal embryo and gives rise to parts such as bone and connective tissue.

Mesophile–A micro-organism able to grow in the temperature range 20 to 50°C; optimal growth often occurs at about 37°C.

Mesophyll (Gr. *mesos,* middle + *phyllon,* leaf)–Parenchyma tissue of leaf occurring between epidermal layers.

Messenger RNA–*Search in* mRNA.

Metabolic cell–A cell that is not dividing.

Metabolism (M.L. from the Gr. *metobolos,* to change)–In an organism or a single cell, the biochemical process by which nutritive material is built up into living matter, or aids in building living matter, or by which complex substances and food are broken down into simple substances.

Metabolite–1. A low-molecular-weight biological compound that is usually synthesized by an enzyme.

2. A compound that is essential for a metabolic process. A substance synthesized by the organism, or taken in from the environment. Autotrophic organisms take in inorganic metabolites, such as water, CO_2, nitrates and some trace elements.

Metacentric chromosome–A chromosome with the centromere near the middle and, consequently, two arms of about equal length.

Metallothionein–A protective protein that binds heavy metals such as cadmium and lead.

Metaphase (Gr. *meta*, after + *phasis*, appearance)–Stage of mitosis during which the chromosomes, or at least the kinetochores, lie in the central plane of the spindle. It is the stage following prophase and preceding anaphase.

Metastasis–The spread of cancer cells to previously unaffected organs.

Methionine–A sulphur-containing amino acid.

Methylation–The addition of a methyl group ($-CH_3$) to a macromolecule, such as the addition of a methyl group to specific cytosine and, occasionally, adenine residues in DNA.

Michaelis constant–*Search in* K_m.

Microalgal culture–Culture in bioreactors of microalgae; microalgae include seaweeds.

Micro-array, DNA–*Search in* DNA microarray.

Microbe–A general term for a microorganism.

Microbial mats–Layered groups or communities of microbial populations.

Microbody (Gr. *mikros*, small + body)–A cellular organelle always bound by a single membrane, frequently spherical, from 20 to 60 nm in diameter, containing a variety of enzymes.

Micro-carriers–Small particles used as a support material for cells, and particularly mammalian cells, which are too fragile to be pumped and stirred as bacterial cells are in a large-scale culture.

Microdroplet array; multiple drop array (MDA); hanging droplet technique–Introduced by Kao and Konstabel (1970), this technique is used to evaluate large numbers of media modifications, employing small quantities of medium into which are placed small numbers of cells. Droplets of liquid are arranged on the lid of a Petri dish, inverted over the bottom half of the dish containing a solution with a lower osmotic pressure, and the dish is sealed. The cells or protoplasts form a monolayer at the droplet meniscus and can easily be examined.

Micro-element–An element required in very small quantities, calcium, copper, zinc, etc.

Micro-encapsulation–A process of enclosing a substance in very small sealed capsules from which material is released by heat, solution or other means.

Micro-environment (Gr. *mikros*, small + O.F. *environ*, about)–The environment close enough to the surface of a living or non-living object to be influenced by it.

Microfibrils (Gr. *mikros*, small + *fibrils*, diminutive of fibre)–Microfibrils are exceedingly small fibres visible only at the high magnification of the electron microscope.

Microgametophytes–*Search in* anther.

Micrograft–*Search in* shoot-tip graft.

Micro-injection–The introduction of small amounts of material (DNA, RNA, enzymes, cytotoxic agents) into a single eukaryotic cell with a fine, microscopic needle, penetrating the cell membrane.

Micro-isolating system–Mechanical separation of single cells or protoplasts thus allowing them to proliferate individually.

Micron; micrometre (Gr. *mikros*, small)–A unit of distance: 10^{-6} m; 0.001 mm. Symbol: µm.

2-micron plasmid–*Search in* 2µm plasmid.

Micronutrient (Gr. *mikros*, small + L. *nutrire*, to nourish)–For growth media: An essential element normally required in concentrations < 0.5 millimole/litre.

Micro-organism–Organism visible only under magnification.

Microplasts–Vesicles produced by subdivision and fragmentation of protoplasts or thin-walled cells.

Micropyle–1. A small opening in the surface of a plant ovule through which the pollen tube passes prior to fertilization.

2. A small pore in some animal cells or tissues.

Microprojectile bombardment–A procedure for modifying cells by shooting DNA-coated metal (tungsten or gold) particles into them. *Search in* biolistics.

Micropropagation–Miniaturized *in vitro* multiplication and/or regeneration of plant material under aseptic and controlled environmental conditions on specially prepared media that contain substances necessary for growth; used for three general types of tissue: excised embryos (= embryo culture); shoot-tips (= meristem culture or mericloning); and pieces of tissue that range from bits of stems to roots. Four stages of plant tissue culture have been defined by Murashige:

Stage I. Establishment of an aseptic culture.

Stage II. The multiplication of propagules.

Stage III. Preparation of propagules for successful transfer to soil (rooting and hardening).

Stage IV. Establishment in soil.

Microsatellite–A form of VNTR. Specifically, a segment of DNA characterized by the occurrence of a variable number of copies (from a few up to 30 or so) of a sequence of around 5 or fewer bases (called a repeat unit, q.v.). A typical microsatellite is the repeat unit AC, which occurs at approximately 100 000 different sites in a typical mammalian genome. At any one site (locus), there are usually several different "alleles," each identifiable according to the number of repeat units. These alleles can be detected by PCR, using primers designed from the unique sequence that is located on either side of the microsatellite. When the PCR product is run on an electrophoretic gel, alleles are seen to differ in length in units equal to the size of the repeat unit, *e.g.*, if the primers correspond to the unique sequence immediately on either side of the microsatellite and are each 20 bases long, and an individual is heterozygous for an AC microsatellite with one allele comprising 5 repeats and the other comprising 6 repeats, the heterozygote will exhibit two bands on the gel, one band being 20 + (2 × 5) + 20 = 50 bases long, and the other allele being 20 + (2 × 6) + 20 = 52 bases long. Microsatellites have been the standard DNA marker: they are easily detectable by PCR, and they tend to be evenly located throughout the genome. Thousands have been mapped in many different species.

Microspore–The smaller of the two kinds of meiospores produced by heterosporous plants in the course of microsporogenesis; in seed plants, microspores give rise to the pollen grain, the male gametophyte.

Microtuber–Cultured tissue capable of growing into tuberous plant.

Microtubules–A minute filament in living cells that is composed of the protein tubulin and occurs singly, in pairs, triplets or bundles. Microtubules help cells to maintain their shape; they also occur in cilia, flagella and the centrioles, and form the spindle during nuclear division.

Middle lamella (L. *lamella*, a thin plate or scale)–Original thin membrane separating two adjacent protoplasts and remaining as a distinct cementing layer between adjacent cell walls.

Mid–parent value–In quantitative genetics, the average of the phenotypes of two mates.

Minimum effective cell density–The inoculum density below which the culture fails to give reproducible cell growth. The minimum density is a function of the tissue (species, explant, cell line) and the culture phase of the inoculum suspension. Minimum density decreases inversely to the aggregate size and division rate of the stock culture.

Minimum inoculum size–The critical volume of inoculum required to initiate culture growth, due to the diffusive loss of cell materials into the medium. The subsequent culture growth cycle is dependent on the inoculum size, which is determined by the volume of medium and size of the culture vessel.

Mini–prep–A small-scale (mini-) preparation of plasmid or phage DNA. Used to analyse DNA in a cloning vector after a cloning experiment.

Minisatellite–A form of VNTR in which the repeat units typically range from 10 to 100 bases. They are usually detected by Southern hybridization, using a probe comprising a clone of the repeat unit. The first DNA fingerprints were minisatellites detected in this way. Minisatellites tend to be located at the ends of chromosomes and in regions with a high frequency of recombination.

Minitubers–Small tubers (5-15 mm in diameter) formed on shoot cultures or cuttings of tuber-forming crops, such as potato.

Mismatch–The lack of a complementary pair of bases in a double helix of DNA, *e.g.*, A:C, G:T.

Mismatch repair–DNA repair processes that correct mismatched base pairs.

Missense mutation–A mutation that changes a codon for one amino acid into a codon specifying another amino acid.

Mist propagation–Application of fine droplets of water to leafy cuttings in the rooting stage to reduce transpiration. *The opposite word is* fog.

Mites–Free-living and parasitic animals belonging to the order Acarina, class Arachnida (with spiders). Mites may infest plant crops, reducing their harvest. They may also infest plant tissue culture work areas and incubation facilities in search of sugars, and so contaminate culture vessels and spread bacteria and fungi.

Mitochondrial DNA (mtDNA)–A circular ring of DNA found in mitochondria. In mammals, mtDNA makes up less than 1 per cent of the total cellular DNA, but in plants the amount is variable. It codes for ribosomal RNA and transfer RNA, but only some mitochondrial proteins (up to 30 proteins in animals), the nuclear DNA being required for encoding most of these.

Mitochondrion (Gr. *mitos*, thread + *chondrion*, a grain; pl: mitochondria)–A small cytoplasmic organelle that carries out aerobic respiration. Oxidative phosphorylation takes place to produce ATP.

Mitosis (Gr. *mitos*, a thread; adj: mitotic; pl: mitoses)–Disjunction of replicated chromosomes and division of the cytoplasm to produce two genetically identical daughter cells. The division involves the appearance of chromosomes, their longitudinal duplication, and equal distribution of newly formed parts to daughter nuclei. It is separated into five stages: interphase, prophase, metaphase, anaphase and telophase.

Mixed bud–A bud containing both rudimentary leaves and flowers.

Mixoploid–Cells with variable (euploid, aneuploid) chromosome numbers. Mosaics or chimeras differ in chromosome number as a result of a variety of mitotic irregularities.

Mobilization–1. The transfer between bacteria of a non-conjugative plasmid by a conjugative plasmid.

2. The transfer between bacteria of chromosomal genes by a conjugative plasmid.

Mobilizing functions–The genes on a plasmid that give it the ability to facilitate the transfer of either a non-conjugative or a conjugative plasmid from one bacterium to another.

Modal class; mode–In a frequency distribution, the class having the greatest frequency.

Model–A mathematical description of a biological phenomenon.

Modification–1. Enzymatic methylation of a restriction enzyme DNA recognition site.

2. Specific nucleotide changes in DNA or RNA molecules.

Modifier; modifying gene–A gene that affects the expression of some other gene.

MOET–*Search in* multiple ovulation and embryo transfer.

Molality–The number of moles of solute per litre of solvent.

Molarity–The number of moles of a substance contained in a kilogram of solution. *Search in* mole.

Mole (symbol: M)–Amount of substance that has a weight in grams numerically equal to the molecular weight of the substance. Also called gram molecular weight. A mole contains 6.023×10^{23} molecules or atoms of a substance.

Molecular biology–The area of knowledge concerned with the molecular aspects of organisms and their cells.

Molecular cloning–The biological amplification of a specific DNA sequence through mitotic division of a host cell into which it has been transformed or transfected. *Search in* cloning.

Molecular genetics–The area of knowledge concerned with the genetic aspects of molecular biology, especially with DNA, RNA and protein molecules.

Molecule (L. diminutive of *moles* a little mass)–A unit of matter, the smallest portion of an element or a compound that retains chemical identity with the substance in mass. The molecule usually consist of a union of two or more atoms; some organic molecules containing a very large number of atoms.

Monoclonal antibody (mAB)–A single type of antibody that is directed against a specific epitope (antigen, antigenic determinant) and is produced by a single clone of B cells or a single hybridoma cell line, which is formed by the fusion of a lymphocyte cell with a myeloma cell. Some myeloma cells synthesize single antibodies naturally.

Monocot–*Search in* monocotyledon.

Monocotyledon (Gr. *monos*, solitary + *kotyledon*, a cup-shaped hollow)–A plant whose embryo has one seed leaf (cotyledon). Examples are cereal grains (corn, wheat and rice), asparagus, and lily. Colloquially called a monocot. *The opposite word is* dicotyledon.

Monoculture–The agricultural practice of cultivating a single crop on a whole farm or area.

Monoecious–Denoting plant species that have separate male and female flowers on the same plant (*e.g.* maize).

Monogastric animals–Animals with simple stomachs that do not ruminate. *The opposite word is* ruminant animals.

Monophyletic–Describing any group of organisms that are assumed to have originated from the same ancestor.

Monogenic–Controlled by a single gene, as opposed to multigenic.

Monohybrid(Gr. *monos*, solitary + L. *hybrida*, a mongrel)–The offspring of two homozygous parents that differ from one another by the alleles present at only one locus.

Monohybrid cross–A cross between parents differing in only one trait or in which only one trait is being considered.

Monolayer–A single layer of cells growing on a surface.

Monomer–A single molecular entity that may combine with others to form more complex structures.

Monoploid–*Search in* haploid.

Monosaccharide–A single sugar. *The opposite word is* polysaccharide.

Monosomic (n: monosomy)–describing a diploid organism lacking one chromosome (2n–1) of its proper (disomic) complement; a form of aneuploidy. *Search in also* disomy.

Mono-unsaturates–Oils containing mono-unsaturated fatty acids.

Monozygotic twins–One-egg or identical twins; twins derived from the splitting of a single fertilized ovum.

Morphogen–A substance that stimulates the development of form or structure in an organism.

Morphogenesis–The development, through growth and differentiation, of form and structure in an organism.

Morphogenic response–The effect on the developmental history of a plant or its parts exposed to a given set of growth conditions or to a change in the environment.

Morphology (Gr. *morphe*, form + *logos*, discourse)–1.The science of studying form and its development.

2. General: Shape, form, external structure or arrangement.

Mosaic–An organism or part of an organism that is composed of cells with different origin.

Mother plant–*Search in* donor plant.

Movable genetic element–*Search in* transposon.

mRNA; messenger RNA–The RNA transcript of a protein-encoding gene. The information encoded in the mRNA molecule is translated into a polypeptide of specific amino acid sequence by the ribosomes. In eukaryotes, mRNAs transfer genetic information from the DNA to ribosomes, where it is translated into protein.

MRUs–Minimum recognition units. *Search in* dabs.

mtDNA–*Search in* mitochondrial DNA.

Multi–copy–Describing plasmids which replicate to produce many plasmid molecules per host genome, *e.g.*, pBR322 is a multi-copy plasmid, there are usually 50 pBR322 molecules (or copies) per *E. coli* genome.

Multigene family–A group of genes that are similar in nucleotide sequence or that produce polypeptides with similar amino acid sequences.

Multigenic–Controlled by several genes, as opposed to monogenic.

Multi-locus probe–A probe that hybridizes to a number of different sites in the genome of an organism. *Search in* probe.

Multimer; multimeric–A protein made up of more than one peptide chain.

Multiple alleles–The existence of more than two alleles at a locus in a population.

Multiple cloning site–*Search in* polylinker.

Multiple drop array (MDA)–*Search in* microdroplet array.

Multiple ovulation and embryo transfer (MOET)–A technology by which a single female that usually produces only one or two offspring can produce a litter of offspring. Involves stimulation of a female to shed large numbers of ova; natural mating or artificial insemination; collection of fertilized ova (either surgically, or non-surgically through the cervix); and transfer (usually non-surgical, through the cervix) of these fertilized ova to recipient females.

Multivalent vaccine–A single vaccine that is designed to elicit an immune response either to more than one infectious agent or to several different epitopes of a molecule.

Mutable genes–Genes with an unusually high mutation rate.

Mutagen–An agent or process which is capable of inducing a mutation, such as UV light. *The opposite word is* mutation.

Mutagenesis–Change(s) in the genetic constitution of a cell through alterations to its DNA.

Mutant–An organism or an allele that differs from the wild type because it carries one or more genetic changes in its DNA. A mutant organism may carry mutated gene(s) (= gene mutation); mutated chromosome(s) (= chromosome mutation); or mutated genome(s) (= genome mutation). a.k.a. a variant.

Mutation (L. *mutare,* to change)–A sudden, heritable change appearing in an individual as the result of a change in the structure of a gene (= gene mutation); changes in the structure of chromosomes (= chromosome mutation); or in the number of chromosomes (= genome mutation). cf genetic diversity; genetic drift.

Mutation pressure–A constant mutation rate that adds mutant genes to a population; repeated occurrences of mutations in a population.

Mutualism–*Search in* symbiosis.

Mycelium (pl: mycelia)–Threadlike filament making up the vegetative portion of thallus fungi.

Mycoprotein–Fungal protein.

Mycotoxin–Toxic substance of fungal origin, such as aflatoxin.

Mycorrhiza (Gr. *mykos,* fungus + *riza,* root)-Fungi that form an association with or have a symbiotic relationship with roots of more developed plants.

Myeloma–A plasma cell cancer.

Myo inositol–*Search in* inositol.

Myosin–*Search in* actin.

N_2-Free nitrogen gas. In liquid form, used as a cryopreservant. *Search in* cryobiological preservation.

Naked bud–A bud not protected by bud scales.

Nanometre 1×10^{-9} m–One millionth of a millimetre, a.k.a. a millimicron; equals ten angstroms.

Narrow-host-range plasmid–A plasmid that can replicate in one, or at most a few, different bacterial species.

Narrow-sense heritability–In quantitative genetics, the proportion of the phenotypic variance that is due to variation in breeding values.

Native protein–The naturally occurring form of a protein.

Natural selection–The differential survival and reproduction of organisms because of differences in characteristics that affect their ability to utilize environmental resources.

NDP–Ribonucleoside diphosphate. *Search in* nucleotide; ribonucleoside triphosphate (NTP).

Necrosis–Death associated with discoloration and dehydration of all or some parts of organs.

Negative autogenous regulation; negative self-regulation–Inhibition of the expression of a gene or set of coordinately regulated genes by the product of the gene or the product of one of the genes.

Negative control system–A mechanism in which the regulatory protein(s) is required to turn off gene expression.

Negative selection–Selection of individuals that do not possess a certain character. A method by which growing cells that do not carry a DNA insert integrated at a specific chromosomal location are selected. *Search in* positive selection.

Negative self-regulation–*Search in* negative autogenous regulation.

Nematodes–A class of slender, unsegmented worms, often parasitic.

Neo-formation–Organogenesis; production of newly formed structures, such as tissues, meristems and embryos.

Neoplasm–Localized cell multiplication. Generally it designates a collection of cells which have undergone genetic transformation, forming a tumour. Neoplasmic cells differ in structure and function from the original cell type.

Neoteny–The retention of juvenile body characters in the adult state, or the occurrence of adult characters in the juvenile state.

Net photosynthesis–Photosynthetic activity minus respiratory activity, as measured by carbon dioxide exchange.

Neutral mutation–A mutation that changes the nucleotide sequence of a gene but has negligible effect on the fitness of the organism.

Neutral theory–The theory that much of evolution has been primarily due to random drift of neutral mutations.

NFT–*Search in* nutrient film technique.

Nick–Verb: To break a phospho-diester bond in the backbone of one of the strands of a duplex DNA molecule. *The opposite word is* cleave; cut.

Nicked circle; relaxed circle–During extraction of plasmid DNA from the bacterial cell, one strand of the DNA becomes nicked. This relaxes the torsional strain needed to maintain supercoiling, producing the familiar form of plasmid. *Search in* plasmid.

Nick translation–A procedure for labelling DNA. A DNA fragment is treated with DNase to produce single-stranded nicks. The nick is moved along the DNA molecule in the presence of labelled deoxyribonucleoside triphosphates by the concerted action of the 5′-> 3′ exonuclease and 5′-> 3′ polymerase activities of *E. coli* DNA polymerase I.

Nitrification–A chemical process in which nitrogen in plant and animal wastes and dead remains is oxidized, first to nitrites and then to nitrates.

Nitrate–The only form in which nitrogen can be used directly by plants; a component of chemical fertilizers.

Nitrite–A salt or ester of nitrous acid.

Nitrocellulose; cellulose nitrate–A nitrated derivative of cellulose. It is made into membrane filters of defined porosity, used to immobilize DNA, RNA or protein, which can then be probed with a labelled sequence or antibody. These filters have a variety of uses in molecular biology, particularly in nucleic acid hybridization experiments. Used extensively in the Southern and northern blotting procedures involving DNA and RNA.

Nitrogen assimilation–The incorporation of nitrogen into organic cell substances by living organisms.

Nitrogen fixation–The conversion of atmospheric nitrogen (N_2) into oxidized forms that can be assimilated by plants. Biological nitrogen fixation is catalysed by the enzyme nitrogenase, which is found only in prokaryotes. Certain blue-green algae and some genera of bacteria (*e.g.*, *Rhizobium* spp.; *Azotobacter* spp.) are capable of biochemically fixing nitrogen. Such bacteria are very important symbionts for plants growing in nitrogen-poor soils.

Nitrogenous bases–The purines (adenine and guanine) and pyrimidines bases (thymine, cytosine and uracil) that form DNA and RNA molecules.

NMP–Ribonucleoside monophosphate. *Search in* nucleotide.

NO–*Search in* nucleolar organizer.

Nod box–A DNA sequence that controls the transcriptional regulation of *Rhizobium* nodulation genes.

Nodal culture–The culture of a lateral bud and a section of adjacent stem tissue.

Node (L. *nodus*, a knot)–Slightly enlarged portion of the stem where leaves and buds arise and where branches originate. Stems have nodes but roots do not.

Nodular–Term commonly used to describe a pebbly (rough) texture of a callus.

Nodulation–The formation of nodules by symbiotic bacteria on the roots of plants.

Nodule–The enlargement or swelling on roots of nitrogen-fixing plants. The nodules contain symbiotic nitrogen-fixing bacteria. *Search in* nitrogen fixation.

Non-autonomous–A term referring to biological units that cannot function by themselves; such units require the assistance of another unit, or "helper". *Search in* autonomous.

Non-disjunction–Failure of disjunction or separation of homologous chromosomes or chromatids in mitosis or meiosis, resulting in too many chromosomes in some daughter cells and too few in others. Examples: In meiosis, both members of a pair of chromosomes go to one pole so that the other pole does not receive either of them; in mitosis, both sister chromatids go to the same pole.

Non-histone chromosomal proteins—In chromosomes, all of the proteins except the histones.

Nonsense mutation–A mutation which converts an amino-acid-specifying codon into a stop codon, *e.g.*, a change from UAU (tyr) to UAG (amber) would lead to the premature termination of a polypeptide chain at the place where a tyrosine was inserted in the wild-type. *Search in* stop codon; suppressor.

Non-target organism–An organism which is affected by an interaction for which it was not the intended recipient.

Non-template strand–In transcription, the non-transcribed strand of DNA. a.k.a. sense strand or coding strand. It will have the same sequence as the RNA transcript, except that 'T' is present at positions where U is present in the RNA transcript.

Non-virulent agent–*Search in* attenuated vaccine.

NOR–*Search in* nucleolar organizer region.

Northern blot–A cellulose or nylon membrane to which RNA molecules have been attached by capillary action. The transferred RNA is hybridized to single-stranded DNA probes. Northern blot technique is often used to measure expression (transcription) of a gene for which a specific cDNA is available for use as a probe. *Search in* Southern blot, western blot.

Northern blotting–Similar to Southern blotting, except that RNA is transferred onto a matrix and the presence of a specific RNA molecule is detected by DNA-RNA hybridization. *Search in* northern hybridization.

Northern hybridization–Hybridization of a labelled DNA probe to RNA fragments that have been transferred from an agarose gel to a nitrocellulose filter. *Search in* northern blotting; hybridization (3).

NTP–Ribonucleoside triphosphate *Search in* nucleotide.

Nucellar embryo–An embryo which has developed vegetatively from somatic tissue surrounding the embryo sac, rather than by fertilization of the egg cell.

Nucellus (L. *nucella*, a small nut)–Tissue composing the chief part of the young ovule in which the embryo sac develops: megasporangium.

Nuclear transfer–A technology by which animals are created by cloning a single diploid somatic cell. It involves taking a single diploid cell from a culture of cells, and inserting it into an enucleated ovum, *i.e.*, an ovum from which the haploid nucleus has been removed. The resultant diploid ovum develops into an embryo that is placed in a recipient female, which gives birth to the cloned animal in the normal manner. Note that the term is somewhat of a misnomer, since it is a whole cell that is transferred, not just the nucleus.

Nuclease–A class of enzymes that degrade DNA or RNA molecules by cleaving the phospho-diester bonds that link adjacent nucleotides. In deoxyribonuclease (DNase), the substrate is DNA. In endonuclease, it cleaves at internal sites in the substrate molecule. Exonuclease progressively cleaves from the end of the substrate molecule. In ribonuclease (RNase), the substrate is RNA. In the S1 nuclease, the substrate is single-stranded DNA or RNA. Nucleases have varying degrees of base-sequence specificity, the most specific being the restriction endonucleases.

Nucleic acid–A macromolecule composed of phosphoric acid, pentose sugar, and organic bases. The two nucleic acids, deoxyribonucleic acid (DNA) and ribonucleic acid (RNA), are made up of long chains of molecules called nucleotides. They were first isolated as part of a protein complex in 1871, and were separated from the protein moiety in 1889. *Search in* DNA; RNA; nucleotides.

Nucleic acid probe–*Search in* DNA probe.

Nuclein–The term used by Friedrich Miescher to describe the nuclear material he discovered in 1869, which today is known as DNA.

Nucleo-cytoplasmic ratio–In a cell, the ratio of nuclear to cytoplasmic volume. This ratio is high in meristematic cells and low in differentiated cells.

Nucleolar organizer (NO); nucleolar organizer region (NOR)–A chromosomal segment containing genes that encode ribosomal RNA; located at the secondary constriction of some chromosomes.

Nucleolar zone–Any chromosome region, irrespective of whether or not it is a secondary-constriction, that is associated with the formation of the nucleolus during telophase.

Nucleolus (L. *nucleolus*, a small nucleus)–An RNA-rich intranuclear organelle in the nucleus of eukaryotic cells, produced by a nucleolar organizer. It represents the storage place for ribosomes and ribosome precursors. The nucleolus consists primarily of ribosomal precursor RNA, ribosomal RNA, their associated proteins, and some, perhaps all, of the enzymatic equipment (RNA polymerase, RNA methylase, RNA cleavage enzymes) required for synthesis, conversion and assembly of ribosomes. Subsequently the ribosomes are transported to the cytoplasm.

Nucleoplasm–The non-staining or slightly chromophilic, liquid or semi-liquid, ground substance of the interphase nucleus and which fills the nuclear space around the chromosomes and the nucleoli.

Little is known of the chemical composition of this ground substance, which is not easily defined. It may be called "karyoplasm" when it is gel-like, and "karyolymph" when it is a colloidal fluid, but generally the terms are synonymous.

Nucleoprotein–Conjugated protein composed of nucleic acid and protein; the material of which the chromosomes are made.

Nucleoside–A base (purine or pyrimidine) that is covalently linked to a 5-carbon (pentose) sugar. When the sugar is ribose, the nucleoside is a ribonucleoside; when it is deoxyribose, the nucleoside is a deoxyribonucleoside. Adenine, guanine and cytosine occur in both DNA and RNA; thymine occurs in DNA; and uracil in RNA. They are the building blocks of DNA and RNA. *Search in* nucleoside analogue.

Nucleoside analogue–A synthetic molecule that resembles a naturally occurring nucleoside, but that lacks the bond site needed to link it to an adjacent nucleotide. *Search in* nucleoside.

Nucleosome–Spherical sub-units of eukaryotic chromatin that are composed of a core particle consisting of an octamer of histones (two molecules each of histones H_{2a}, H_{2b}, H_3 and H_4) and 146 nucleotide pairs.

Nucleotide–A nucleoside with one or more phosphate groups linked to the 5′ carbon of the pentose sugar. Ribose-containing nucleosides include ribonucleoside monophosphate (NMP), ribonucleoside diphosphate (NDP), and ribonucleoside triphosphate (NTP). When the nucleoside contains the sugar deoxyribose, the nucleotides are called deoxyribonucleoside mono-, di-, or tri-phosphates (dNMP, dNDP, or dNTP). A building block of DNA and RNA. *Search in* chromosome; codon; complementary nucleotides; dideoxynucleotide; DNA; gene; oligonucleotide; RNA.

Nucleus (L. *nucleus*, kernel of a nut)–A dense protoplasmic-membrane-bound region of a eukaryotic cell that contains the chromosomes separated from the cytoplasm by a membrane; present in all eukaryotic cells except mature sieve-tube elements.

Null mutation–*Search in* amorph.

Nullisomy (adj: nullisomic)–An otherwise diploid cell or organism lacking both members of a chromosome pair (chromosome formula $2n-2$). *Search in also* disomy.

Nurse culture–Planting a cell from a suspension culture on a raft of filter paper above a callus tissue piece (nurse tissue). The filter paper serves to prevent tissue union but allows the flow of essential substances from the nurse to the isolated cell. In such culture, a piece of callus is first placed on nutrient agar. Over this tissue is laid a strip of filter paper.

Nutrient cycle–The passage of a nutrient or element through an ecosystem, including its assimilation and release by various organisms and its transformation into various organic or inorganic chemical forms.

Nutrient deficiency–Absence or insufficiency of some factor needed for normal growth and development.

Nutrient film technique (NFT)–Hydroponic technique used to grow plants. NFT delivers a film of water or nutrient solution either continuously or through on-off cycles (*e.g.*, on 8 minutes and off 7 minutes).

Nutrient gradient–A diffusion gradient of nutrients and gases that develops in tissues where only a portion of the tissue is in contact with the medium. Gradients are less likely to form in liquid media than in callus cultures.

Nutrient medium (pl: nutrient media)–A solid, semi-solid or liquid combination of: major and minor salts; an energy source (sucrose); vitamins; plant growth regulators; and occasionally other defined or undefined supplements. Often made from stock solutions, then sterilized by autoclaving or filtering through a micropore filter.

Octoploid–Cell or organism with eight sets of chromosomes, *i.e.*, chromosome number 2n = 8*x*.

Oestrogen; estrogen–The generic term for a group of female sex hormones which control the development of sexual characteristics and control oestrus.

Oestrous cycle (from oestrus)–The cycle of reproductive activity shown by most sexually mature non-pregnant female mammals.

Oestrus (adj: oestrous)–In female mammals, the period of sexual excitement and acceptance of the male.

Offset–Young plant produced at the base of a mature plant.

Offshoot–Short, usually horizontal, stem produced near the crown of a plant.

Offspring; progeny (both same in plural)–New individual organisms that result from the process of sexual or asexual reproduction.

Okazaki fragment–Since DNA can be replicated in only one direction, *i.e.*, nucleotides can be added only at the 3′ end, only one of the two strands of a double helix can be replicated continuously. The other strand is replicated in small segments (Okazaki fragments) that are subsequently joined together by DNA ligase. *Search in* primosome.

OLA–*Search in* oligonucleotide ligation assay.

Oligomer–A molecule formed from a small number of monomers.

Oligonucleotide–A short molecule (usually 6 to 100 nucleotides) of single-stranded DNA.

Oligonucleotide ligation assay (OLA)–A diagnostic technique for determining the presence or absence of a specific nucleotide pair within a target gene, often indicating whether the gene is wild type (normal) or mutant (defective).

Oligonucleotide-directed mutagenesis; oligonucleotide-directed site-specific mutagenesis–*Search in* site-specific mutagenesis.

Oligosaccharide–Carbohydrate consisting of several linked sugar units.

Oncogene–A gene that causes cells to grow in an uncontrolled manner, *i.e.*, that causes cancer. Oncogenes are mutant forms of normal functional genes (called proto-oncogenes) that have a role in normal cell proliferation. Oncogenes are found in tumours and in retroviruses; in the latter case, having been picked up along with retroviral genes during retroviral replication in the host. Once incorporated as part of the retroviral genome, they are activated at inappropriate times and places by retroviral promoters, thereby becoming oncogenes.

Search in cellular oncogene; dominant oncogene; immortalizing oncogene; recessive oncogene.

Oncogenesis–The progression of cytological, genetic and cellular changes that culminate in a tumour.

Oncogenic–Oncogenic genes are responsible for the transformation of normal cells into tumour cells. *Search in Agrobacterium.*

Onco-mouse–A mouse that has been genetically modified to incorporate an oncogene; oncogenes cause cells to undergo cancerous transformation.

Ontogeny–Developmental life history of an organism.

Oocyte–The egg mother cell; it undergoes two meiotic divisions (oogenesis) to form the egg cell. The primary oocyte is before completion of the first meiotic division; the secondary oocyte is after completion of the first meiotic division.

Oogenesis–The formation and growth of the egg or ovum in an animal ovary.

Oogonium (pl: oogonia)–1. A germ cell of the female animal, that gives rise to oocytes by mitotic division.

2. The female sex organ of algae and fungi.

Oospore (Gr. *oion*, an egg + spore)–1. A resistant spore developing from a zygote, resulting from the fusion of heterogametes in certain algae and fungi.

2. The non-motile female gamete in plants and some algae which is rich in nutrient.

Open continuous culture–A continuous culture in which inflow of fresh medium is balanced by outflow of a corresponding volume of culture. Cells are constantly washed out with the outflowing liquid. In a steady state, the rate of cell wash-out equals the rate of formation of new cells in the system. *Search in* continuous culture; batch culture; closed continuous culture.

Open pollination–Pollination by wind, insects or other natural mechanisms.

Open reading frame (ORF)–A sequence of nucleotides in a DNA molecule that has the potential to encode a peptide or protein: it starts with a start triplet (ATG), is followed by a string of triplets each of which encodes an amino acid, and ends with a stop triplet (TAA, TAG or TGA). This term is often used when, after the sequence of a DNA fragment has been determined, the function of the encoded protein is not known. The existence of open reading frames is usually inferred from the DNA (rather than the RNA) sequence.

Operational definition–An operation or procedure that can be carried out to define or delimit something.

Operator–The region of DNA that is upstream from a gene or genes and to which one or more regulatory proteins (repressor or activator) binds to control the expression of the gene(s).

Operon–A functionally integrated genetic unit for the control of gene expression in bacteria. It consists of one or more genes that encode one or more polypeptide(s) and the adjacent site (promoter and operator) that controls their expression by regulating the transcription of the structural genes.

Opine–The condensation product of an amino acid with either a keto-acid or a sugar, produced by the crown gall tissues. Opine synthesis is a unique characteristic of tumour cells.

OPU–*Search in* ovum pickup.

Ordinate–The vertical axis of a graph. Opposite: absciss; abscissa.

ORF–*Search in* open reading frame.

Organ–A tissue or group of tissues that constitute a morphologically and functionally distinct part of an organism.

Organ culture–The aseptic culture of complete living organs of animals and plants outside the body in a suitable culture medium. Animal organs must be small enough to allow the nutrients in the culture medium to penetrate all the cells.

Organellar genes–Genes located on organelles outside the nucleus.

Organelle–A membrane-bounded specialized region within a cell, such as the mitochondrion or dictyosome, that carries out a specialized function in the life of a cell.

Organic–Referring in chemistry to compounds containing carbon, many of which have been in some manner associated with living organisms.

Organic complex–A chemically undefined compound added to nutrient media to stimulate growth, *e.g.*, coconut milk; yeast extract; casein hydrolysate.

Organic co-solvent–A compound used to dissolve some neutral organic substances, such as in media preparation. Organic co-solvents include alcohols (usually ethanol), acetone and dimethylsulphoxide (DMSO).

Organic evolution–The process by which changes in the genetic composition of populations of organisms occur in response to environmental changes.

Organized growth–The development under tissue culture conditions of organized explants (meristem tips or shoot tips, floral buds or organ primordia). *Search in* unorganized growth.

Organized tissue–Composed of regularly differentiated cells.

Organizer–An inductor; a chemical substance in a living system that determines the fate in development of certain cells or groups of cells.

Organogenesis–The initiation of adventitious or *de novo* shoots or roots from callus, meristem or suspension cultures. *Search in* micropropagation; regeneration.

Organoid–An organ-like structure produced in culture, such as leaves, roots or callus.

Organoleptic–Having an effect on one of the organs of sense, such as taste or smell.

Organism–An individual living system, such as animal, plant or micro-organism, that is capable of reproduction, growth and maintenance.

Origin of replication–The nucleotide sequence at which DNA synthesis (replication) is initiated.

Ortet–The plant from which a clone is obtained.

Osmic acid (= osmium tetroxide) (OsO_4; f.w. 254.20)–A fixing agent commonly used to prepare tissue samples for electron microscopy.

Osmolarity–The total molar concentration of the solutes. Osmolarity affects the osmotic potential of solution or nutrient medium.

Osmosis (Gr. *osmos*, a pushing)–Diffusion from areas of high concentration to areas of lower concentration of a solvent through a differentially permeable membrane.

Osmotic potential [value]–Potential brought about by dissolving a substance, especially in water.

Osmoticum–An agent, such as PEG, mannitol, glucose or sucrose, employed to maintain the osmotic potential of a nutrient medium equivalent to that of the cultured cells (isotonic). Because of this osmotic equilibrium, cells are not damaged *in vitro*.

Outbreeding–A mating system characterized by the breeding of genetically unrelated or dissimilar individuals. Since genetic diversity tends to be enhanced and since vigour or fitness of individuals can be increased by this process, it is often used to counter the detrimental effects of continuous inbreeding.

Outflow–The volume of growing cells that is removed from a bioreactor during a continuous fermentation process.

Oxygen-electrode-based sensor–Sensor in which an oxygen electrode–a standard electrochemical cell which measures the amount of oxygen in a solution–is coated with a biological material which generates or absorbs oxygen. When the biological coating is active, the amount of oxygen next to the electrode changes and the signal from the electrode changes.

Ovary–1. Enlarged basal portion of the pistil which becomes the fruit.

2. The reproduction organ in female animals in which eggs (ova) are produced.

Overdominance–A condition in which heterozygotes are superior (on some scale of measurement) to either of the associated homozygotes.

Overhang–*Search in* extension.

Overlapping reading frames–Start triplets in different reading frames generate different polypeptides from the same DNA sequence. *Search in* reading frame.

Ovulation–In mammals, the process of escape of the ovum (egg cell) from the ovary.

Ovule (L. *ovulum*, diminutive of *ovum*, egg)–The part of the reproductive organs in seed plants that consists of the nucellus, the embryo sac and integuments.

Ovum (L. *ovum*, egg; pl: ova)–1. A gamete of female animals, produced by the ovary.

2. The oosphere in plants.

Ovum pickup (OPU)–The non-surgical collection of ova from a female.

P_1–Symbol for the parental generation or parents of a given individual.

Pachynema (adj: pachytene)–A mid-prophase stage in meiosis, immediately following zygonema and preceding diplonema. In microscopic preparations, the chromosomes are visible as long, paired threads. Rarely, four chromatids are detectable.

Packaging cell line–A cell line that is designed to produce viral particles that do not contain nucleic acid. After transfection of these cells with a full-size viral genome, fully infective viral particles are assembled and released.

Packed cell volume (PCV)–The volume of cells in a set volume of culture expressed as a percentage of that set volume after sedimentation (packing) by means of low speed centrifugation.

PAGE–*Search in* polyacrylamide gel electrophoresis.

Pairing; synapsis–The pairing of homologous chromosomes during the prophase of the first meiotic division, when crossing over occurs.

Pair-rule gene–A gene that influences the formation of body segments in *Drosophila*.

Palaeontology–The study of the fossil record of past geological periods and of the phylogenetic relationships between extinct and contemporary plant and animal species.

Palindrome (Gr. *palindromos*, running back again)–Originally a word, sentence or verse that reads the same from right to left as it does from left to right. In biotechnology, the term is applied to a segment of DNA in which the base-pair sequence reads the same in both directions from a central point of symmetry. *Search in* the entry for *inverted repeat* for an example of a palindrome. *Search in also* palindromic sequence.

Palindromic sequence–A segment of duplex DNA whose 5′-to-3′ sequence is identical on each DNA strand. The sequence is the same when one strand is read left-to-right and the other strand is read right-to-left. Typically, recognition sites for type II restriction endonucleases are palindromes. *Search in* the entry for *inverted repeat* for an example of a palindrome.

Palisade parenchyma–Elongated cells found just beneath the upper epidermis of leaves, and containing many chloroplasts.

pAMP–Ampicillin-resistant plasmid. *Search in* plasmid.

Panicle (L. *panicula*, a tuft)–An inflorescence, the main axis of which is branched; the branches bear loose racemose flower clusters.

Panicle culture–Aseptic culture of grain panicle segments to induce microspore germination and development.

Panmictic population–A population in which mating occurs at random.

Panmixis–Random mating in a population. *The opposite word is* apomixis.

Paper raft technique–*Search in* nurse culture.

PAR–*Search in* photosynthetically active radiation.

Paracentric inversion–An inversion that is entirely within one arm of a chromosome and does not include the centromere.

Paraffin [wax]–A translucent, white, solid hydrocarbon with a low melting point. Paraffin is used as an embedding medium to support tissue for sectioning for light microscopy observation.

Parallel evolution–The development of different organisms along similar evolutionary paths due to similar selection pressures acting on them.

ParafilmTM–A stretchable film based on paraffin wax; used to seal tubes and Petri dishes. ParafilmTM is a proprietary name which is applied colloquially to similar products.

Parahormone–A substance with hormone-like properties that is not a secretory product (*e.g.*, ethylene; carbon dioxide).

Parameter–A value or constant pertaining to an entire population. *The opposite word is* statistic.

Parasexual cycle–A sexual cycle involving changes in chromosome number but differing in time and place from the usual sexual cycle; occurring in those fungi in which the normal cycle is suppressed or apparently absent.

Parasexual hybridization–*Search in* somatic hybridization.

Parasite (Gr. *parasites*, one who eats at the table of another)–An organism deriving its food from the living body of another organism.

Parasitism–The close association of two or more dissimilar organisms, where the association is harmful to at least one. *The opposite word is* commensalism; symbiosis.

Parasporal crystal–Tightly packaged insect pro-toxin molecules that are produced by strains of *Bacillus thuringiensis* during the formation of resting spores.

Parenchyma–1. A plant tissue consisting of spherical, undifferentiated cells, frequently with air spaces between them.

2. Loose, spongy connective tissue formed of large cells.

Parenchymatous–Adjective used to describe spherical and

undifferentiated cells with primary cell walls, capable of both cell division and differentiation.

Par gene–A gene found in bacterial and plant cells, and involved in partition of plasmids in cells.

Parthenocarpy (Gr. *parthenos*, virgin + *karpos*, fruit)–The development of fruit without fertilization.

Parthenogenesis (Gr. *parthenos*, virgin + *genesis*, origin)–Production of an embryo from an unfertilized egg. *The opposite word is* androgenesis; apomixis; gynogenesis.

Partial digest–Addition of a restriction enzyme to a DNA sample under particular conditions or for a limited period, such that only a proportion of the target sites in any individual molecule are cleaved. Partial digests are often performed to give an overlapping collection of DNA fragments for use in the construction of a gene bank. *Search in* complete digest; library.

Particle radiation–Refers to gamma (γ) particles (positively charged) and beta (ß) particles (negatively charged), electrons, protons and neutrons. In plant tissue culture, these particles are used to produce mutant cells or organisms.

Parts per million (ppm)–Units of any given substance per one million equivalent units, such as the weight units of solute per million weight units of solution (*i.e.*, 1 ppm = 1 mg/l).

Parturition–The process of giving birth.

Passage–The transfer or transplantation of cells from one culture vessel to another. *The opposite word is* sub-culture.

Passage number–The number of times the cells in the culture have been sub-cultured. In descriptions of this process, the dilution ratio of the cells should be stated so that the relative cultural "age" can be ascertained.

Passage time–Interval between successive sub-cultures.

Passive immunity–1. Natural acquisition of antibodies by the foetus or neonate (newborn) from the mother.

2. The artificial introduction of specific antibodies by the injection of serum from an immune animal.

In both cases, temporary protection is conferred on the recipient.

Patent–A government-issued document that assigns the holder the exclusive right–for a defined period of time–to manufacture, use or sell an invention.

Paternal–Pertaining to the father.

Pathogen (Gr. *pathos*, suffering + *genesis*, beginning)–An organism that causes a disease in another organism.

Pathogen-free–Freedom from disease-causing organisms (bacteria, fungi, viruses, etc.).

Pathotoxin–Very dilute substance synthesized and released by some pathogen, and which interacts with the host metabolism.

Pathovar–In plant pathology, strains of bacteria causing disease in specific plant cultivars.

pBR322–One of the first plasmid vectors widely used; especially used for cloning DNA in *E. coli*. *Search in* plasmid.

PCR–*Search in* polymerase chain reaction.

PCV–*Search in* packed cell volume.

Pectin (Gr. *pektos*, congealed)–A white amorphous substance which, when combined with acid and sugar, yields a jelly substance cementing cells together (the middle lamella).

Pectinase (Gr. *pektos*, congealed)–Enzyme catalysing the hydrolysis of pectins.

Pedicel (L. *pediculus*, a little foot)–Stalk or stem of the individual flowers of an inflorescence.

Pedigree–A table, chart or diagram recording the ancestry of an individual.

Peduncle (L. *pedunculus*, a late form of *pediculus*, a little foot)–Stalk or stem of a flower that is born singly; the main stem of an inflorescence.

PEG–*Search in* polyethylene glycol.

Penetrance–The percentage of individuals in population that show a particular phenotype among those capable of showing it, *i.e.*, among those that have the genotype normally associated with that phenotype.

Peptide–A sequence of amino acids linked by peptide bonds; a breakdown or build-up unit in protein metabolism and it is a primary component of a protein molecule.

Peptide bond–A chemical bond holding amino acid sub-units together in proteins.

Peptide vaccine–A short chain of amino acids that can induce antibodies against a specific infectious agent.

Peptidyl-tRNA binding site; P–site–The site on a ribosome that hosts the tRNA to which an amino acid for the growing polypeptide chain is attached.

Peptidyl transferase–An enzyme bound tightly to the large sub-unit of the ribosome that catalyses the formation of peptide bonds between amino acids during translation.

Perennial (L. *perennis*, lasting years)–A plant that grows more or less indefinitely from year to year and, once mature, usually produces seed each year.

Pericentric inversion–An inversion that includes the centromere, hence involving both arms of a chromosome.

Periclinal–The plane of cell wall orientation or cell division parallel to the surface of the organ. *The opposite word is* anticlinal.

Periclinal chimera–Genotypically or cytoplasmically different tissues arranged in concentric layers.

Pericycle (Gr. *peri*, around + *kyklos*, circle)–Region of the plant bounded externally by the endodermis and internally by the phloem. Most roots originate from the pericycle.

Periplasm–The space (periplasmic space) between the cell (cytoplasmic) membrane of a bacterium or fungus and the outer membrane or cell wall.

Permanent wilting point (PWP)–The moisture content of soil at which plants wilt to such an extent that they fail to recover even when placed in a humid atmosphere.

Permeable (L. *permeabilis*, that which can be penetrated)–Used of a membrane, cell or cell system through which substances may diffuse.

Persistence–Ability of an organism to remain in a particular setting for a period of time after it is introduced.

Persistent–1. Continues to exist or to remain attached.

2. Chemicals with a long inactivation time, such as some pesticides, which may accumulate in the food chain.

PERV–*Search in* porcine endogenous retrovirus.

Pesticide–A toxic chemical product that kills harmful organisms (*e.g.*, insecticides, fungicide, weedicides, rodenticides).

Petal–One of the parts of the flower that make up the corolla.

Petiole (L. *petiolus*, a little foot or leg)–Stalk of leaf. *Search in* pedicel; peduncle.

Petite mutant–A respiration-deficient yeast mutant that produces small colonies when grown on glucose-containing medium.

Petri dish–Flat round dish with a matching lid, made of glass or plastic material, and used for culturing organisms. a.k.a. plates, hence the term to "plate" a culture.

PFGE–*Search in* pulsed-field [gel] electrophoresis.

pH–A measure of acidity and alkalinity. Equal to the log of the reciprocal of the hydrogen ion concentration of a solution, expressed in gram per litre. A reading of 7 is neutral (*e.g.*, pure water), whereas below 7 is acidic and above 7 is alkaline.

Phage–*Search in* bacteriophage.

Phagemids–Cloning vectors that contain components derived from both phage chromosomes and plasmids.

Phagocytes–Immune system cells that ingest and destroy viruses, bacteria, fungi and other foreign substances or cells.

Phagocytosis–The process by which minute food particles or foreign particles invading the body are engulfed and broken down by certain animal cells.

Pharmaceutical agent–*Search in* therapeutic agent.

Phase state–The coupling or repulsion of two linked genes.

pH-electrode-based sensor–Sensor in which a standard electrochemical pH electrode is coated with a biological material. Many biological processes raise or lower pH, and the changes can be detected by the pH electrode.

Phenocopy–An organism whose phenotype (but not genotype) has been changed by the environment to resemble the phenotype usually associated with a mutant organism.

Phenolic oxidation–Many plant species contain phenolic compounds that blacken through oxidation. The process is initiated after plants are wounded. Phenolic oxidation may lead to growth inhibition or, in severe cases, to tissue necrosis and death. Antioxidants are incorporated into the sterilizing solution or isolation medium to prevent or reduce oxidative browning. *The opposite word is* activated charcoal.

Phenols; phenolics–Compounds with hydroxyl group(s) attached to the benzene ring, forming esters, ethers and salts. Phenolic substances are produced from newly explanted tissues, oxidizing to form coloured compounds visible in nutrient media.

Phenotype (Gr. *phaneros*, showing + type)–The visible appearance or set of traits of an organism resulting from the combined action of genotype and environment. *The opposite word is* genotype.

Phenylalanine–*Search in* amino acid.

Pheromone–A hormone-like substance that is secreted by an organism into the environment as a specific signal (chemical biocommunication) to another organism, usually of the same species.

Phloem (Gr. *phloos*, bark)–Food-conducting tissue in plants, consisting of sieve tubes, companion cells, phloem parenchyma and fibres.

Phosphatase–An enzyme that hydrolyses esters of phosphoric acid, removing a phosphate group from an organic compound.

Phospho-diester bond–A bond in which a phosphate group joins adjacent carbons through ester linkages. A condensation reaction between adjacent nucleotides results in a phospho-diester bond between 3′ and 5′ carbons in DNA and RNA.

Phospholipase A2–An enzyme which degrades type A2 phospholipids.

Phospholipid–A class of lipid molecules in which a phosphate group is linked to glycerol and two fatty acyl groups. A major component of biological membranes. *Search in* inositol lipid.

Phosphorolysis–The cleavage of a bond by orthophosphate; analogous to hydrolysis referring to cleavage by water.

Phosphorylation–The addition of a phosphate group to a compound.

Photoautotroph–*Search in* autotroph; heterotroph.

Photo-bioreactor–Bioreactor dependent on sunlight, which is taken up by its content of plant material, usually algae.

Photoheterotroph–*Search in* heterotroph.

Photon–A quantum of light; the energy of a photon is proportional to its frequency; E = hv, where E is energy; h is Planck's constant, 6.62 × 10^{-27} erg-second; and v is the frequency.

Photoperiod (Gr. *photos*, light + period)–The length of day or period of daily illumination provided or required by plants for reaching the reproductive stage.

Photoperiodism–The response of an organism to changes in day length.

Photophosphorylation–The formation of ATP from ADP and inorganic phosphate using light energy in photosynthesis.

Photoreactivation–A DNA repair process that is light dependent.

Photosynthate–The carbohydrates and other compounds produced in photosynthesis.

Photosynthesis–A chemical process by which green plants synthesize organic compounds from carbon dioxide and water in the presence of sunlight.

Photosynthetic–Able to use sunlight energy to convert atmospheric carbon dioxide into organic compounds. Nearly all plants, most algae and some bacteria are photosynthetic.

Photosynthetic efficiency–Efficiency of converting light energy into organic compounds.

Photosynthetically active radiation (PAR)–Radiant energy captured by the photosynthetic system in the light reactions, usually taken to be the wavelengths between 400 and 700 nm.

Phototropism (Gr. *photos*, light + *tropos*, turning)–A growth curvature in which light is the stimulus. It is positive if movement is towards light source and negative if movement is away from light source.

Phylogeny–A diagram illustrating the deduced evolutionary history of populations of related organisms.

Physical map–A map showing physical locations on a DNA sequence, such as restriction sites and sequence-tagged sites. Also a diagram of a chromosome or a karyotype, showing the location of loci (genes and markers) *Search in* mapping.

Phytochrome–A reversible pigment system of a protein nature, found in the cytoplasm of green plants. Phytochrome is associated with the absorption of light that affects growth, development and differentiation of a plant, independent of photosynthesis, *e.g.*, in the photoperiodic response.

Phytohormone–A substance that stimulates growth or other processes in plants. They include auxins, abscissic acid, cytokinins, gibberellins and ethylene. Phytohormones are chemical messengers that may pass through cells, tissues and organs and stimulate biochemical, physiological and morphological responses.

Phytokinin–*Search in* cytokinin.

Phytoparasite (adj: phytoparasitic)–Parasite on plants.

Phytopathogen–An organism that causes disease in plants.

Phytosanitary–Plant health, including quarantine.

Phytostat–The name adopted by Tulecke in 1965 for an apparatus designed for the semi-continuous chemostat culture of plant cells.

Pigments–Molecules that are coloured by the light they absorb. Some plant pigments are water soluble and are found mainly in the cell vacuole.

Pinocytosis–The process by which a living cell engulfs a minute droplet of liquid.

Pipette–A slender graduated tube into which small amounts of liquids are taken up by suction, for measuring and transferring.

Pistil (L. *pistillum*, a pestle)–Central organ of the flower, typically consisting of ovary, style and stigma. The pistil is usually referred to as the female part of a perfect flower.

Plantlet–A small rooted shoot developed from seed or from cultured cells either by embryogenesis or organogenesis.

Plant cell culture–Growth of plant cells or roots of plants in bioreactors.

Plant cell immobilization–Entrapment of plant cells in gel matrices; the cells are suspended in small drops of the material, which then is set or allowed to harden to make little carriers. Materials such as alginates, agar or polyacrylamide can be used.

Plant genetic resources (PGR)–Defined in the International Undertaking on Plant Genetic Resources (FAO, 1983) to mean the reproductive or vegetative propagating material of the following categories of plants:

(i) cultivated varieties (cultivars) in current use and newly developed varieties;

(ii) obsolete cultivars;

(iii) primitive cultivars (landraces);

(iv) wild and weed species, near relatives of cultivated varieties; and

(v) special genetic stocks (including elite and current breeder's lines and mutants).

Plaque–A clear spot on an otherwise opaque culture plate of bacteria or cultured bacteria cells, showing where cells have been lysed by viral infection.

Plasma cells–Antibody-producing white blood cells derived from B lymphocytes.

Plasma membrane–*Search in* cell membrane; plasmalemma.

Plasmalemma (Gr. *plasma*, anything formed + *lema*, a husk or shell of a fruit)–1. A delicate cytoplasmic double membrane found on the outside of the protoplast, adjacent to the cell wall. *Search in* cell membrane.

2. **2μm plasmid** A naturally occurring, double-stranded, circular DNA plasmid (6 318 bp) found in the nuclei of yeast. Many yeast plasmid vectors are derived from the 2μm plasmid. a.k.a. 2μm circle; 2 μ-plasmid; 2-micron plasmid.

Plasmid–An extrachromosomal, autonomous circular DNA molecule found in certain bacteria, capable of autonomous replication. Plasmids can transfer genes between bacteria and are important tools of transformation in genetic engineering. They exist in an autonomous state and are transferred independently of chromosomes. *Search in* nicked circle; pAMP; relaxed plasmid; stringent plasmid; supercoiled plasmid.

Plasmodesma (Gr. *plasma*, something formed + *demos*, a bond, a band; pl: plasmodesmata)–Fine protoplasmic thread passing through the plant cell wall that separates two protoplasts.

Plasmolysis (Gr. *plasma*, something formed + *lysis*, a loosening)–Separation of the cytoplasm from the cell wall, due to removal of water from the protoplast.

Plastid (Gr. *plastis,* a builder)–A cytoplasmic body found in the cells of plants and some protozoa. Chloroplastids, for example, produce chlorophyll that is involved in photosynthesis.

Plastoquinone–A quinone which is one of a group of compounds involved in the transport of electrons in photosynthesis in chloroplasts.

Plate–1. *Verb*: To distribute a thin film of something. Hence micro-organisms or plant cells are plated onto nutrient agar.

2. *Noun*: Refers to the two segments of a Petri dish or a similar-shaped item.

Platform shaker–*Search in* shaker.

Plating efficiency–The percentage of inoculated cells which give rise to cell colonies when seeded into culture vessels.

Pleiotropy (adj: pleiotropic)–The situation in which a particular gene has an effect on several different traits.

Ploidy–The number of sets of chromosomes per cell, *e.g.*, haploid, diploid and polyploid.

Plumule (L. *plumula,* a small feather)–The first bud of an embryo or that portion of the young shoot above the cotyledons.

Pluripotent–*Search in* totipotent.

Pneumatic reactor–*Search in* airlift fermenter.

Point mutation–A change in DNA at a specific site in a chromosome. Includes nucleotide substitutions and the insertion or deletion of one or a few nucleotide pairs. *Search in* mutation.

Polar bodies–In female animals, the products of a meiotic division that do not develop into an ovum. The first polar body comprises one of the two products of meiosis I, and it may not go through meiosis II. The second polar body comprises one of the two products of meiosis II.

Polar mutation–A mutation that influences the functioning of genes that are downstream in the same transcription unit.

Polar nuclei–Two centrally located nuclei in the embryo sac that unite with a second sperm cell in a triple fusion. In certain seeds, the product of this fusion develops into the endosperm.

Polar transport–The directed movement within plants of compounds (usually endogenous plant growth regulators) mostly in one direction; polar transport overcomes the tendency for diffusion in all directions.

Polarity (Gr. *pol,* an axis)–The observed differentiation of an organism, tissue or cell into parts having opposed or contrasted properties or form.

Pole cells–A group of cells in the posterior of *Drosophila* embryos that are precursors to the adult germ line.

Pollen (L. *pollen,* fine flour)–The mass of germinated microspores or partially developed male gametophytes of seed plants.

Pollen culture–The *in vitro* culture and germination of pollen grains. Callus cultures thus obtained will form shoots or embryoids which develop into monoploid plants. *Search in* anther culture; microspore culture.

Pollen grain–A microspore produced in the pollen sac of angiosperms or the microsporangium of gymnosperms. Unicellular, with variable shape and size, and usually ovoid from 25 to 250μm.

Pollination–Transfer of pollen from anther to stigma in the process of fertilization in angiosperms; transfer of pollen from male to female cone in the process of fertilization in gymnosperms.

Poly (A) polymerase–Enzyme that catalyses the addition of adenine residues to the 3′ end of pre-mRNAs to form the poly-(A) tail. *Search in* polyadenylation; polymerase.

Poly-(A) tail–*Search in* polyadenylation.

Polyacrylamide gel electrophoresis (PAGE)–A method for separating nucleic acid or protein molecules according to their molecular size. The molecules migrate through the inert gel matrix under the influence of an electric field. In the case of protein PAGE, detergents such as sodium dodecyl sulphate (SDS) are often added to ensure that all molecules have a uniform charge. Secondary structure can often lead to the anomalous migration of molecules. Therefore it is common to denature protein samples by boiling them prior to PAGE. In the case of nucleic acids, denaturing agents such as formamide, urea or methyl mercuric hydroxide are often incorporated into the gel itself, which may also be run at high temperature. PAGE is used to separate the products of DNA-sequencing reactions and the gels employed are highly denaturing, since molecules differing in size by a single nucleotide must be resolved.

Polyacrylamide gels–Often referred to incorrectly as acrylamide gels. These gels are made by cross-linking acrylamide with *N N′*-methylene-*bis*-acrylamide. Polyacrylamide gels are used for the electrophoretic separation of proteins, DNA and RNA molecules. Polyacrylamide beads are also used as molecular sieves in gel chromatography, marketed as Bio-gelTM.

Polyadenylation–Post-transcriptional addition of a polyadenylic acid tail to the 3′ end of eukaryotic mRNAs. Also called poly-(A) tailing. The adenine-rich 3′ terminal segments is called a poly (A) tail.

Polyclonal antibody–A serum sample that contains a mixture of immunoglobulin molecules secreted against a specific antigen, each recognizing a different epitope, some antibodies of which bind to different antigenic determinants of one antigen.

Polycloning site–*Search in* polylinker.

Polyembryony–In the ordinary course of events, one embryo is formed in each ovule and is derived from the fertilization of the ovum in the solitary embryo sac. However, two or more embryos could start development even though only one may reach maturity.

Polyethylene glycol (PEG); carbowax–A polymer having the general formula: $HOCH_2(CH_2OCH_2)\,XCH_2$ OH and available in a range of molecular weights from *ca* 1000 to *ca* 6000. PEG 4000 and PEG 6000 are commonly used to promote cell or protoplast fusion, and to facilitate DNA uptake in the transformation of organisms such as yeast. PEG is also used to concentrate solutions by withdrawing water from them.

Polygene–One of many genes of small effect influencing the development of a quantitative trait; results in continuous variation and in quantitative inheritance. *Search in* gene.

Polygenic–Controlled by many genes of small effect.

Polylinker–A segment of DNA that contains a number of different restriction endonuclease sites. a.k.a. multiple cloning site (MCS).

Polymer–A compound composed of many identical smaller sub-units; resulting from the process of polymerization.

Polymerase chain reaction (PCR)–A procedure that amplifies a particular DNA sequence. It involves multiple cycles of ***denaturation, annealing*** to oligonucleotide primers, and ***extension*** (polynucleotide synthesis), using a thermostable DNA polymerase, deoxyribonucleotides, and primer sequences in multiple cycles of denaturation-renaturation-DNA synthesis. *Search in* polymerase.

Polymerase–An enzyme that catalyses the formation of polymeric molecules from monomers. A DNA polymerase synthesizes DNA from deoxynucleoside triphosphates using a complementary DNA strand and a primer. An RNA polymerase synthesizes RNA from monoribonucleoside triphosphates and a complementary DNA strand. *Search in* poly-(A) polymerase; polymerase chain reaction; RNA polymerase; *taq* polymerase.

Polymerization–Chemical union of two or more molecules of the same kind such as glucose or nucleotides to form a new compound (starch or nucleic acid) having the same elements in the same proportions but a higher molecular weight and different physical properties.

Polymery–The phenomenon whereby a number of genes at different loci (which may be polygenes) can act together to produce a single effect.

Polymorphism–The occurrence of two or more alleles at a locus in a population.

Polynucleotide–A chain of nucleotides in which each nucleotide is linked by a single phospho-diester bond to the next nucleotide in the chain. They can be double–or single-stranded. The term is used to describe DNA or RNA. *Search in* nucleotide.

Polypeptide–A linear molecule composed of two or more amino acids linked by covalent (peptide) bonds. They are called dipeptides, tripeptides and so forth, according to the number of amino acids present.

Polyploid (Gr. *polys*, many + *ploid*, fold)–Tissue or cells with more than two complete sets of chromosomes, that results from chromosome replication without nuclear division or from union of gametes with different number of chromosome sets, hence triploid (3*x*), tetraploid (4*x*), pentaploid (5*x*), hexaploid (6*x*), heptaploid (7*x*), octoploid (8*x*).

Polysaccharide (Gr. *polys*, many + Gr. *sakcharon*, sugar)-Long-chain molecules, such as starch and cellulose, composed of multiple units of a monosaccharide.

Polysaccharide capsules–Carbohydrate coverings with antigenic specificity that are present on some types of bacteria.

Polytene chromosomes–Giant chromosomes produced by interphase replication without division, and consisting of many identical chromatids arranged side by side in a cablelike pattern.

Polyunsaturates–Oils containing polyunsaturated fatty acids.

Polyspermy–The entry of several sperm into the egg during fertilization, although only one sperm nucleus actually fuses with the egg nucleus.

Polyvalent vaccine–A recombinant organism into which antigenic determinants have been cloned from a number of different disease-causing organisms, and used as a vaccine. *Search in* vaccine.

Polyvinylpyrrolidone (PVP) (C_6H_9NO) n)–An occasional constituent of plant tissue culture isolation media. PVP is of variable molecular weight and has antioxidant properties, so is used to prevent oxidative browning of explanted tissues. PVP is less frequently used as an osmoticum in culture media.

Population density–Number of cells or individuals per unit. The unit could be an area or volume of medium.

Population genetics–The branch of genetics that deals with frequencies of alleles and genotypes in breeding populations. *The opposite word is.* quantitative genetics.

Population–A local group of organisms belonging to the same species and interbreeding.

Porcine endogenous retrovirus (PERV)–The provirus of a porcine retrovirus. With the increasing interest in the use of pig organs for xenotransplantation to humans, there has been concern that PERVs could be activated after transplantation, creating an infection in the human recipient.

Positional candidate gene–A gene known to be located in the same region as a DNA marker that has been shown to be linked to a single-locus trait or to a quantitative trait locus (QTL) and whose function suggests that it could be the source of genetic variation in the trait in question.

Positional cloning–The process that commences with searching for markers linked to a particular inherited trait, then using those markers to identify the approximate location of the gene responsible for the trait, and then using various cloning strategies (including chromosome walking, q.v.) to identify, isolate and characterize the gene. Originally, the strategy was called *reverse genetics*, a term that some consider to be a misnomer and misleading.

Position effect–The situation in which a change in phenotype results from the change of the position of a gene or group of genes.

Positive control system–A mechanism in which the regulatory protein(s) is required to turn on gene expression.

Positive selectable marker–*Search in* dominant marker selection.

Positive selection–1. *Search in* dominant marker selection.

2. A method by which cells that carry a DNA insert integrated at a specific chromosomal location are selected.

Post-replication repair–A recombination-dependent mechanism for repairing damaged DNA.

Post-translational modification–The specific addition of phosphate groups, sugars (glycosylation), or other molecules to a protein after it has been synthesized.

Potentiometric–*Search in* enzyme electrode.

ppm–*Search in* parts per million.

Precocious germination–Premature germination of the embryo, prior to completion of embryogenesis.

Pre-filter–A coarse filter used to screen out large particles before air is forced through a much finer filter. *Search in* HEPA filter; laminar air-flow cabinet.

Pre-mRNA–*Search in* primary transcript.

Pressure potential–The pressure generated within a cell; it is the difference between the osmotic potential within the cell and the water potential of the external environment, provided the cell volume is constant.

Pre-transplant–Stage III in tissue culture micropropagation; the rooting, hardening stage prior to transfer to soil. *Search in* micropropagation.

Preventive immunization; vaccination–Infection with an antigen to elicit an antibody response that will protect the organism against future infections.

Pribnow box–Consensus sequence near the mRNA start-point of prokaryotic genes. *Search in* TATA box.

Primary (L. *primus*, first)–First in order of time or development.

Primary antibody–In an ELISA or other immunological assay, the antibody that binds to the target molecule.

Primary cell–A cell or cell line taken directly from a living organism, which is not immortalized.

Primary cell wall–The cell wall layer formed during cell expansion. Plant cells possessing only primary walls may divide or undergo differentiation.

Primary culture–A culture started from cells, tissues or organs taken directly from organisms. A primary culture may be regarded as such until it is sub-cultured for the first time. It then becomes a cell line.

Primary germ layers–*Search in* germ layers.

Primary growth–1. Refers to apical meristem-derived growth; the tissues of a young plant.

2. Refers to explant growth during the initial culture period; such as primary callus growth.

Primary immune response–The immune response that occurs during the first encounter of a mammal with a given antigen. *The opposite word is* secondary immune response.

Primary meristem–Meristem of the shoot or root tip giving rise to the primary plant body.

Primary tissue–A tissue that has differentiated from a primary meristem.

Primary transcript–The RNA molecule produced by transcription prior to any post-transcriptional modifications; also called a pre-mRNA in eukaryotes.

Primer–A short DNA or RNA fragment annealed to a template of single-stranded DNA, providing a 3' hydroxyl end from which DNA polymerase extends a new DNA strand to produce a duplex molecule.

Primer DNA polymerase–DNA polymerase that provides primers for the DNA polymerization. Unlike RNA polymerase, DNA polymerase is unable to initiate the *de novo* synthesis of a polynucleotide chain. DNA polymerase can only add nucleotides to a free 3′ hydroxyl group at the end of a pre-existing chain. A short oligonucleotide, known as a primer, is therefore needed to supply such a hydroxyl group for the initiation of DNA synthesis.

Primer walking–A method for sequencing long (>1 kb) cloned pieces of DNA. The initial sequencing reaction reveals the sequence of the first few hundred nucleotides of the cloned DNA. On the basis of these data, a primer containing about 20 nucleotides and complementary to a sequence near the end of sequenced DNA is synthesized, and is then used for sequencing the next few hundred nucleotides of the cloned DNA. This procedure is repeated until the complete nucleotide sequence of the cloned DNA is determined.

Primordium–A group of cells which gives rise to an organ.

Primosome–A protein-replication complex that catalyses the initiation of synthesis of Okazaki fragments during discontinuous replication of DNA. It involves DNA primase and DNA helicase activities.

Prion–*Search in* proteinaceous infectious particle.

Probability–The frequency of occurrence of an event.

Proband–The individual in a family in whom an inherited trait is first identified.

Probe–1. For diagnostic tests, the agent that is used to detect the presence of a molecule in a sample.

2. A DNA or RNA sequence labelled or marked with a radioactive isotope or that is used to detect the presence of a complementary sequence by hybridization with a nucleic acid sample.

Probe DNA–A labelled DNA molecule used to detect complementary-sequence nucleic acid molecules by molecular hybridization. To localize the probe DNA sequence and reveal the complementary hybridization sequence, autoradiography or fluorescence is used. *Search in* multilocus probe; nucleotide.

Procambium (L. *pro*, before + cambium)–A primary meristem that gives rise to primary vascular tissues and, in most woody plants, to the vascular cambium.

Procaryote; prokaryotic–*Search in* prokaryote; prokaryotic.

Processed pseudo-gene–A copy of a functional gene which has no promoter, no introns and which, consequently, is not itself transcribed. Pseudo-genes are thought to originate from the integration into the genome of cDNA copies synthesized from mRNA molecules by reverse transcriptase. Pseudo-genes therefore have a poly (dA) sequence at their 5′ ends. Because they are not subject to any evolutionary pressure to maintain their coding potential, pseudo-genes accumulate mutations and often have stop codons in all three reading frames.

Production environment–All input-output relationships, over time, at a particular location. The relationships will include biological, climatic, economic, social, cultural and political factors, which combine to determine the productive potential of a particular livestock enterprise. (Source: FAO, 1999)

high-input production environment–A production environment where all rate-limiting inputs to animal production can be managed to ensure high levels of survival, reproduction and output. Output and production risks are constrained primarily by managerial decisions. (*Source*: FAO, 1999)

medium-input production environment–A production environment where management of the available resources has the scope to overcome the negative effects of the environment on animal production, although it is common for one or more factors to limit output, survival or reproduction in a serious fashion. (*Source*: FAO, 1999)

low-input production environment–A production environment where one or more rate-limiting inputs impose continuous or variable severe pressure on livestock, resulting in low survival, reproductive rate or output. Output and production risks are exposed to major influences which may go beyond human management capacity. (*Source*: FAO, 1999)

Production traits–Characteristics of animals, such as the quantity or quality of the milk, meat, fibre, eggs, draught, etc., they (or their progeny) producc, which contribute directly to the value of the animals for the farmer, and that are identifiable or measurable at the individual level. Production traits of farm animals ate generally quantitatively inherited, *i.e.*, they are influenced by many genes whose expression in a particular animal also reflects environmental influences. (*Source*: FAO, 1999)

Productivity–The amount of product that is produced within a given period of time from a specified quantity of resource.

Pro-embryo (L. *pro*, before + *embryon*, embryo)–A group of cells arising from the division of the fertilized egg cell or somatic embryo before those cells which are to become the embryo are recognizable.

Progeny–*Search in* offspring.

Progeny testing–For a single locus: the practice of ascertaining the genotype of an individual from its offspring, such as by mating it to other individuals and examining the progeny. For a quantitative trait: the collection and use of progeny performance as a clue to the breeding value of the parent of those progeny.

Progesterone–A hormone produced primarily by the corpus luteum of the ovary, but also by the placenta, that prepares the inner lining of the uterus for implantation of a fertilized egg cells.

Prokaryote (L. *pro*, before + Gr. *karyon*, a nut, referring in modern biology to the nucleus)–A member of a large group of organisms, including bacteria and blue-green algae, which do not have the DNA separated from the cytoplasm by a membrane in their cells. The DNA is usually in one long strand. Prokaryotes do not undergo meiosis and do not have functional organelles such as mitochondria and chloroplasts. *The opposite word is* eukaryote.

Prolactin–A hormone, produced by the anterior pituitary gland, that stimulates and controls lactation in mammals.

Proliferation (L. *proles*, offspring + *ferre*, to bear)–Increase by frequent and repeated reproduction; growth by cell division.

Pro-meristem–The embryonic meristem that is the source of organ initials or foundation cells.

Promoter–1. A nucleotide sequence of DNA to which RNA polymerase binds and initiates transcription. It usually lies upstream of (5′ to) a coding sequence. A promoter sequence aligns the RNA polymerase so that transcription will initiate at a specific site.

2. A chemical substance that enhances the transformation of benign cells into cancerous cells. *Search in* constitutive promoter.

Promoter sequence–A regulatory DNA sequence that initiates the expression of a gene.

Pro-nuclear micro-injection–The initial method of transgenesis in animals, involving injecting many copies of a particular gene into one of the two pro-nuclei of a fertilized ova. Characterised by a very low success rate. As animal cloning becomes more common, pro-nuclear micro-injection will be replaced by micro-injection of cloned genes into a culture of cloned embryos produced by nuclear transfer, which can be tested for expression of the transgene before transfer to recipient females. *Search in* transgenesis.

Pro-nucleus–Either of the two haploid gamete nuclei just prior to their fusion in the fertilized ovum.

Proofreading–The enzymatic scanning of newly-synthesized DNA for structural defects, such as mismatched base pairs. It is effected by DNA polymerase.

Propagation (L. *propagare*, to propagate)–The multiplication of plants by numerous types of vegetative material; an ancient practice dating from the dawn of agriculture, carried out in a nursery or directly in the field (vegetative propagation), and now in *in vitro* culture (micropropagation).

Propagule–Any structure capable of giving rise to a new plant by asexual or sexual reproduction, including bulbils, leafbuds, etc.

Pro-phage–The genome of a temperate bacteriophage integrated into the chromosome of a lysogenic bacterium and replicated along with the host chromosome.

Prophase (Gr. *pro*, before + *phasis*, appearance)–An early stage in nuclear division, characterized by the shortening and thickening of the chromosomes and their movement to the metaphase plate. It occurs between interphase and metaphase. During this phase, the centriole divides and the two daughter centrioles move apart. Each sister DNA strand from interphase replication becomes coiled, and the chromosome is longitudinally double except in the region of the centromere. Each partially separated chromosome is called a chromatid. The two chromatids of a chromosome are sister chromatids.

Protamines–Small basic proteins that replace the histones in the chromosomes of some sperm cells.

Protease–An enzyme that hydrolyzes proteins, cleaving the peptide bonds that link amino acids in protein molecules.

Protein (Gr. *proteios*, of the first rank)–A macromolecule composed of one to several polypeptides. Each polypeptide consists of a chain of amino acids linked together by covalent (peptide) bonds. They are naturally-occurring complex organic substances (egg albumen, meat) composed essentially of carbon, hydrogen, oxygen and nitrogen, plus sulphur or phosphorus, which are so associated as to form sub-microscopic chains, spirals or plates and to which are attached other atoms and groups of atoms in a variety of ways. The word was coined by Jöns J. Berzelius (1838) to emphasize the importance of this group of molecules. *Search in* polypeptide.

Protein crystallization–Making crystals of a protein, as a key part of most methods used for determining a protein's three-dimensional structure.

Protein drug–*Search in* therapeutic agent.

Protein engineering–Generating proteins with modified structures that confer properties such as higher catalytic specificity or thermal stability.

Protein kinase–An enzyme that adds phosphate groups to a protein molecule at the sites of serine, threonine or tyrosine residues.

Protein metabolic step–One step in the chain of reactions that take place in an organism and dictate the composition of that organism.

Protein sequencing–The process of determining the amino-acid sequence of a protein or its component polypeptides.

Protein synthesis–The creation of proteins from their constituent amino acids, in accordance with the genetic information carried in the DNA of the chromosomes.

Proteinaceous infectious particle; prion–An abnormal form of a normal cell protein, with no detectable nucleic acid, found in the brain of mammals and believed to be the agent responsible for the class of diseases called spongiform encephalopathies, including scrapie in sheep and bovine spongiform encephalopathy (BSE; mad cow disease) in cattle.

Proteolysis (Protein degradation)–Enzymatic degradation of a protein.

Proteolytic–The ability to break down protein molecules.

Protoclone–Regenerated plant derived from protoplast culture or a single colony derived from protoplasts in culture.

Protocol (Gr. *protokollon*, first leaf glued to a manuscript and describing the contents)–The step-by-step experiments proposed to describe or solve a scientific problem, or the defined steps of a specific procedure.

Protocorm–In orchids, seeds contain an unorganized embryo comprising only a few hundred cells. During seed germination this embryo forms a tuberous structure, called a protocorm, from which develops a complete plant. In cultures, vegetative parts of several orchids form round, smooth protocorms which may be multiplied indefinitely or induced to regenerate a whole plant.

Protoderm (Gr. *protos*, first + *derma*, skin)–A primary meristem tissue that gives rise to epidermis.

Protogyny–The condition in which the female reproductive organs (carpels) of a flower mature before the male ones (stamens), thereby ensuring that self-fertilization does not occur.

Protomeristem–*Search in* promeristem.

Proto–oncogene–A normal cellular gene that can be changed to an oncogene by mutation or by being incorporated into a retrovirus and then being transcribed at inappropriate times and/or in inappropriate tissues. *Search in* cellular oncogene; oncogene.

Protoplasm–The essential, complex living substance of cells, upon which all vital functions of nutrition, secretion, growth and reproduction depend.

Protoplast (Gr. *protoplastos*, formed first)–A bacterial or plant cell for which the relatively rigid wall has been removed either chemically or enzymatically, leaving its cytoplasm enveloped by only a delicate peripheral membrane. Protoplasts are spherical and smaller than the elongate, angular shaped and often vacuolated cells from which they have been released.

Protoplast culture–The isolation and culture of plant protoplasts by mechanical means or by enzymatic digestion of plant tissues or organs, or cultures derived from these. Protoplasts are utilized for selection or hybridization at the cellular level and for a variety of other purposes.

Protoplast fusion–The coalescence of the plasmalemma and cytoplasm of two or more protoplasts in contact with one another. Initial adhesion of two protoplasts is a random process but coalescence can be induced.

Prototroph–An organism such as a bacterium that will grow on a minimal medium.

Protozoan (pl: protozoa)–A microscopic, single-cell organism.

Pro-toxin–A latent, non-active precursor form of a toxin.

Protruding end–*Search in* extension.

Provirus–A retrovirus in which the single RNA strand has been converted into a double-strand DNA, which has been integrated into a host genome.

Pseudo-affinity chromatography–*Search in* affinity chromatography.

Pseudo-autosomal region–A section at one end of the X and Y chromosomes for which there is sufficient homology that there is pairing (synapsis) during meiosis I.

Pseudo-gene–A copy (duplicate) of a functional gene that, as a result of mutation following duplication, can no longer function.

Pseudocarp; false fruit–A fruit that incorporates, in addition to the ovary wall, other parts of the flower, such as the receptacle (*e.g.*, strawberry).

***Pseudomonas* spp.**–A common Gram-negative bacterial genus that is widely distributed. Many of the soil forms produce a pigment that fluoresces under ultraviolet light, hence the descriptive term fluorescent pseudomonas.

P-site–*Search in* peptidyl-tRNA site.

Psychrophile–A micro-organism that can grow at temperatures below 30°C and as low as 0°C.

PUC–A widely used expression plasmid, containing a galactosidase gene. *Search in* plasmid.

Pulsed-field gel electrophoresis (PFGE)–A procedure used to separate very large DNA molecules by alternating the direction of electric current in a pulsed manner across a semisolid gel.

Punctuated equilibrium–The occurrence of speciation events in bursts, separated by long intervals of species stability.

Pure culture–Axenic culture.

Pure line–A strain in which all members have descended by self-fertilization or close inbreeding. A pure line is genetically uniform.

Purification tag–*Search in* affinity tag.

Purine–A double-ring, nitrogen-containing base present in nucleic acids; adenine (A) and guanine (G) are the two purines present in most DNA and RNA molecules.

PVP–*Search in* polyvinylpyrrolidone.

PWP–*Search in* permanent wilting point.

Pyrethrins–Active constituents of pyrethrum (*Tanacetum cinerariifolium*) flowers, used as insecticides.

Pyrimidine–A single-ring, nitrogen-containing base present in nucleic acids; cytosine (C) and thymine (T) are commonly present in DNA, whereas uracil usually replaces thymine in RNA.

Pyrogen–Bacterial substance that causes fever in mammals released by activated white blood cells.

Quadrivalent–*Search in* tetrad (second definition).

Quantitative genetics–The area of genetics concerned with the inheritance of continuously-varying traits. Most practical improvement programs involve the application of quantitative genetics.

Quantitative inheritance–Inheritance of measurable traits (height, weight, colour intensity, etc.) that depend on the cumulative action of many genes.

Quantitative trait–A measurable trait that shows continuous variation; a trait that can not be classified into a few discrete classes.

Quantitative trait locus (QTL)–A locus that affects a quantitative trait. The plural form (quantitative trait loci) is also abbreviated as QTL.

Quantum (L. *quantum*, how much)–An elemental unit of energy. Its energy value is hv, where h, Planck's constant, is 6.62×10^{-27} erg-second and v is the frequency of the vibrations or waves with which the energy is associated. *The opposite word is* photon.

Quantum speciation–The rapid formation of new species, primarily by genetic drift.

Quarantine (It. *quarantina*, from *quaranta* forty)–Originally, keeping a person or living organism in isolation for a period (originally 40 days) after arrival to allow disease symptoms to appear, if there was any disease present. Now used for regulations restricting the sale or shipment of living organisms, usually to prevent disease or pest invasion of an area.

Quiescent–Quiet, at rest, but not necessarily dormant, and having the potential for resumed activity; can apply to non-meristematic cells.

QTL–*Search in* quantitative trait locus.

Race–A distinguishable group of organisms of a particular species, that are geographically, ecologically, physiologically, physically and/or chromosomically distinct from other members of the species.

Raceme–An inflorescence in which the main axis is elongated but the flowers are borne on pedicels that are about equal in length.

Rachilla (Gr. *rhachis*, a backbone + L. diminutive suffix–*illa*)–Shortened axis of a spikelet.

Rachis (Gr. *rhachis*, a backbone)–Main axis of a spike; axis of fern leaf (frond) from which pinnae arise; in compound leaves, the extension of the petiole corresponding to the midrib of an entire leaf.

Radicle (L. *radix*, root)–That portion of the plant embryo which develops into the primary or seed root.

Radioactive isotope; radioisotope–An unstable isotope that emits ionizing radiation.

Raft culture–*Search in* nurse culture.

Ramet–An individual member of a clone. *The opposite word is* ortet.

Random amplified polymorphic DNA (RAPD; pronounced 'rapid')–A technique using single, short (usually 10-mer) synthetic oligonucleotide primers for PCR. The primer, whose sequence has been chosen at random, initiates replication at its complementary sites on the DNA, producing fragments up to about 2 kb long, which can be separated by electrophoresis and stained with ethidium bromide. A primer can exhibit polymorphism between individuals, and polymorphic fragments can be used as markers.

Random genetic drift–*Search in* genetic drift..

Random mutagenesis–A non-directed change of one or more nucleotide pairs in a DNA molecule.

Random primer method–A protocol for labelling DNA *in vitro*. A sample of random oligonucleotides (6 or 14 nucleotides long) containing all possible combinations of nucleotide sequences is hybridized to a DNA probe. Then, in the presence of a DNA polymerase and the four deoxyribonucleotides–one of which is labelled–the 3´ hydroxy ends of the hybridized oligonucleotides provide initiation sites for DNA synthesis that uses the separated strands of the probe DNA as a template. This reaction produces labelled copies of portions of the probe DNA.

RAPD–*Search in* random amplified polymorphic DNA.

Reading frame–A series of triplets beginning from a specific nucleotide. Each triplet is represented by a single amino acid in the protein synthesized. The reading frame defines which sets of three nucleotides are read as triplets in the DNA, and hence as codons in the corresponding mRNA; this is determined by the initiation codon, AUG. Thus the sequence

AUGGCAAAAUUUCCC would read as

AUG/GCA/AAA/UUU/CCC/ and not as

A/UGC/CAA/AAU/UUC/CC.

Depending on where one begins, each DNA strand contains three different reading frames. *Search in* open reading frame, overlapping reading frames.

Read-through–Transcription or translation that proceeds beyond the normal stopping point because of the absence of the transcription or translation termination signal of a gene.

Reca–A protein in most bacteria, and that is essential for DNA repair and DNA recombination.

Recalcitrant–Of seeds: unable to survive drying and subsequent storage at low temperature. *Search in* field gene bank.

Receptacle (L. *receptaculum*, a reservoir)–Enlarged end of the pedicel or peduncle, to which other flower parts are attached.

Receptor–A molecule that can accept the binding of a ligand.

Receptor–binding screening–One of the biotechnology-based methods for discovering conventional drugs. The method relies on the fact that many drugs act by binding to specific proteins (receptors) on or in cells: these proteins usually bind to hormones or to other cells, and control the cell's behaviour, although they may be enzymes or structural elements of the cell. The drug interferes with the normal role of the protein.

Recessive–Describing an allele whose effect with respect to a particular trait is not evident in heterozygotes. Opposite to *dominant*.

Recessive-acting oncogene–*Search in* recessive oncogene.

Recessive oncogene; recessive-acting oncogene; anti-oncogene–A single copy of this gene is sufficient to suppress cell proliferation; the loss of both copies of the gene contributes to cancer formation. *Search in* oncogene.

Reciprocal crosses–Crosses between the same two strains, but with the sexes reversed; *e.g.*, female A × male B and male A × female B.

Reciprocating shaker–A platform shaker used for agitating culture flasks, with a back and forth action at variable speeds.

Recognition sequence–*Search in* recognition site.

Recognition site–A nucleotide sequence-composed typically of 4, 6 or 8 nucleotides–that is recognized by and to which a restriction endonuclease (restriction enzyme) binds. For type II restriction enzymes (those used in gene-cloning experiments) it is also the sequence within which the enzyme specifically cuts (and their corresponding enzymes methylate) the DNA, *i.e.*, for type II enzymes, the recognition site and the target site are the same sequence. Type-I enzymes bind to their recognition site and then cleave the DNA at some more or less random position outside that recognition site. *The opposite word is* restriction site.

Recombinant–A term used in both classical and molecular genetics.

1. In classical genetics: An organism or cell that is the result of recombination (crossing-over), *e.g.*, Parents: *AB/ab* and *ab/ab*; recombinant offspring: *Ab/ab*.

2. In molecular genetics: A molecule containing DNA from different sources. The word is typically used as an adjective, *e.g.*, recombinant DNA.

Recombinant DNA–The result of combining DNA fragments from different sources.

Recombinant DNA technology–A set of techniques which enable one to manipulate DNA. One of the main techniques is DNA cloning (because it produces an unlimited number of copies of a particular DNA segment), and the result is sometimes called a DNA clone or gene clone (if the segment is a gene), or simply a clone. An organism manipulated using recombinant DNA techniques is called a genetically modified organism (GMO).

Among other things, recombinant DNA technology involves:

–identifying genes;

–cloning genes;

–studying the expression of cloned genes; and

–producing large quantities of the gene product.

Recombinant protein–A protein whose amino acid sequence is encoded by a cloned gene.

Recombinant RNA–A term used to describe RNA molecules joined *in vitro* by T4 RNA ligase.

Recombinant toxin–A single multifunctional toxic protein that has been created by combining the coding regions of various genes.

Recombinant vaccine–A vaccine produced from a cloned gene.

Recombination–The process of crossing over, which occurs during meiosis I. It involves breakage in the same position of each of a pair of non-sister chromatids from homologous chromosomes, followed by joining of non-sister fragments, resulting in a reciprocal exchange of DNA between non-sister chromatids within an homologous pair of chromosomes.

Recombination fraction; recombination frequency–The proportion of gametes that have arisen from recombination between two loci. It is estimated as the number of recombinant individuals among a set of offspring of a particular mating, divided by the total number of offspring from that mating. Represented by the Greek letter theta (θ). Linkage maps are created from estimates of recombination fraction between all pair-wise combinations of loci. *Search in* map distance.

Reconstructed cell–A viable transformed cell resulting from genetic engineering.

Re-differentiation–Cell or tissue reversal from one differentiated type to another differentiated type of cell or tissue. *Search in* de-differentiation.

Reduction division–Phase of meiosis in which the maternal and paternal chromosomes of the bivalent separate. *Search in* equational division.

Regeneration (L. *re*, again + *generate*, to beget)–The growth of new tissues or organs to replace those injured or lost. In tissue culture, regeneration is used to define the development of organs or plantlets from a tissue, callus culture or from a bud. *Search in* conversion; micropropagation; organogenesis.

Regulator–Substance regulating growth and development of cells, organs, etc.

Regulatory gene–A gene whose protein controls the activity of other genes or metabolic pathways.

Rejuvenation–Reversion from adult to juvenile stage.

Relaxed circle–*Search in* nicked circle.

Relaxed circle plasmid–*Search in* plasmid.

Relaxed plasmid–A plasmid that replicates independently of the main bacterial chromosome and is present in 10-500 copies per cell. *Search in* plasmid.

Release factors–1. Soluble protein that recognizes termination codons in mRNAs and terminate translation in response to these codons.

2. A hormone that is produced by the hypothalamus and stimulates the release of a hormone from the anterior pituitary gland into the bloodstream.

Re-naturation–The re-association of two nucleic acid strands after denaturation. The restoration of a molecule to its native form. In nucleic acid biochemistry, this term usually refers to the formation of a double-stranded helix from complementary single-stranded molecules. Some simple proteins can also be re-natured and regain their function.

Rennin–An enzyme secreted by cells lining the stomach in mammals, and that is responsible for clotting milk.

Repeat unit–A sequence of bases that occurs repeatedly in the genome, often end-on-end, *i.e.*, tandemly.

Repetitive DNA–DNA sequences that are present in a genome in multiple copies, sometimes a million times or more.

Replacement; gene replacement–A method of substituting a cloned gene, or part of a gene, which may have been mutated *in vitro*, for the wild-type copy of the gene within the host's chromosome. *Search in* homogenotization.

Replacement therapy–The administration of metabolites, co-factors or hormones that are deficient as the result of a genetic disease.

Replica plating–A procedure for duplicating the bacterial colonies growing on agar medium in one Petri plate to agar medium in another Petri plate.

Replication–The synthesis of duplex (double-stranded) DNA by copying from a single-stranded template.

Replicative form (RF)–The molecular configuration of viral nucleic acid that is the template for replication in the host cell.

Replicon–The portion of a DNA molecule which is replicable from a single origin. Plasmids and the chromosomes of bacteria, phages and other viruses usually have a single origin of replication and, in these cases, the entire DNA molecule constitutes a single replicon. Eukaryotic chromosomes have multiple internal origins and thus contain several replicons. The word is often used in the sense of a DNA molecule capable of independent replication, *e.g.*, "The shuttle vector pJDB219 is a replicon in both yeast and *E. coli*."

Replisome–The complete replication apparatus present at a replication fork that carries out the semi-conservative replication of DNA.

Reporter gene–A gene that encodes a product that can readily be assayed. Thus reporter genes are used to determinate whether a particular DNA construct has been successfully introduced into a cell, organ or tissue.

Repressible enzyme–An enzyme whose synthesis is diminished by a regulatory molecule.

Repression–Inhibition of transcription by preventing RNA polymerase from binding to the transcription initiation site: a repressed gene is "turned off."

Repressor–A protein which binds to a specific DNA sequence (the operator) upstream from the transcription initiation site of a gene or operon and prevents RNA polymerase from commencing mRNA synthesis. Examples of repressors are the C_1 protein of bacteriophage and the *lac*1 protein of the *lac* operon.

Reproduction–The production of an organism, cell or organelle like itself (self propagation).

1. Sexual reproduction: the regular alternation (in the life-cycle of haplontic, diplontic and diplohaplontic organisms) of meiosis and fertilization (karyogamy) which provides for the production of offspring. The main biological significance of sexual reproduction lies in the fact that it achieves genetic recombination.

2. Asexual or agamic reproduction: the development of a new individual from either a single cell (agamospermy) or from a group of cells (vegetative reproduction) in the absence of any sexual process. *Search in also* apomixis.

Repulsion–The phase state in which a dominant (or wild-type) allele at one locus and a recessive (or mutant) allele at a second locus occur on the same chromosome. Also called trans configuration. *Search in* coupling.

Residues–The components of macromolecules, *e.g.*, amino acids, nucleotides.

Resistance–Term commonly used to describe the ability of an organism to withstand a stress, a force or an effect of a disease, or its agent or a toxic substance.

Resistance factor–A plasmid that confers antibiotic resistance to a bacterium.

Rest period–An endogenous physiological condition of viable seeds, buds or bulbs that prevents growth even in the presence of otherwise favourable environmental conditions. By some seed physiologists, this is referred to as dormancy.

Restitution nucleus–A nucleus with unreduced or doubled chromosome number that results from the failure of a meiotic or mitotic division.

Restriction endonuclease [enzyme]–A class of endonucleases that cleaves DNA after recognizing a specific sequence, *e.g.*, *Bam*H1 (5´GGAT CC3´), *Eco*RI (5´GAATTC3´), and *Hind*III (5´AAGCTT3´). There are three types of restriction endonuclease enzymes:

Type I: Cuts non-specifically a distance greater than 1000 bp from its recognition sequence and contains both restriction and methylation activities.

Type II: Cuts at or near a short, and often palindromic, recognition sequence. A separate enzyme methylates the same recognition sequence. They may make the cuts in the two DNA strands exactly opposite one another and generate blunt ends, or they may make staggered cuts to generate sticky ends. The type II restriction enzymes are the ones commonly exploited in recombinant DNA technology.

Type III: Cuts 24-26 bp downstream from a short, asymmetrical recognition sequence. Requires ATP and contains both restriction and methylation activities.

Restriction enzyme–*Search in* restriction nuclease.

Restriction exonuclease [enzyme]–A class of nucleases that degrades DNA or RNA, starting from an end either 5´ or 3´.

Restriction fragment–A fragment of DNA produced by cleaving (digesting, cutting) a DNA molecule with one or more restriction endonucleases.

Restriction fragment length polymorphism (RFLP)–The occurrence of variation in the length of DNA fragments that are produced after cleavage with a type II restriction endonuclease. The differences in DNA lengths are due to the presence or absence of recognition site(s) for that particular restriction enzyme. RFLPs were initially detected using hybridization with DNA probes after separation of digested genomic DNA by gel electrophoresis (Southern analysis). Now they are typically detected by electrophoresis of digested PCR product.

Restriction map–The linear array of restriction endonuclease sites on a DNA molecule. *Search in* mapping.

Restriction nuclease–A bacterial enzyme that cuts DNA at a specific site.

Restriction site–The specific nucleotide sequence in DNA that is recognized by a type II restriction endonuclease and within which it makes a double-stranded cut. Restriction sites usually comprise four or six base pairs that typically are palindromic,

e.g., 5´GGCC3´

3´CCGG5´

The two strands may be cut either opposite to one another, to create blunt ends, or in a staggered manner, giving sticky ends, depending on the enzyme involved. *Search in* restriction endonuclease.

Reticulocyte–A young red blood cell.

Retro-element–Any of the integrated retroviruses or the transposable elements that resemble them.

Retro-poson; retro-transposon–A transposable element that moves via reverse transcription (*i.e.*, from DNA to RNA to DNA) but lacks the long terminal repeat sequences.

Retroviral vectors–Gene transfer systems based on viruses that have RNA as their genetic material.

Retrovirus–A class of eukaryotic RNA viruses that can form double-stranded DNA copies of their genomes by using reverse transcription; the double-stranded forms integrate into chromosomes of an infected cell. Many naturally occurring cancers of vertebrate animals are caused by retroviruses. Also, the AIDS virus is a retrovirus.

Reversal transfer–Transfer of a culture from a callus-supporting medium to a shoot-inducing medium.

Reverse genetics–*Use* positional cloning.

Reverse mutation–*Search in* reversion.

Reverse transcriptase; RNA–dependent DNA polymerase–An enzyme that uses RNA molecule as a template for the synthesis of a complementary DNA strand.

Reverse transcription–The synthesis of DNA on a template of RNA, accomplished by reverse transcriptase.

Reversion; reverse mutation–Restitution of a mutant gene to the wild-type condition, or at least to a form that gives the wild phenotype; more generally, the appearance of a trait expressed by a remote ancestor.

RF–*Search in* replicative form.

RFLP–*Search in* restriction fragment length polymorphism.

Rhizobacterium–A micro-organism whose natural habitat is near, on or in plant roots.

Rhizobium (pl: rhizobia)–Prokaryote able to establish symbiotic relationship with leguminous plants, as a result of which elemental nitrogen is fixed or converted to ammonia. *Search in* nitrogen fixation.

Rhizosphere–The portion of soil in the immediate vicinity of growing plant roots.

Ribose–A monosaccharide ($C_5H_{10}O_5$) rarely occurring free in nature, but important as a component of RNA.

Ribonuclease–Any enzyme that hydrolyses RNA.

Ribonucleic acid–*Search in* RNA.

Ribosomal binding site–A sequence of nucleotides near the 5′ end of a bacterial mRNA molecule that facilitates the binding of the mRNA to the small ribosomal sub-unit. Also called the Shine-Delgarno sequence.

Ribosomal RNA–*Search in* rRNA.

Ribosome (*ribo*, from RNA + Gr. *soma*, body)–The sub-cellular structure that contains both RNA and protein molecules and mediates the translation of mRNA into protein. Ribosomes comprise large and small sub-units. *Search in* organelle; translation.

Ribozyme; gene shears–RNA molecule that can catalyse chemical reactions, often cutting other RNAs.

Ribulose–A keto-pentose sugar ($C_5H_{11}O_5$) that is involved in carbon dioxide fixation in photosynthesis.

Ribulose biphosphate (RuBP)–A five-carbon sugar that is combined with carbon dioxide to form a six-carbon intermediate in the first stage of the dark reaction of photosynthesis.

Rinderpest–Cattle plague; a viral infection of cattle, sheep and goats.

Ri plasmid–A class of large conjugative plasmids found in the soil bacterium *Agrobacterium rhizogenes*. Ri plasmids are responsible for hairy root disease of certain plants. A segment of the Ri plasmid is found in the genome of tumour tissue from plants with hairy root disease.

R-loops–Single-stranded DNA regions in RNA-DNA hybrids formed *in vitro* under conditions where RNA-DNA duplexes are more stable than DNA-DNA duplexes.

RNA Ribonucleic acid–An organic acid composed of repeating nucleotide units of adenine, guanine, cytosine and uracil, whose ribose components are linked by phosphodiester bonds. The information-carrying material in some viruses. More generally, a molecule derived from DNA by transcription that may carry information (messenger RNA (mRNA)), provide sub-cellular structure (ribosomal RNA (rRNA)), transport amino acids (transfer RNA (tRNA)) or facilitate the biochemical modification of itself or other RNA molecules. *Search in* antigen RNA; gene splicing; heterogeneous nuclear RNA (hnRNA); mRNA; ribosomal RNA; RNA polymerase; small nuclear RNA; transfer RNA.

RNA-dependent DNA polymerase–*Search in* reverse transcriptase.

RNA editing–Post-trancriptional processes that alter the information encoded in gene transcripts (RNAs).

RNA polymerase–An enzyme that catalyses the synthesis of RNA from a DNA template. *Search in* polymerase; RNA.

RNAase–*Search in* RNase.

RNase–Ribonuclease. A group of enzymes that catalyse the cleavage of nucleotides in RNA.

Roentgen (symbol: r)–Obsolete unit of ionizing radiation. The SI unit is the sievert (symbol: Sv; 1 Sv 8.4 r)

Root–The descending axis of a plant, normally below ground, which serves to anchor the plant and to absorb and conduct water and mineral nutrients.

Root apex–The apical meristem of a root; very similar to the shoot apical meristem in that it forms the three meristematic areas: the protoderm (developing into the epidermis); the procambium (which develops into the stele); and the growth meristem (which forms the cortex).

Root cap–A thimblelike mass of cells covering and protecting the apical meristem of a root.

Root culture–The culture of isolated root tips of apical or lateral origin to produce *in vitro* root systems with indeterminate growth habits. Root culture was among the first kinds of plant tissue cultures, and is still largely used in the study of developmental phenomena, and mycorrhizal, symbiotic and plant-parasitic relationships.

Root cutting–Cutting made from sections of roots alone.

Root hairs–Outgrowths from epidermal cell walls of the root specialized for water and nutrient absorption.

Root nodule–A small round mass of cells that is located on the roots of plants and contains nitrogen-fixing bacteria.

Root tuber–Thickened root that stores carbohydrates.

Root zone–The volume of soil or growing medium containing the roots of a plant. In soil science, the depth of the soil profile in which roots are normally found. *Search in* rhizosphere.

Rootstock–The trunk or root material to which buds or scions are inserted in grafting. *Search in* stock.

Rotary shaker–Rotating apparatus with a platform on which, containers can be shaken, such as Erlenmeyer flasks containing cells in liquid nutrient medium.

rRNA; ribosomal RNA–The RNA molecules which are essential structural and functional components of ribosomes, the organelles responsible for protein synthesis. The different rRNA molecules are known by their sedimentation (Svedberg; symbol S) values. *E. coli* ribosomes contain one 16S rRNA molecule (1541 nucleotides long) in the same (small) sub-unit and a 23S rRNA (2904 nucleotides) and a 5S rRNA (120 nucleotides) in the large sub-unit. These three rRNA molecules are synthesized as part of a large precursor molecule which also contains the sequences of a number of tRNAs. Special processing enzymes cleave this large precursor to generate the functional moieties. *Search in* RNA.

RuBP–*Search in* ribulose biphosphate.

Ruminant animals; ruminants–Animals having a rumen–a large digestive vat in which fibrous plant material is partially broken down by microbial fermentation, prior to digestion in a "true" stomach (the abomasum). There are also two other stomachs–the reticulum and the omasum. Typical ruminants are cattle and sheep.

Runner–A lateral stem that grows horizontally along the ground surface and gives rise to new plants either from axillary or terminal buds. *The opposite word is* stolon.

Rust–A generic descriptor for various plant diseases, especially those caused by a group of parasitic fungi of the phylum *Basidiomycota*, that attack the leaves and stems of crops.

S phase–The phase in the cell cycle during which DNA synthesis occurs.

S_1 mapping–*Search in* S_1 nuclease.

S_1 nuclease–An enzyme that specifically degrades RNA or single-stranded DNA to 5′ mononucleotides. Purified from the filamentous fungus *Aspergillus oryzae* S_1 nuclease is used in assessing the extent of a hybridization reaction by removing unpaired regions. It is also used to remove the sticky ends of restriction fragments. In S_1 mapping, the coding region of a gene is detected by performing mRNA-DNA hybridization and removing unpaired DNA with S_1 nuclease.

Saccharifaction–Following liquefaction, the hydrolysis of polysaccharides by glucoamylase to maltose and glucose.

Saline resistance–*Search in* salt tolerance.

Salmonella–A genus of rod-shaped, Gram-negative bacteria that are a common cause of food poisoning.

Salt tolerance; saline resistance–The ability to withstand a concentration of sodium (Na^+ ion), or of any other salt, in the soil (or in culture), which is damaging or lethal to other plants. Breeding and selection for increased tolerance and resistance in crop plants is of great current interest.

Sap–Fluid content of the xylem and phloem cells of plants. Fluid contents of the vacuole are referred to as cell sap.

Saprophyte–A vegetable organism that derives its nutriment from decaying organic matter.

Satellite DNA–That portion of the DNA in plant and animal cells consisting of highly repetitious sequences (millions of copies) typically in the range from 5 to 500 bases. Thousands of copies occur tandemly (end-on-end) at each of many sites. It can be isolated from the rest of the DNA by density gradient centrifugation.

Satellite RNA–A small, self-splicing RNA molecule that accompanies several plant viruses, including tobacco ringspot virus. a.k.a. viroid.

Saturates–Noun: Oils containing saturated fatty acids.

SCA (single chain antigen)–Antibody-binding domains in which the two chains are produced by a gene and linked by a short peptide. *Search in* dabs.

Scaffold–The central core structure of condensed chromosomes. The scaffold is composed of non-histone chromosomal proteins.

Scale up–Conversion of a process, such as fermentation of a micro-organism, from a small scale to a larger scale.

Scanning electron microscope (SEM)–An electron-beam based microscope used to examine, in a three dimensional screen image, the surface structure of prepared specimens.

Scarification–The chemical or physical treatment given to some seeds (where the seed coats are very hard or contain germination inhibitors) in order to break or weaken the seed coat sufficiently to permit germination.

Scientific name–A unique identifier consisting of a genus and a species name (the specific epithet) in Latin, assigned to each recognized and described species of organism. Based on the Linnean system of classification. Susceptible to considerable blurring at the edges because, as Darwin came to realize, nature has not packaged living things into neatly discrete entities. *The opposite word is* cultivar; pathovar.

Scion–The twig or bud to be grafted onto another plant, the root stock, in a budding or grafting operation.

Scion-stock interaction–The effect of a rootstock on a scion (and vice versa) in which a scion on one kind of rootstock performs differently than it would on its own roots or on a different rootstock.

Sclerenchyma (Gr. *skleros*, hard + *echyma*, a suffix denoting tissue)–A strengthening tissue in plants, composed of cells with heavily lignified cell walls.

SCP–*Search in* single-cell protein.

Scrapie–A disease of sheep; a spongiform encephalopathy. *Search in* proteinaceous infectious particle.

Screen–To separate by exclusion or collection on the basis of a set of criteria (biochemical, anatomical, physiological, etc.). Screening is often applied to the process of selection for specific purposes, such as disease resistance or improved agronomic qualities in plants, improved performance in animals, specific enzyme properties in micro-organisms, etc.

SDS (Sodium dodecyl sulphate)–*Search in* gel electrophoresis.

Secondary antibody–In an ELISA or other immunological assay system, the antibody that binds to the primary antibody. The secondary antibody is often conjugated with an enzyme such as alkaline phosphatase.

Secondary cell wall–The innermost layer of cell wall, with a highly organized microfibrillar structure, which is formed in certain cells after cell elongation has ceased. It gives rigidity to the cells.

Secondary growth–Type of growth characterized by an increase in thickness of stem and root and resulting from formation of secondary vascular tissues by the vascular cambium.

Secondary immune response–The rapid immune response that occurs during the second (and subsequent) encounters of the immune system of a mammal with a specific antigen. *The opposite word is* primary immune response.

Secondary metabolite–A compound that is not necessary for growth or maintenance of cellular functions but is synthesized, generally, for the protection of a cell or micro-organism, during the stationary phase of the growth cycle.

Secondary oocyte–*Search in* oocyte.

Secondary phloem–Phloem tissue formed by the vascular cambium during secondary growth in a vascular plant.

Secondary plant products–Metabolic products not having a known functional or structural use in plant cells. They have been extracted from plant tissue cultures for pharmaceutical and food processing purposes (*e.g.*, essential oils, food additives, flavours).

Secondary messenger–A chemical compound within a cell that is responsible for initiating the response to a signal from a chemical messenger (such as a hormone) that cannot enter the target cell itself.

Secondary root–A branch or lateral root.

Secondary spermatocyte–*Search in* spermatocyte.

Secondary thickening–Deposition of secondary cell wall materials which result in an increase in thickness in stems and roots.

Secondary vascular tissue–Vascular tissue (xylem and phloem) formed by the vascular cambium during secondary growth in a vascular plant.

Secondary xylem–Xylem tissue formed by the vascular cambium during secondary growth in a vascular plant.

Secretion–The passage of a molecule from the inside of a cell through a membrane into the periplasmic space, or the extracellular medium.

Seed–Botanically, a seed is the matured ovule without accessory parts.

Colloquially, a seed is anything which may be sown; *i.e.*, seed potatoes (which are vegetative tubers); seed of corn, sunflower, etc.

Seed storage proteins–Proteins accumulated in large amounts in seeds not because of their enzymatic or structural properties but simply as a convenient source of amino acid for use when the seed germinates. They are of interest to biotechnologists:

1. As a source of protein. Much of the world's food comes from plant seeds or fruits, and much of the protein in those seeds is storage protein. Thus a substantial amount of the world's food protein comes from plant storage protein. Any improvement of the nutritional content of those proteins could correspondingly improve human diet.

2. As expression systems. Storage proteins are produced in very large amounts relative to other proteins, and are stored in stable, compact bodies in the plant seed. Several workers are seeking to make the plants produce other proteins in similarly large amounts and in as convenient a form, by splicing the gene for a desired protein into the middle of a plant storage protein gene.

Segment-polarity gene–A gene that functions to define the anterior and posterior components of body segments in *Drosophila*.

Segregant–A hybrid resulting from the crossing of two genetically unlike individuals.

Segregation–The separation of the two members of a chromosome pair from each other at meiosis; the result is seen as the separation of alleles from each other in the gametes of heterozygotes; the occurrence of different phenotypes among offspring, resulting from chromosome or allele separation in their heterozygous parents. Mendel's first principle of inheritance (the Law of Segregation) predicts that heterozygotes will produce equal numbers of gametes containing each allele.

Selectable–Having a gene product that, when present, enables a researcher to identify and preferentially propagate a particular organism or cell type. *Search in* reporter gene.

Selectable marker–A gene whose expression allows the identification of:

1. A specific trait or gene in an organism.

2. Cells that have been transformed or transfected with a vector containing the marker gene. *Search in* β-lactamase, kanr.

Selection–1. Differential survival and reproduction phenotypes.

2. A system for either isolating or identifying specific organisms in a mixed culture.

Selection coefficient, *s*–The proportion by which the fitness of a genotype is less than the fitness of a standard genotype, which is usually the genotype with the highest fitness. In general, relative fitness = 1–*s*.

Selection culture–A selection based on difference(s) in environmental conditions or in culture medium composition, such that preferred variant cells or cell lines (presumptive or putative mutants) are favoured over other variants or the wild-type.

Selection differential, *S*–The difference between the mean of the individuals selected to be parents and the mean of the overall population.

Selection pressure–The intensity of selection acting on a population of organisms or cells in culture. Its effectiveness is measured in terms of differential survival and reproduction, and consequently in change in the frequency of alleles in a population.

Selection response–The difference between the mean of the individuals selected to be parents and the mean of their offspring. Predicted response = heritability (narrow-sense) × selection differential.

Selection unit–The minimum number of organisms or cells effective in the screening process.

Selective agent–An environmental or chemical agent characterized by its lethal or sub-lethal stress on growing plants, or portion thereof in culture. A selective agent is mainly used when selection of resistant or tolerant individuals is the research aim. *Search in* single-cell line.

Self-fertilization–The process by which pollen of a given plant fertilizes the ovules of the same plant. Plants fertilized in this way are said to have been selfed. An analogous process occurs in some animals, such as nematodes and molluscs.

Self-incompatibility–In plants, the inability of the pollen to fertilize ovules (female gametes) of the same plant.

Self-pollination–Pollen of a plant is transferred to the female part of the same plant or another plant with the same genetic makeup. Opposite: cross-pollination.

Self-replicating elements–Extra-chromosomal DNA elements that have origins of replication for the initiation of their own DNA synthesis.

Self-sterility–*Search in* self-incompatibility.

SEM–*Search in* scanning electron microscope.

Semen sexing–*Search in* sperm sexing.

Semi-conservative replication–During DNA duplication, each strand of a parent DNA molecule is a template for the synthesis of its new complementary strand. Thus, one half of a pre-existing DNA molecule is conserved during each round of replication. *The opposite word is* replisome.

Semi-continuous culture–Cells in an actively dividing state are maintained in culture by periodically draining off the medium and replenishing it with fresh medium.

Semi-liquid–*Search in* semi-solid.

Semi-permeable membrane–A cell or plasma membrane that is partially permeable; certain ions or molecules (water, solvents) can pass through it but others cannot (such as certain solutes).

Semi-solid–Gelled but not firmly so; small amounts of a gelling agent are used to obtain a semi-solid medium; called also semi-liquid.

Semi-sterility–A condition of only partial fertility in plant zygotes; usually associated with translocations.

Senescence–The last stage in the post-embryonic development of multicellular organisms, during which loss of functions and degradation of biological components occur. A physiological ageing process in which cells and tissues deteriorate and finally die.

Sense RNA–A primary transcript (RNA) that contains a coding region (contiguous sequence of codons) that is translated to produce a polypeptide.

Sensitivity–For diagnostic tests, the smallest amount of the target molecule that the assay can detect.

Sepsis–Destruction of tissue by pathogenic micro-organisms or their toxins, especially through infection of a wound. *The opposite word is* aseptic; axenic; sterile.

Septate (L. *septum*, fence)- Divided by cross walls into cells or compartments.

Septum (L. *septum*, fence)–Any dividing wall or partition; frequently a cross wall in a fungal or algal filament that separates two cavities or structures.

Sequence hypothesis–Francis Crick's seminal concept that genetic information exists as a linear DNA code; DNA and protein sequence are collinear.

Sequence-tagged site (STS)–Short, unique DNA sequence (usually 200 to 500 bp) that, by being able to be amplified by PCR, is uniquely "tagged" to the site on the chromosome from which it was amplified.

Sequencing–The determination of the order of nucleotides in a DNA or RNA molecule, or that of amino acids in a polypeptide chain. *Search in* DNA sequencing.

Serial divisions–Splitting at about monthly intervals of excised shoot-tip material growing on culture medium, in order to induce additional plantlets.

Serial float culture–A technique of floating anthers on liquid medium developed by Sunderland. Anther dehiscence, pollen release and development occur at intervals of several days, and in different nutrient media.

Serology (adj: serological)–The study of serum reactions between an antigen and its antibody. Serology is mainly used to identify and distinguish between antigens, such as those specific to micro-organisms or viruses. Serology is also employed as an indicator technique to assay plants suspected of being virus-infected.

Serum albumin–A globular protein obtained from blood and body fluids.

Sewage treatment–Sewage treatment is one of the most widespread biotechnological processes in Western societies, to deal with the huge amounts of human and animal waste that such societies produce. Sewage treatment methods vary widely, but all have a biological basis to break down the organic material in sewage and convert it into something that can be safely discharged into the environment (usually rivers or seas).

Sex chromosomes–Chromosomes that are connected with the determination of sex: X and Y chromosomes in mammals; W and Z chromosomes in birds.

Sex determination–The method by which the distinction between males and females is established in a species.

Sexduction–The incorporation of bacterial genes into F factors and their subsequent transfer, by conjugation, to a recipient cell.

Sex hormones–Hormones secreted in the body that influence several characters in the individuals.

Sex linkage–The location of a gene on a sex chromosome, typically on the X chromosome.

Sex mosaic–*Search in* gynandromorph.

Sexed embryos–Embryos separated according to sex.

Sex factor–A bacterial episome (*e.g.*, the F plasmid in *E. coli*) that enables the cell to be a donor of genetic material. The sex factor may be propagated in the cytoplasm, or it may be integrated into the bacterial chromosome.

Sex-influenced dominance–The tendency for the type of gene action to vary between the sexes within a species. Thus the presence of horns in some breeds of sheep appears to be dominant in males and recessive in females.

Sex-limited–Expression of a trait in only one sex; *e.g.*, milk production in mammals; egg production in chickens.

Sexual reproduction–The process where two cells (gametes) fuse to form one fertilized cell or zygote. *The opposite word is* asexual reproduction; gamete; hybrid.

Shake culture–An agitated suspension in culture providing adequate aeration for cells in the liquid medium. Usually an Erlenmeyer flask containing the culture is attached to a horizontal or platform shaker, or agitated with a magnetic stirrer.

Shaker; platform shaker–A platform fitted with clips for grasping Erlenmeyer flasks, with set or variable speed control. Shaking speed must be adjusted for gentle and even agitation of suspension cultures. *The opposite word is* reciprocating shaker.

Shear–1. The sliding of one layer across another, with deformation and fracturing in the direction parallel to the movement. This term usually refers to the forces that cells are subjected to in a bioreactor or a mechanical device used for cell breakage.

2. To fragment DNA molecules into smaller pieces. DNA, as a very long

and fairly stiff molecule, is very susceptible to hydrodynamic shear forces. Forcing a DNA solution through a hypodermic needle will fragment it into small pieces. The size of the fragments obtained is inversely proportional to the diameter of the needle's bore. The actual sites at which the shear force breaks a DNA molecule are approximately random. Therefore DNA fragments may be generated by random shear and then cloned (by either tailing their ends or using linkers) so as to create a complete gene library of an organism. This method is little used now, having been replaced by the use of partial digests with four-base-pair cutters, such as *Sau*3A, as a means of generating random DNA fragments.

Shine-Dalgarno sequence–A conserved sequence in prokaryotic mRNAs that is complementary to a sequence near the 5′ terminus of the 16S ribosomal RNA and is involved in the initiation of translation. *Search in* ribosomal binding site.

Shoot–A young branch that grows out from the main stock of a tree, or the young main portion of a plant growing above ground.

Shoot differentiation–The development of growing points, leaf primordia and finally shoots from a shoot tip, axial bud, or even a callus surface.

Shoot-tip graft; micrograft–A shoot tip or meristem tip is grafted onto a prepared seedling or micro-propagated rootstock in culture. Meristem tip grafting is mainly used for *in vitro* virus elimination with *Citrus* spp. and other plants.

Shoot tip; shoot apex–The terminal bud (0.1–1.0 mm) of a plant, which consists of the apical meristem (0.05–0.1 mm) and the immediate surrounding leaf primordia and developing leaves and adjacent stem tissue.

Short template–A DNA strand that is synthesized during the polymerase chain reaction and has a primer sequence at one end and a sequence complementary to the second primer at the other end.

Short-day plant–Plant that requires a night (or dark period) longer than its critical dark period to induce flower formation.

Shuttle vector; bifunctional vector–A plasmid capable of replicating in two different host organisms because it carries two different origins of replication and can therefore be used to 'shuttle' genes from one to the other. For example, the YEp, pJDB219, is a shuttle vector able to replicate in *E. coli* from its pMB9 origin and in *Saccaromyces cerevisiae* from its 2 μm-plasmid origin.

Sib-mating–Crossing of siblings. Matings involving two individuals of the same parentage; brother-sister matings.

Siderophore–A low molecular weight substance that binds very tightly to iron. Siderophores are synthesized by a variety of soil micro-organisms to ensure that the organism is able to obtain sufficient amounts of iron from the environment.

Sieve cell–In vascular plants, a long and slender sieve element with relatively unspecialized sieve areas and with tapering end walls that lack sieve plates.

Sieve plate–Perforated wall area in a sieve-tube element through which strands connecting sieve-tube protoplasts pass.

Sievert (symbol: Sv)–The SI unit of ionizing radiation.

Sieve tube–A tube within the phloem tissue of a plant, and composed of joined sieve elements which is specialized for the conduction of food materials.

Sigma factor–The sub-unit of prokaryotic RNA polymerases that is responsible for the initiation of transcription at specific initiation sequences.

Signal peptide–*Search in* signal sequence.

Signal sequence–A segment of about 15 to 30 amino acids at the N terminus of a protein, that enables the protein to be secreted (pass through a cell membrane). The signal sequence is removed as the protein is secreted. Also called signal peptide, leader peptide.

Signal transduction–The biochemical events that conduct the signal of a hormone or growth factor from the cell exterior, through the cell membrane, and into the cytoplasm. This involves a number of molecules, including receptors, ligands and messengers.

Signal-to-noise ratio–A specifically produced response compared to the response level when no specific stimulus (activity) is present.

Silencer–A DNA sequence that helps to reduce or shut off the expression of a nearby gene.

Simple sequence repeat (SSR)–*Search in* microsatellite.

Simplicity–For diagnostic tests, the ease with which an assay can be implemented.

SINEs–Short interspersed nuclear elements. Families of short (150 to 300 bp), moderately repetitive elements of eukaryotes, occurring about 100,000 times in a genome. SINES appear to be DNA copies of certain tRNA molecules, created presumably by the unintended action of reverse transcriptase during retroviral infection.

Single-cell line; cell strain–A culture initiated from a single cell, usually from suspension cultures of single cells or small aggregates plated on solidified medium. The latter may incorporate a selective agent, from which tolerant or resistant individual cell lines or cell clones can be selected. *Search in* selective agent.

Single-cell protein (SCP)–Protein produced by micro-organisms. The dried mass of a pure sample of a protein-rich-micro-organism, which may be used either as feed (for animals) or as a food (for humans).

Single-chain antigen–*Search in* SCA.

Single copy–A gene or DNA sequence which occurs only once per (haploid) genome. Most structural genes, those encoding functional proteins, are single-copy genes.

Single-node culture–Culture of separate lateral buds with each carrying a piece of stem tissue.

Single-nucleotide polymorphism (**SNP**; pronounced "**snip**")–A polymorphism at a particular base site in a coding sequence, *e.g.*, at base 306 in a particular gene, one individual could be heterozygous for A and G: the maternal allele could have an A at this site, while the paternal allele has a G at this site. This type of polymorphism is extensive throughout the genome, and has the great advantage of being detectable without the need for gel electrophoresis, which opens the way for large-scale automation of genotyping.

Single-strand-DNA-binding protein–A protein that coats DNA single strands, keeping them in an extended state.

Single-stranded–A term used to describe nucleic acid molecules consisting of only one polynucleotide chain. The genomes of certain phages, *e.g.*, MI3, are single-stranded DNA molecules; rRNA, mRNA and tRNA are all single-stranded nucleic acids, but they all contain double-stranded regions formed by the intra-strand base-pairing of self-complementary sequences.

Sires–Male animals used for breeding.

Site-directed mutagenesis–The introduction of base changes–mutations–into a piece of DNA at a specific site, using recombinant DNA methods.

Site-specific–A term used to describe any process or enzyme which acts at a defined sequence within a DNA or RNA molecule. Type II restriction enzymes are site-specific endonucleases and the recombination systems encoded by some transposons are site-specific, such as is the integration of phage into the *E. coli* chromosome.

Site-specific mutagenesis–A technique to change one or more specific nucleotides within a cloned gene in order to create an altered form of a protein with one or more specific amino acid changes. Oligonucleotide-directed muta-genesis; oligonucleotide-directed site-specific mutagenesis.

Six-base cutter–A type II restriction endonuclease that binds (and subsequently cleaves) DNA at sites that contain a sequence of six nucleotide pairs that is uniquely recognized by that enzyme. Because any sequence of six bases occurs less frequently by chance than any sequence of four bases, six-base cutters cleave less frequently than do four-base cutters. Thus, six-base cutters create larger fragments than four-base cutters.

Small nuclear ribonucleoprotein [**snRNP**; pronounced **snurp(s)**]–A compound comprising small nuclear RNA and nuclear protein, that is heavily involved in the post-transcriptional processing of mRNA, especially the removal of introns. snRNPs are a major component of spliceosomes.

Small nuclear RNA (snRNA)–Short RNA transcripts of 100-300 bp that associate with proteins to form small nuclear ribonucleoprotein particles (snRNPs) ; most snRNAs are components of the spliceosomes that excise introns from pre-mRNAs in RNA processing. *Search in* RNA.

SNP–*Search in* single nucleotide polymorphism.

snRNA–*Search in* small nuclear RNA.

snRNP–*Search in* small nuclear ribonucleoprotein.

Sodium dodecyl sulphate (SDS)–*Search in* gel electrophoresis.

Soil amelioration–The improvement of poor soils, usually using bacteria or fungi. This contrasts with bioremediation, which is the cleaning up of toxins, usually in soils. Amelioration includes breaking down organic matter; forming humus; by solubilizing them, making minerals–such as phosphates–in the soil available to plants; fixing nitrogen; and sometimes an element of bioremediation as well.

Soilless culture or soil-free culture–Tissue culture and hydroponics. Growing plants in nutrient solution without soil.

Solid media–Nutrient media that has been solidified, such as by addition of agar.

Somaclonal variation–Epigenetic or genetic changes, sometimes expressed as a new trait, resulting from *in vitro* culture of higher plants.

Somatic–Referring to vegetative or non-sexual stages of a life-cycle.

Somatic cell–Any cell of a multicellular organism that composes the body of that organism but does not produce gametes. *The opposite word is* gamete; somatic cell gene therapy.

Somatic cell embryogenesis–Embryos are produced either from somatic cells of explants (direct embryogenesis) or by induction on callus formed by explants (indirect embryogenesis).

Somatic cell gene therapy–The delivery of a gene or genes to a tissue other than reproductive cells of an individual, with the aim of correcting a genetic defect. *The opposite word is* somatic cell.

Somatic cell variant–A somatic cell with unique characters not present in the other cells, such as might be selected for in a screening trial that following a mutation event.

Somatic embryo; somatic embryoid–An organized embryonic structure morphologically similar to a zygotic embryo but initiated from somatic (non-zygotic) cells. Under *in vitro* conditions, somatic embryos go through developmental processes similar to embryos of zygotic origin.

Somatic hybridization–1. Asexual fusion of protoplasts from somatic cells of genetically different parents.

2. Hybridization by induced fusion of cells (protoplasts) from two contrasting genotypes for production of hybrids or cybrids which contain various mixtures of nuclear and/or cytoplasmic genomes, respectively.

Somatic hypermutation–A high frequency of mutation that occurs in the gene segments encoding the variable regions of antibodies during the differentiation of B lymphocytes into antibody producing plasma cells.

Somatic reduction–Halving of the chromosomal number of somatic cells; a possible method of producing "haploids" from somatic cells and calluses by artificial means.

Somatocrinin–Growth-hormone-releasing hormone. *Search in* growth hormone.

Somatostatin–Growth-hormone-inhibiting hormone. *Search in* growth hormone.

Somatotrophin; somatotropin–*Search in* growth hormone.

Sonication–Disruption of cells or DNA molecules by high frequency sound waves. a.k.a. ultrasonication.

SOS response–The synthesis of a whole set of DNA repair, recombination and replication proteins in bacteria containing severely damaged DNA (*e.g.*, following exposure to UV light).

Source DNA–The DNA from an organism that contains a target gene; this DNA is used as starting material in a cloning experiment.

Source organism–A bacterium, plant or animal from which DNA is purified and used in a cloning experiment.

Southern blot–A cellulose or nylon membrane to which DNA fragments previously separated by gel electrophoresis have been transferred by capillary action. Named after Ed Southern.

Southern blotting–A technique for transferring denatured DNA molecules that have been separated electrophoretically, from a gel to a matrix (such as a nitrocellulose membrane) on which a hybridization assay can be performed.

Southern hybridization–A procedure in which a cloned, labelled segment of DNA is hybridized to DNA restriction fragments on a Southern blot.

Southern transfer–*Search in* Southern blotting.

Speciation–The development of one or more species from an existing species.

Sparger–A device that introduces into a bioreactor air in the form of separate fine streams.

Specialized–Anatomically or physiologically adapted for particular functions or habitats.

Species (L. *species*, appearance, form, kind)–A class of potentially interbreeding individuals that are reproductively isolated from other such groups having many characteristics in common. A somewhat arbitrary and sometimes blurred classification; but still quite useful in many situations.

Specificity–For diagnostic tests, the ability of a probe to react precisely with a specific target molecule.

Spent medium–After each sub-culture, the medium is discarded because it has been depleted of nutrients, dehydrated or accumulated toxic metabolic products.

Sperm–Abbreviation of spermatozoon.

Spermatid–One of the four cells formed by the meiotic divisions in spermatogenesis. Spermatids become mature spermatozoa (sperm).

Spermatocyte–Sperm mother cell. The cell that undergoes two meiotic divisions (spermatogenesis) to form four spermatids; the *primary* spermatocyte before completion of the first meiotic division; the *secondary* spermatocyte after completion of the first meiotic division.

Spermatogenesis–The series of cell divisions in the testis by which maturation of the gametes (sperm) of the male takes place.

Spermatogonium (pl: spermatogonia)–Primordial male germ cell that may divide by mitosis to produce more spermatogonia. A spermatogonium may enter a growth phase and give rise to a primary spermatocyte.

Spermatozoon (pl: spermatozoa; abbr: sperm)–The mature, mobile reproductive cell of male animals, produced by the testis.

Sperm competition–Competition between different spermatozoa to reach and fertilize the egg cell of a single female.

Sperm sexing–The separation of sperm into those bearing an X chromosome and those bearing a Y chromosome, in order to be able to produce, via artificial insemination or *in vitro* fertilization, animals of a specified sex. Achieved by means of inactivation of X-bearing or Y-bearing sperm via antibodies directed against sex-specific peptides on the surface of sperm cells, or fluorescence-activated cell sorting (FACS), in which sperm that have been pre-treated with a fluorescent dye that binds to DNA are separated according to the quantity of fluorescence detected by a laser beam, based on the principle that X-bearing sperm contain more DNA than Y-bearing sperm.

Spharoblast–Nodule of wood which can give rise to adventitious shoots with juvenile characteristics.

Spheroplast (formerly also sphaeroplast)–A microbial or plant cell from which most of the cell wall has been removed, usually by enzymic treatment. Strictly, in a spheroplast, some of the wall remains, while in a protoplast the wall has been completely removed. In practice, the two words are often used interchangeably. *Search in* protoplast.

Spike (L. *spica*, an ear of grain)–An inflorescence in which the main axis is elongated and the flowers are sessile.

Spikelet (L. *spica*, an ear of grain + diminutive ending–let)–The unit of inflorescence in grasses; a small group of grass flowers.

Spindle (A.S. *spinel*, a tool for spinning thread by hand)–In mitosis and meiosis, refers to the spindle-shaped intracellular structure in which the chromosomes move.

Spine–Hard, sharp structure on the surface of a plant; usually a modified leaf.

Spliceosomes Organelles responsible for the removal of introns from mRNA by means of splicing.

Splicing–During the maturation of eukaryotic mRNA, the process that eliminates intervening intron sequences and covalently joins exon sequences of RNA. *The opposite word is* split gene; exon; guide sequence. In recombinant DNA technology, the term refers to the latter of the two processes just described, namely joining fragments of DNA together. *Search in* gene splicing.

Split genes–In eukaryotes, structural genes are typically divided up (split) by a number of non-coding regions called introns. *The opposite word is* exon; guide sequence; splicing.

Spore–1. A reproductive cell that develops into an individual without union with other cells; some spores such as meiospores occur at a critical stage in the sexual cycle, but others are asexual in nature.

2. A small, protected reproductive form of a micro-organism, often synthesized when nutrient levels are low.

Spririllum–A rigid, spiral-shaped bacterium.

Spirochaete–A non-rigid, corkscrew-shaped bacterium that moves by means of muscular flexions of the cell.

Spore mother cell–*Search in* sporocyte.

Sporocyte–A diploid cell that gives rise to four haploid spores by meiosis.

Sporophyte–The diploid generation in the life cycle of a plant, and that produces haploid spores by meiosis.

Sport–An individual or portion thereof distinguished by a spontaneous mutation or new form due to mutation. Sports are sometimes of great agricultural worth, but alternatively, they may be disadvantageous and may be rogued during agricultural production.

SSR simple sequence repeat–*Search in* microsatellite.

Stages of culture (I-IV)–*Search in* micropropagation.

Staggered cuts–Symmetrically cleaved phospho-diester bonds that lie on both strands of duplex DNA but are not opposite one another.

Stamen (L. *stamen*, the standing-up things or a tuft of thready things)–Flower structure made up of an anther (pollen-bearing portion) and a stalk or filament. The stamen is the male part of the flower.

Standard deviation–A statistical measure of variability in a population of individuals or in a set of data; the square root of the variance.

Standard error–A statistical measure of variation in a population of means, used to indicate how well sample estimates represent population parameters.

Starch (M.E. *strechen*, to stiffen)–A complex insoluble carbohydrate, consisting of various proportions of two glucose polymers, amylose and amylopectin; the chief food storage substance of plants, which is composed of several hundred hexose sugar units and which easily breaks down on hydrolysis into these separate units.

Sporophyll–A leaf that bears spore producing structures (sporangia).

Start codon; initiator codon–The set of three nucleotides in an mRNA molecule with which the ribosome starts the process of translation. The start codon sets the reading frame for translation. The most commonly used start codon is AUG, which is decoded as methionine in eukaryotes and as *N*-formyl-methionine in prokaryotes. AUG appears to be the only start codon used by eukaryotes, while in bacteria, GUG (valine) may sometimes be employed. *Search in* initiation codon; initiator.

Stationary culture–A culture maintained in the growth chamber with no agitation movement. The antonym is shake culture.

Stationary phase–The plateau of the growth curve after log growth, during which cell number remains constant. New cells are produced at the same rate as older cells die. *Search in* growth phases.

Statistic–An estimate based on a sample or samples of a population, providing an indication of the true population parameter.

Steady state–In a continuous fermentation process, the condition when the number of cells that are removed with the outflow is exactly balanced by newly synthesized cells.

Stele (Gr. *stele*, a post)–The central cylinder, inside the cortex, of roots and stems of vascular plants. *Search in* meristele.

Stem (O.F. *stemn*)–The main body of the above-ground portion of a tree, shrub, herb or other plant; the ascending axis, whether above or below ground, of a plant.

Stem cell–An undifferentiated active somatic cell that undergoes division and gives rise to other stem cells or to cells that differentiate to form specialized cells.

Sterile–1. Medium or object with no perceptible or viable micro-organisms.

2. Incapable of fertilizing or being infertile.

Sterile room–Operation room for performing inoculations under aseptic conditions; usually now replaced by use of laminar air-flow cabinets, in which filtered air is blown from the inside to the outside. *Search in* laminar air-flow cabinet.

Sterility–Inability to produce offspring.

Sterilize 1. The process of elimination of micro-organisms, such as by chemicals, heat, irradiation or filtration.

2. The operation of making an animal incapable of producing offspring.

Sterilization–The act of sterilizing.

Steward bottle–Flask developed by Steward for the growth of cells and tissues in a liquid medium, in which they can be periodically submerged during rotation.

Sticky ends; cohesive ends–The single-stranded nucleotide sequence left on a restriction fragment by type II restriction enzymes that cut each strand at a separate location. These unpaired regions are available for hybridization with complementary ends on other fragments during the creation of recombinant DNA. a.k.a. protruding end; overhang; cohesive end. *Search in* cohesive ends.

Stigma(L. *stigma*, a prick, a spot, a mark)–Receptive portion of the style, to which pollen adheres.

Stirred-tank fermenter–A growth vessel in which cells or micro-organisms are mixed by mechanically-driven impellers.

Stock–The lower portion of a graft. *Search in* rootstock.

Stock plant–The source plant from which cuttings or explants are made. Stock plants are usually maintained carefully in an optimum state for (sometimes prolonged) explant use. Preferably they are certified-pathogen-free plants.

Stock solutions–Pre-prepared solutions of individual components and used to prepare many different types of media. Certain substances, including Ca and Mg, sulphates and phosphates must not be combined until actual medium preparation because insoluble combinations are formed and precipitate.

Stolon (L. *stolo*, a shoot)–A lateral stem that grows horizontally along the ground surface. The runners of white clover, strawberry and bermuda grass are examples of stolons. *The opposite word is* runner.

Stoma (Gr. *stoma*, mouth; pl: stomata)–1. Any of various small openings or pores in an animal body, especially an opening resembling a mouth in various invertebrates.

2. Botany: A minute pore in the epidermis of the leaf or stem of a plant, forming a slit of variable width between two specialized cells (guard-cells), which allows movement of gases, including water vapour, to and from the intercellular spaces. Also, the whole pore with its associated guard-cells.

Stomatal complex–Includes the stoma, together with its guard cells and, when present, the subsidiary cells.

Stomatal index–(number of stomata per mm^2 × 100)/(number of stomata per mm^2 + number of epidermal cells per mm^2). This value has been found useful in comparing leaves of different sizes. Relative humidity and light intensity during leaf development affect the value of stomata index.

Stop codon; termination codon–A set of three nulecotides for which there is no corresponding tRNA molecule to insert an amino acid into the polypeptide chain. Protein synthesis is hence terminated and the completed polypeptide released from the ribosome. Three stop codons are found: UAA (ochre), UAG (amber) and UGA (opal). Mutations which generate any of these three codons in a position which normally contains a codon specifying an amino acid are known as *nonsense mutations*. Stop codons can also be called nonsense codons. *Search in* chain terminator; nonsense mutation; suppressor.

Strain–A group of individuals from a common origin within a species.

Strain isolation–Isolation of any bacterium, animal or plant from the outside world.

Stratification–Treatment of moist seeds at low temperature (+2°–+4°C) to break dormancy.

Stress–*Search in* water stress.

Stringency–Reaction conditions–notably temperature, salt concentration(s) and pH–that dictate the annealing of single-stranded DNA/DNA, DNA/RNA and RNA/RNA hybrids. At high stringency, duplexes form only between strands with perfect one-to-one complementarity; lower stringency allows annealing between strands with some degree of mismatch between bases.

Stringent plasmid–A plasmid that only replicates along with the main bacterial chromosome and is present as a single copy, or at most several copies, per cell. *Search in* plasmid.

Stroma–Tissue that forms the framework of an organ.

Structural gene–A DNA sequence that forms the blueprint for the synthesis of a polypeptide.

Structure-functionalism–The scientific tradition that stresses the relationship between a physical structure and its function, *e.g.*, the related disciplines of anatomy and physiology.

STS–*Search in* sequence-tagged site.

Style (Gr. *stylos*, a column)–Slender column of tissue that arises from the top of the ovary and through which the pollen tube grows.

Sub-clone–A method in which smaller DNA fragments are cloned from a large insert which has already been cloned in a vector.

Sub-cloning–1. Splicing part of a cloned DNA molecule into a different cloning vector.

2. The process of transferring a cloned DNA fragment from one vector to another. *Search in* cloning.

Sub-culture–Division and transfer of a portion or inoculum of a culture to fresh medium. Sometimes used to denote the adding of fresh liquid to a suspension culture. *The opposite word is* inoculum; passage.

Sub-culture interval–The time between subsequent sub-cultures of cells. Sub-culture interval has no relationship to the term *cell generation time*.

Sub-culture number–The number of times cells, etc., have been sub-cultured, *i.e.*, transplanted by inoculation from one culture vessel to another.

Sub-species–Population(s) of organisms sharing certain characteristics that are not present in other populations of the same species.

Sub-strain–Derived from a strain by isolating a single cell or groups of cells having properties or markers not shared by all cells of the strain.

Substrate–1. A compound that is altered by an enzyme.

2. Food source for growing cells or micro-organisms.

3. Material on which a sedentary organism lives and grows.

Sub-unit vaccine–One or more immunogenic proteins either purified from the disease-causing organism or produced from a cloned gene. A vaccine composed of a purified antigenic determinant that is separated from the virulent organism. *The opposite word is* vaccine; enzyme.

Sucker–A shoot that arises from an underground root or stem and grows at the expense of the parent plant. *The opposite word is* turion.

Suckering–Type of vegetative propagation where lateral buds grow out to produce an individual that is a clone of the parent.

Sucrose density gradient centrifugation–A procedure used to fractionate mRNAs or DNA fragments on the basis of size.

Substitution–A point mutation in which one base pair in the DNA sequence is replaced by another.

Superbug–Jargon for the bacterial strain of *Pseudomonas* developed by Chakrabarty, who combined hydrocarbon-degrading genes carried on different plasmids into one organism. Although this genetically engineered microorganism is neither "super" nor a "bug", it represents a landmark example because it showed how genetically modified microbial strains could be used in a novel way and because it was the basis for the precedent-setting legal decision that declared that genetically engineered organisms were patentable.

Supercoil–A DNA molecule that contains extra twists as a result of overwinding (positive supercoils) or underwinding (negative supercoils).

Supercoiled plasmid–The predominant *in vivo* form of plasmid, in which the plasmid is coiled around histone-like proteins. Supporting proteins are stripped away during extraction from the bacterial cell, causing the plasmid molecule to supercoil around itself *in vitro. Search in* Plasmid.

Supergene–A group of neighbouring genes on a chromosome and that tend to be inherited together and sometimes are functionally related.

Supernatant–The soluble liquid fraction of a sample after centrifugation or precipitation of insoluble solids.

Suppressor–Mutations in suppressor genes are able to overcome (suppress) the effects of mutations in other, unlinked, genes. A common, and very useful, kind of suppressor mutation occurs within the gene encoding a tRNA molecule and results in a change in the tRNA's anticodon. Such a mutant tRNA can reverse the effect of chain-terminating mutations, such as amber or ochre, in protein-encoding genes. *Search in* stop codon; nonsense mutation.

Suppressor mutation–A mutation that partially or completely cancels the phenotypic effect of another mutation.

Suppressor-sensitive mutant–An organism that can grow when a second genetic factor–a suppressor–is present, but not in the absence of this factor.

Surfactant–A surface-active agent or wetting agent, such as Tween 20TM or Tween 80TM, TeepolTM, Lissapol FTM, AlconoxTM, etc. Surfactants act by lowering the surface tension and are common addenda to solutions used to surface sterilize materials prior to aseptic excision of explants. *Search in* detergent; wetting agent.

Susceptible–The characteristic of a host organism such that it is incapable of suppressing or retarding an injurious pathogen or other factor.

Suspension culture–A type of culture in which (single) cells and/or clumps of cells grow and multiply while suspended in a liquid medium. *Search in* cell suspension.

Sustainable intensification of animal production systems–The manipulation of inputs to, and outputs from, livestock production systems aimed at increasing production and/or productivity and/or changing product quality, while maintaining the long-term integrity of the systems and their surrounding environment, so as to meet the needs of both present and future human generations. (Source: FAO, 1999)

Sustainable agricultural intensification respects the needs and aspirations of local and indigenous people, takes into account the roles and values of their locally adapted genetic resources, and considers the need to achieve long-term environmental sustainability within and beyond the agro-ecosystem. (*Source*: FAO, 1999)

Symbiont–An organism living in symbiosis with another, dissimilar organism.

Symbiosis (Gr. *syn*, with + *bios*, life)–The beneficial association of two different kinds of living organisms where there is benefit to both or where both receive an advantage from the association. An example is the association of the mycelium of mycorrhizal fungi with roots of seed plants. *Search in* commensalism; parasitism.

Symbiotic association–An intimate partnership between two organisms, in which the mutual advantages normally outweigh the disadvantages.

Sympatric speciation–The formation of new species by populations that inhabit the same or overlapping geographic regions.

Sympodial–A type of plant development in which the terminal bud of the stem stops growing due to either its abortion or its development into a flower or an inflorescence, and the uppermost lateral bud takes over the further axial growth of the stem.

Symplast–The system of protoplasts in plants, that are interconnected by plasmodesmata.

Synapsis–*Search in* pairing.

Synaptonemal complex–A ribbonlike protein structure formed between synapsed homologues at the end of the first meiotic prophase, binding the chromatids along their length and facilitating chromatid exchange.

Synchronous culture–A culture in which the majority of the cells are dividing at the same time or are at a specific phase of the cell cycle.

Syncytium–A group of cells in which cytoplasmic continuity is maintained. *The opposite word is* acellular.

Syndrome–A group of symptoms that occur together and represent a particular disease.

Syngamy–Fertilization.

Synergids–The two nuclei within the embryo sac at the upper end in the ovule of the flower, which, with the third (the egg), constitute the egg apparatus.

Synkaryon–A nucleus formed by the fusion of nuclei from two different somatic cells during somatic-cell hybridization. *Search in* hybrid cell.

Synteny–The occurrence of two or more loci on the same chromosome, without regard to the distance between them.

T–The thymine residue in DNA.

Tag–*Search in* label.

Tailing–The *in vitro* addition of the same nucleotide by the enzyme terminal transferase, to the 3´-hydroxyl ends of a duplex DNA molecule. a.k.a. homopolymeric trailing.

Tandem array–The existence of two or more identical DNA sequences in series, *i.e.*, end–to-end.

Tank bioreactor–Vessel in which fermentation takes place. A tank bioreactor is a vessel in which a micro-organism is grown in a large volume of liquid. This contrasts with fibre or membrane bioreactors and immobilized cell reactors. The large majority of bioreactors used in biotechnology are tank bioreactors, and most tank bioreactors are stirred-tank bioreactors, because stirring helps to distribute effectively gas and nutrients to the growing organism.

Tap root–Root system in which the primary root has a much larger diameter than the lateral roots. Opposite: fibrous root.

***Taq* polymerase**–A heat-stable DNA polymerase isolated from the thermophilic bacterium *Thermus aquaticus* and used in PCR. *Search in* polymerase.

Target–For diagnostic tests, the molecule or nucleic acid sequence that is being sought in a sample.

Target site duplication–A sequence of DNA that is duplicated when a transposable element inserts; usually found at each end the insertion.

Targeted drug delivery–A method of delivering a drug to the site in the body where it is needed, rather than allowing it to diffuse into many sites.

Targeting vector–A cloning vector carrying a DNA sequence capable of participating in a crossing-over event at a specified chromosomal location in the host cell.

TATA box–A conserved adenine- and thymine-rich promoter sequence located 25-30 bp upstream of a gene, which is the binding site of RNA polymerase. *Search in* Pribnow box.

Tautomeric shift–The transfer of a hydrogen atom from one position in an organic molecule to another position.

Tautomerism–A type of isomerism in which the two isomes are in equilibrium.

T cells; T lymphocyte–Lymphocyte that pass through the thymus gland during maturation. Different kinds of T cells play important roles in the immune response, being primarily responsible for the T cell-mediated response or cellular immune response.

T-cell receptor–An antigen-binding protein that is located on the surfaces of killer T cells and mediates the cellular immune response of mammals. The genes that encode T cell antigens are assembled from gene segments by somatic recombination processes that occur during T lymphocyte differentiation.

T-cell-mediated (cellular) immune response–The synthesis of antigen-specific T cell receptors and the development of killer T cells in response to an encounter of immune system cells with a foreign immunogen.

T-DNA–The segment of DNA in the Ti plasmid of *Agrobacterium tumefaciens* that is transferred to plant cells and inserted into the chromosomes of the plant.

T4 DNA ligase–An enzyme from bacteriophage-T4-infected cells, and that catalyses the joining of duplex DNA molecules and repairs nicks in DNA molecules. The enzyme requires that one of the DNA molecules has a 5´-phosphate group and that the other has a free 3´-hydroxyl group.

tRNA; transfer RNA–The class of small RNA molecules that transfer amino acids to the ribosome during protein synthesis. Transfer RNA molecules are folded into a 'clover-leaf' secondary structure by intrastrand base pairing. The anticodon loop contains a nucleotide-triplet complementary to a specific codon within the mRNA molecule. Each tRNA is 'charged' with the correct amino acid molecule, via its 3´ adenosine moiety, by an enzyme called aminoacyl-tRNA synthetase.

Telomerase–An enzyme that adds telomeric sequences to the ends of eukaryotic chromosomes.

Telemeter–The unique structure found at the end of eukaryotic chromosomes containing specialized sequences of DNA that assures the completion of a cycle of DNA replication.

Telophase (Gr. *telos*, end + phase)–The last stage in each mitotic or meiotic division, in which the chromosomes are assembled at the poles of the division spindle.

TEM–*Search in* transmission electron microscope.

Temperate phage–A phage (virus) that invades but may not destroy (lyse) the host (bacterial cell). However, it may subsequently enter the lytic cycle.

Temperature-sensitive mutant–An organism that can grow at one temperature but not at another.

Temperature-sensitive protein–A protein that is functional at one temperature but loses function at another (usually higher) temperature.

Template–An RNA or single-stranded DNA molecule upon which a complementary nucleotide strand is synthesized. A pattern or mould. DNA stores coded information and acts as a model or template from which information is copied into complementary strands of DNA or transcribed into mRNA.

Template strand–The polynucleotide strand that a polymerase uses for determining the sequence of nucleotides during the synthesis of a new nucleic acid strand.

Term finalization–Repelling movement of the centromeres of bivalents in the diplotene stage of the meiotic prophase, that tends to move the visible chiasmata toward the ends of the bivalents.

Terminal bud–A branch tip, an undeveloped shoot containing rudimentary floral buds or leaves, enclosed within protective bud scales.

Terminal transferase–An enzyme that adds nucleotides to the 3′ terminus of DNA molecules.

Termination codon–*Search in* stop codon.

Termination signal–In transcription, a nucleotide sequence that specifies RNA chain termination.

Terminator–(of transcription)

1. A DNA sequence just downstream of the coding segment of a gene, which is recognized by RNA polymerase as a signal to stop synthesizing mRNA. In prokaryotes, terminators usually have an inverted repeat followed by a short stretch of Us at the very end of the transcribed portion. There may also be sequences beyond the transcribed part of the gene which influence the termination of transcription.

2. A name given to antisense DNA inserted in plants to make impossible the use of a second generation of seed by a farmer.

Terminator codon–*Search in* stop codon.

Terminator region–A DNA sequence that signals the end of transcription.

Test-tube fertilization–*In vitro* fertilization.

Test tube–Tube in which cells, tissues, etc., can be cultured.

Testcross–Backcross to the recessive parental type, or a cross between genetically unknown individuals and a fully recessive tester to determine whether an individual in question is heterozygous or homozygous for a certain allele. It is also used as a test for linkage, *i.e.*, to estimate recombination fraction.

Tetracycline–An antibiotic that interferes with protein synthesis in prokaryotes.

Tetrad–1. The four cells arising from the second meiotic division in plants (pollen tetrads) or fungi (ascospores).

2. The quadruple group of chromatids that is formed by the association of duplicated homologous chromosomes during synapsis in meiosis I. a.k.a: quadrivalent.

Tetraploid–An organism whose cells contain four haploid ($4x$) sets of chromosomes.

Tetrasomic (noun: tetrasome)–Pertaining to a nucleus or an organism with four members of one of its chromosomes, whereas the remainder of its chromosome complement is diploid. Chromosome formula: $2n + 2$.

Tetratype–In fungi, a tetrad of spores that contains four different types; *e.g.*, *AB aB Ab* and *ab*.

Thallus (Gr. *thallos*, a sprout)–Plant body without true roots, stems, or leaves.

Therapeutic agent–A compound that is used for the treatment of a disease or for improving the well-being of an organism. a.k.a. pharmaceutical agent; drug; protein drug.

Thermic shock–Exposure to reduced or increased temperature for several days.

Thermolabile–Not heatproof, *e.g.*, a substance which disintegrates or is unstable upon heating.

Thermophile–An organism which grows at a higher temperature than most other organisms. In general, a wide range of bacteria, fungi and simple plants and animals can grow at temperature up to 50°C; thermophiles are considered to be organisms which can grow at above 50°C. They can be classified according to their optimal growth temperature, into simple thermophiles (50-65°C), thermophiles (65-85°C), and extreme thermophiles (>85°C). Thermophiles and extreme thermophiles are usually found growing in very hot places, such as hot springs and geysers, smoker vents on the sea floor, and domestic hot water pipes.

Thermosensitivity–Loss of activity of a protein at high temperature.

Thermostability–Retention of activity at high temperature.

Thermotherapy–Technique mainly used for virus or mycoplasma elimination. Plants are exposed to elevated temperatures as a treatment. Thermotherapy is used alone or in combination with meristem culture or meristem tip culture.

Thinning–1. Removal of older stems to promote new growth.

2. Removal of excess fruits to improve the size and quality of the remaining fruits.

3. Removal of seedlings spaced too closely for optimum growth.

Thymidine–A nucleoside present in DNA but absent in RNA.

Thymidine kinase (tk)–An enzyme that allows a cell to utilize an alternate metabolic pathway for incorporating thymidine into DNA. Used as a selectable marker to identify transfected eukaryotic cells.

Thymine–A pyrimidine base found in DNA. The other three organic bases–adenine, cytosine, and guanine–are found in both RNA and DNA; in RNA, thymine is replaced by uracil.

Ti plasmid–Tumour-inducing plasmid. A giant plasmid of *Agrobacterium tumefaciens* that is responsible for the induction of tumours in infected plants. Ti plasmids are used as vectors to introduce foreign DNA into plant cells. *Search in* vector.

Tissue–A group of cells of similar structure which sometimes performs a special function.

Tissue culture–A general term used to describe the culture of cells, tissues or organs in a nutrient medium under sterile conditions.

T lymphocyte–*Search in* T cell.

tk–*Search in* thymidine kinase.

Tolerance–A form of genetic resistance in which an organism attacked or affected by a disease pathogen (or pest) exhibits less reduction in yield or performance in comparison with members of other affected cultivars or breeds.

Tonoplast (Gr. *tonos*, stretching tension + *plastos*, moulded, formed)–The cytoplasmic membrane bordering the vacuole, with a role in regulating the pressure exerted by the cell sap.

Topo-isomerase–An enzyme that introduces or removes supercoils from DNA.

Torr–An obsolete unit of pressure equal to that exerted by a column of mercury 1mm high at 0°C and standard gravity (1mm Hg); named after Evangelista Torricelli (1608-1647), the inventor of the mercury barometer. 1 Torr = 1/760 atm = 133.322 Pa.

Totipotency–Having the potentiality of forming all the types of cells in the body. The property of somatic cells to be induced to undergo regeneration. The diploid zygote formed at fertilization is a single cell which is capable of division and differentiation to give rise to the total range of cell types found in the adult organism.

Totipotent cell; totipotent nucleus–An undifferentiated cell (or nucleus), such as a blastomere, that, when isolated or suitably transplanted, can develop into a complete embryo.

Toxicity–Negative effect of a compound, as shown by altered morphology or physiology. It is meaningful only when the effect itself is also described, such as changes in the rate of cell growth, cell death, etc.

Toxin (L. *toxicum*, poison)–A compound produced by an organism and poisonous to plants or animals.

Trace element–*Search in* microelement.

Tracer–An added or injected substance that can be followed within a reaction or an organism, such as radioactive isotopes and certain dyes.

Tracheid (Gr. *tracheia*, windpipe)–An elongated, tapering xylem cell, with lignified pitted walls, and adapted for conduction and support. Found in conifers, ferns and related plants.

Trait–*Search in* phenotype.

Trans-acting–A term describing substances that are diffusable and that can affect spatially separated entities within cells.

Transcapsidation–The partial or full coating of the nucleic acid of one virus with a coat protein of a differing virus. *Search in* coat protein.

Trans configuration–*Search in* repulsion.

Transcript–An RNA molecule that has been synthesized from a specific DNA template. In eukaryotes, the primary transcript produced by RNA polymerase must often be processed or modified in order to form the mature, functional mRNA, tRNA or rRNA

Transcription–Process through which RNA is formed along a DNA template. The enzyme RNA polymerase catalyses the formation of RNA from ribonucleoside triphosphates.

Transcription factor–A protein that regulates the transcription of genes.

Transcription unit–A segment of DNA that contains signals for the initiation and termination of transcription, and is transcribed into one RNA molecule.

Transcription vector–A cloning vector that allows the foreign gene or DNA sequence to be transcribed *in vitro*.

Transcriptional anti-terminator–A protein that prevents RNA polymerase from terminating transcription at specific transcription termination sequences.

Transducing phage–*Search in* transduction.

Transduction (t)–The transfer of DNA sequences from one bacterium to another via lysogenic infection by a bacteriophage (transducing phage). Genetic recombination in bacteria mediated by bacteriophage. Abortive t: Bacterial DNA is injected by a phage into a bacterium, but unable to replicate.

Transfection–The transfer of DNA to an eukaryotic cell.

Transfer–The process of moving cultured tissue or cells to a fresh medium.

Transferred DNA–*Search in* T-DNA.

Transferase–Enzyme that catalyses the transfer of a group of atoms from one molecule to another.

Transfer RNA (tRNA)–RNA that transports amino acids to the ribosomes, where the amino acids are assembled into proteins.

Transformant–In prokaryotes, a cell that has been genetically altered through the uptake of foreign DNA. In higher eukaryotes, a cultured cell that has acquired a malignant phenotype. *Search in* transformation.

Transformation–1. The uptake and establishment of DNA in a bacterium or yeast cell, in which the introduced DNA often changes the phenotype of the recipient organism.

2. Conversion by various means of animal cells in tissue culture from controlled to uncontrolled cell growth. Typically through infection by a tumour virus or transfection with an oncogene. *Search in* transformant; transformation efficiency.

Transformation efficiency–The number of cells that take up foreign DNA as a function of the amount of added DNA; expressed as transformants per microgram of added DNA. *Search in* transformation.

Transformation frequency–The fraction of a cell population that takes up foreign DNA; expressed as the number of transformed cells divided by the total number of cells in a population.

Transforming oncogene–A gene that upon transfection converts a previously immortalized cell to the malignant phenotype. *Search in* oncogene.

Transgene–A gene from one genome that has been incorporated into the genome of another organism. Often refers to a gene that has been introduced into a multicellular organism.

Transgenesis–The introduction of a gene or genes into animal or plant cells, which leads to the transmission of the input gene (transgene) to successive generations. *The opposite word is* transfection.

Transgenic–An organism in which a foreign gene (a transgene) is incorporated into its genome. The transgene is present in both somatic and germ cells, is expressed in one or more tissues, and is inherited by offspring in a Mendelian fashion. *Search in* transgenic animal; transgenic plant.

Transgressive variation–The appearance in the F_2 (or later) generation of individuals showing more extreme development of a trait than either of the original parents.

Trans heterozygote–A double heterozygote that contains two mutations arranged in the *trans* configuration.

Transient–Of short duration.

Transition–The mutation caused by substitution in DNA or RNA of one purine by another purine, or of one pyrimidine by another pyrimidine.

Transition stage–The integration period of juvenile and reproductive stages of growth.

Transition-state intermediate–In a chemical reaction, an unstable and high energy configuration assumed by reactants on the way to making products. Enzymes are thought to bind and stabilize the transition state, thus lowering the energy of activation needed to drive the reaction to completion.

Translation–The process of polypeptide synthesis in which the amino acid sequence is determined by mRNA, mediated by tRNA molecules, and carried out on ribosomes.

Translational initiation signal–*Search in* initiation codon.

Translational start codon–*Search in* initiation codon.

Translational stop signal–*Search in* termination codon.

Translocation–(L. *trans*, across + *locare*, to place) 1. The movement of nutrients or products of metabolism from one location to another.

2. Change in position of a segment of a chromosome to another, non-homologous chromosome.

Transmission electron microscope (TEM)–A microscope which uses an electron beam to obtain images of objects, with a much greater resolving power than a light microscope. *Search in* electron microscope.

Transplant–1. noun: A plant grown in a cold frame, greenhouse, tissue culture or indoors for later planting outdoors.

2. To dig up and move a plant to another location.

Transposable genetic element–A DNA element that can move from one location in the genome to another. *Search in* transposon.

Transposase–An enzyme encoded by a transposon gene and that catalyses the movement of a DNA sequence to a different site in a DNA molecule, by catalysing the excision of the transposon from one site and its insertion into a new chromosomal site.

Transposition–The process whereby a transposon or insertion sequence inserts itself into a new site on the same or another DNA molecule. The exact mechanism is not fully understood and different transposons may transpose by different mechanisms. Transposition in bacteria does not require extensive DNA homology between the transposon and the target DNA. The phenomenon is therefore descri-bed as illegitimate recombination.

Transposon–A transposable or movable genetic element. A relatively small DNA segment that has the ability to move (mobile genetic element) from one chromosomal position to another, *e.g.*, Tn 5 is a bacterial transposon that carries the genes for resistance to the antibiotics neomycin and kanamycin and the genetic information for insertion and excision of the transposon.

Transposon tagging–The insertion of a transposable element into or nearby a gene, thereby marking that gene with a known DNA sequence.

Trans test–*Search in* complementation test.

Transversion–The substitution of a purine for a pyrimidine, or a pyrimidine for a purine.

Tribrid protein–A fusion protein that has three segments, each encoded by parts of different genes.

Trichome (Gr. *trichoma*, a growth of hair)–A short filament of cells. *Search in* hair.

Tri-hybrid–The offspring from homozygous parents differing in three pairs of genes.

Tri-nucleotide repeats–Tandem repeats of three nucleotides that are present in many genes. In several cases, these trinucleotide repeats have undergone expansions in copy number, and that has resulted in inherited diseases.

Tripartite mating–A process in which conjugation is used to transfer a plasmid vector to a target cell when the plasmid vector is not self-mobilizable. When (1) cells that have a plasmid with conjugative and mobilizing functions are mixed with (2) cells that carry the plasmid vector and (3) target cells, mobilizing plasmids enter the cells with the plasmid vector and mobilize the plasmid vector to enter into the target cells. Following tripartite mating, the target cells with the plasmid vector are separated from the other cell types in the mixture by various selection procedures.

Triplet–A sequence of three nucleotides of DNA which specifies an amino acid. The elucidation of the genetic code involved the binding of charged tRNA species to chemically synthesized ribonucleotide triplets.

Triploid–A cell or organism containing three times the haploid number of chromosomes. Meiosis does not occur normally.

Trisomy (adj: trisomic) (Gr. *treis*, three + *soma*, body)–An otherwise diploid cell or organism that has an extra chromosome of one pair (chromosome formula: 2n + 1). *The opposite word is* disomy; monosomy.

Triticale–The species formed by crossing wheat and rye.

Tropism (Gr. *trope*, a turning)–An involuntary plant response to a stimulus, in which a bending, turning or growth occurs, such as phototropism, geotropism or hydrotropism. The response may be positive (towards) or negative (away from) to the stimulus.

True-to-type–Applied to a plant or propagation source, this term denotes correct cultivar identification and lack of variation in productivity or performance. Verification is determined visually by an expert or through biochemical, serological or other means.

Trypsin–A proteolytic enzyme that hydrolyzes peptide bonds on the carboxyl side of the amino acids arginine and lysine.

Trypsin inhibitors–Substances inactivating the enzyme trypsin, which is needed for digestion of peptides.

Tuber–Food-storing modified roots in plants like potato.

Tubulin–The major protein component of the microtubules of eukaryotic cells.

Tumble tube–A glass tube mainly used *in vitro* to agitate and consequently aerate suspension cultures. The tube, which is commonly attached to a slowly revolving platform, is closed at both ends, with a side-neck opening.

Tumour virus–A virus capable of transforming a cell to a malignant phenotype. *Search in* virus.

Tumour-inducing plasmid–*Search in* Ti plasmid.

Tunica–An outer one-to four-cell layered region of the apical meristem, where cell division is anticlinal, *i.e.*, perpendicular to the surface. *Search in* apical meristem.

Turbidostat–An open continuous culture in which a pre-selected biomass density is uniformly maintained by automatic removal of excess cells. The fresh medium flows in response to an increase in the turbidity (usually cell density) of the culture.

Turgid (L. *turgidus*, swollen, inflated)–Swollen, distended; referring to a cell that is firm due to water uptake.

Turgor potential–*Search in* pressure potential.

Turgor pressure (L. *turgor*, a swelling)–The pressure within the cell resulting from the absorption of water into the vacuole and the imbibition of water by the protoplasm.

Turion–An underground bud or shoot from which an aerial stem arises. *The opposite word is* sucker.

Twins–Two individuals originating from the same zygote.

U–The uracil residue in RNA.

Ultrasonication–*Search in* sonication.

Ultraviolet light; ultraviolet radiation (UV)–The portion of the electromagnetic spectrum with wavelengths from about 100 to 400 nm; between ionizing radiation (X-rays) and visible light. UV is absorbed by DNA and is highly mutagenic to unicellular organisms and to the epidermal cells of multicellular organisms. UV light is used in tissue culture for its mutagenic and bactericidal properties.

Undefined–A medium or substance added to medium in which not all of the constituents or their concentrations are chemically defined, such as media containing coconut milk, malt extract, casein hydrolysate, fish emulsion or other complex compounds. *The opposite word is* organic complex.

Understock–Host plant for a grafted scion, a branch or shoot from another plant; an understock may be a fully grown tree or a stump with a living root system.

Undifferentiated–In a meristematic state or resembling a meristem; lacking the specialized or differential gene expression characteristic of specialized cells.

Unequal crossing over–Crossing over between repeated DNA sequences that have paired out of register, creating duplicated and deficient products.

Unicellular–Describing tissues, organs or organisms consisting of a single cell.

Unisexual–Describing animals and plants possessing either male or female reproductive organs, but not both.

Univalent–An unpaired chromosome at meiosis.

Universal donor cells–Cells that, after introduction into a recipient, will not induce an immune response that leads to their rejection.

Universality–Referring to the genetic code, the codons have the same meaning, with minor exceptions, in virtually all species.

Unorganized growth–*In vitro* formation of tissues with few differentiated cell types and no recognizable structure; many call are unorganized. *The opposite word is* organized growth.

Upstream processing–*Search in* upstream (2).

Upstream–1. In molecular biology, the stretch of DNA base pairs that lie in the 5′ direction from the site of initiation of transcription. Usually the first transcribed base is designated +1 and the upstream nucleotides are marked with minus signs, *e.g.*,–1,–10. Also, to the 5′ side of a particular gene or sequence of nucleotides.

2. In chemical engineering, those phases of a manufacturing process that precede the biotransformation step. Refers to the preparation of raw materials for a fermentation process. Also called upstream processing.

Uracil–A pyrimidine base found in RNA but not in DNA. In DNA, uracil is replaced by thymine. It is analogous to thymine in DNA.

Utilization of farm animal genetic resources–The use and development of animal genetic resources for the production of food and agriculture.
(*Source*: FAO, 1999)

The use in production systems of AnGRs that already possess high levels of adaptive fitness to the environments concerned, and the deployment of sound genetic principles, will facilitate sustainable development of the AnGRs and the sustainable intensification of the production systems themselves. The wise use of AnGRs is possible without depleting domestic animal diversity. Development of AnGRs includes a broad mix of ongoing activities that must be well planned and executed for success, and compounded over time, hence with high value. It requires careful definition of breeding objectives, and the planning, establishment and maintenance of effective and efficient animal recording and breeding strategies. (Source: FAO, 1999)

v/v–On a volume per volume basis; the percent of the volume of a constituent in 100 units of volume, *e.g.*, (ml/100 ml) × 100.

Vaccination–*Search in* preventive immunization.

Vaccine–A preparation of dead or weakened pathogens, or of derived antigenic determinants, that is used to induce formation of antibodies or immunity against the pathogen.

Vaccinia–The cowpox virus used to vaccinate against smallpox and, experimentally, as a carrier of genes for antigenic determinants cloned from other disease organisms.

Vacuole (L. diminutive of *vacuus*, empty)–A cavity in a plant cell, bounded by a membrane; in which various plant products and by-products are stored.

Vacuum–Vacuum is created through use of a vacuum pump or aspirator pump, to facilitate specific biological preparations, such as inclusions or disinfection of material for *in vitro* culture, etc.

Variable domains–Regions of antibody chains that have different amino acid sequences in different antibody molecules. These regions are responsible for the antigen-binding specificity of the antibody molecule.

Variable expressivity–Variation in the phenotype caused by different alleles of the same gene and/or by the action of other genes and/or by the action of non-genetic factors. *The opposite word is* expressivity.

Variable number tandem repeat–*Search in* VNTR.

Variable surface glycoprotein (VSG)–One of a battery of antigenic determinants expressed by a micro-organism to elude immune detection.

Variance–In statistics: The sum of the squared deviations, divided by one less than the number of observations. A statistical measure of variation in a population.

Variant–An organism that is genetically different from the wild type organism.

Variation–Differences between individuals within a population or among populations.

Variegated–Plants having both green and albino tissues. This difference in colour may result from viral infection, nutritional deficiency, or may be under genetic or physiological control.

Variety–A naturally occurring subdivision of a species, with distinct morphological characters and given a Latin name according to the rules of the International Code of Nomenclature. A taxonomic variety is known by the first validly published name applied to it so that nomenclature tends to be stable. *The opposite word is* cultivar; pathovar.

Vascular (L. *vasculum*, a small vessel)–Referring to any plant tissue or region consisting of or giving rise to conducting tissue, *e.g.*, bundle, cambium, ray.

Vascular bundle; fascicle–A strand of tissue containing primary xylem and primary phloem (and procambium if present) and frequently enclosed by a bundle sheath of parenchyma or fibres.

Vascular cambium–In biennials and perennials, cambium giving rise to secondary phloem and secondary xylem.

Vascular plants–Plants possessing organized vascular tissues.

Vascular system–1. A specialized network of vessels for the circulation of fluids throughout the body tissue of an animal.

2. The system of vascular tissue in plants.

Vascular tissue–The tissue that conducts water and nutrients throughout the plant body in higher plants.

Vector (L. *vehere*, to carry)–1. An organism, usually an insect, that carries and transmits disease-causing organisms.

2. A plasmid or phage that is used to deliver selected foreign DNA for cloning and in gene transfer. *Search in* Ti plasmid.

Vegetative propagation–Same as asexual or non-sexual propagation.

Vehicle–The host organism used for the replication or expression of a cloned gene or other sequence The term is little used and is often confused with vector. *Search in* vector.

Velocity density gradient centrifugation–A procedure used to separate macromolecules based on their rate of movement through a density gradient.

Velogenetics–The combined use of marker-assisted selection and embryo technologies such as OPU, IVM and IVF, in order to increase the rate of genetic improvements in animal populations.

Vermiculite–Coarse aggregate material made from expanded mica having a high cation exchange capacity and high water-holding capacity and used as a rooting medium and a soil additive.

Vernalization–The process by which floral induction in some plants is promoted by exposing the plants to chilling for a certain duration.

Vessel (L. *vasculum*, a small vessel)–1. A series of xylem elements whose function is to conduct water and nutrients in plants.

2. A container, such as a Petri dish or test tube, used for tissue culture.

Vessel element–A type of cell occurring within the xylem of flowering plants. Many are water-conducting vessels.

Viability–The capability to live and develop normally.

Viability test–Assay of the number or percent of living cells or plants in a population that have been given a specific treatment; such as ascertaining cell viability after cryopreservation treatment, or for seeds.

Viable–Capable of germinating, living, growing and developing.

Vibrio–Comma-shaped bacterium.

Viral vaccines–Vaccines consisting of live viruses rather than dead ones or separated parts of viruses. However, as the virus itself cannot be used, because that would simply give the patient the disease, the virus is genetically engineered so that it elicits the immune response to the viral pathogen without causing the disease itself.

Two genetic engineering methods can be used. The first is to make the disease virus harmless, but still able to replicate in cultured animal cells. This is similar to producing an 'attenuated' virus, *i.e.*, one which has been grown in the laboratory until it loses its ability to cause disease. However, the genetic engineering route seeks to make sure that the attenuated virus has no chance of mutating back to a wild type, pathogenic virus, by deleting whole genes or replacing key regions of genes with completely different genetic material.

The second approach is to clone the gene for a protein from the pathogenic virus into another, harmless virus, so that the result 'looks' like the pathogenic virus but does not cause disease.

***vir* genes**–A set of genes on a Ti plasmid that prepare the T-DNA segment for transfer into a plant cell.

Viral coat protein–Protein present in the outer layer of a virus.

Viral oncogene–A gene in the viral genome that contributes to malignancies in vertebrate hosts. *Search in* oncogene.

Viral pathogen–A disease-causing virus.

Virion–An infectious virus particle. A plant pathogen that consists of a naked RNA molecule of approximately 250-350 nucleotides, whose extensive base pairing results in a nearly correct double helix. *Search in* satellite RNA.

Viroid–An infectious entity similar to a virus but smaller, consisting only of a strand of nucleic acid without the protein coat characteristic of a virus. *Search in* satellite RNA.

Virulence (L. *virulentia* a stench)–The degree of ability of an organism to cause disease. The relative infectiousness of a bacterium or virus, or its ability to overcome the resistance of the host metabolism.

Virulent phage–A phage (virus) that destroys the host (bacterium).

Virus (L. *virus*, a poisonous or slimy liquid)–An infectious particle composed of a protein capsule and a nucleic acid core (DNA or RNA), which is dependent on a host organism for replication. The DNA or a double-stranded DNA copy of an RNA virus genome is integrated into the host chromosome during lysogenic infection or replicated during the cystic cycle. *Search in* coat protein; DNA; genome; host; nucleic acid; prion; RNA; tumour virus; viroid.

Virus-free–Plant, animal, cell, tissue or meristem which exhibits no viral symptoms or contains no identifiable virus-particles.

Virus-tested–Description of a organism or a cell stock certified as being free of certain specified viruses following recognized procedures of virus diagnosis.

Visible light–The part of the electromagnetic spectrum with wavelengths between 380 nm and 750 nm and perceived by the human eye.

Vitamin B complex–A large group of water soluble vitamins that function as co-enzymes, including thiamine (B_1); riboflavin or vitamin G (B_2); niacin or nicotinic acid (B_3); pantothenic acid (B_5); pyridoxine (B_6); cyanocobalamin (B_{12}); biotin or vitamin H; folic acid or vitamin M (Bc); inositol; choline; and others.

Vitamin C–*Search in* ascorbic acid.

Vitamin H–*Search in* biotin; vitamin B complex.

Vitamins (L. *vita*, life + amine)–Naturally occurring organic substances required by living organisms in relatively small amounts to maintain normal health, and which are added to tissue culture media to enhance growth, usually acting as enzyme co-factors.

Vitrified; water soaked–Cultured tissue having leaves and sometimes stems with a glassy, transparent or wet and often swollen appearance. The process of vitrification is a general term for a variety of physiological disorders that lead to shoot tip and leaf necrosis.

Vivipary–1. A form of reproduction in animals in which the developing embryo obtains its nourishment directly from the mother via a placenta or by other means.

2. A form of asexual reproduction in certain plants, in which the flower develops into a budlike structure that forms a new plant when detached from the parent.

3. The development of young plants in the inflorescence of the parent plant.

V_{max}–The maximal rate of an enzyme-catalysed reaction. V_{max} is the product of E_o (the total amount of enzyme) times the value of K_{eat} (the catalytic rate constant).

VNTR–Variable number tandem repeat. A short DNA sequence that is present as tandem repeats and in highly variable copy number.

V regions–Variable regions in antibodies. *Search in* CDR.

Volatilization–The conversion of a solid or liquid into a gas or vapour.

VSG–*Search in* variable surface glycoprotein.

w/v–Weight per volume; the weight of a constituent in 100 cm^3 of solution, expressed as a percentage.

Walking–A method for cloning large regions of a chromosome. Starting from a known site, a gene library is screened for clones that hybridize to DNA probes taken from the ends of the first clone. These clones are then isolated, and their ends used to screen the library again. These clones are then isolated and their ends used, and so on. *Search in* gene walking; chromosome walking.

Wall pressure–Pressure that a cell wall exerts against the turgor of the cell contents. Wall pressure is equal and opposite to the turgor potential.

Wash-out–The loss of the slower growing micro-organism when two organisms are being grown together.

Water potential–Refers to the difference between the activity of water molecules in pure distilled water at atmospheric pressure and 30°C (standard conditions), and the activity of water molecules in any other system. The activity of these water molecules may be greater (positive) or less (negative) than the activity of the water molecules under standard conditions.

Water soaked–*Search in* vitrified.

Water stress–The condition when plants are unable to absorb enough water to replace that lost by transpiration. The results may be wilting, cessation of growth, or even death of the plant or plant parts.

Wavelength–The distance between two corresponding points on any two consecutive waves. For visible light it is very small and is generally measured in nanometres.

Wax (A.S. *weax*, wax)–Esters of alcohol higher than glycerol, which are insoluble in water and difficult to hydrolyze; wax forms protective waterproof layers on leaves, stems, fruits, animal fur and integuments of insects.

Weed–Simply any plant growing where it is not wanted. In agriculture, used for a plant which has good colonizing capability in a disturbed environment, and can usually compete with a cultivated species therein. Weeds are typically considered as unwanted, economically useless or pest species.

Weediness–In agriculture, the ability of a plant to colonize a disturbed habitat and compete with cultivated species.

Western blot–A technique in which protein is transferred from an electrophoretic gel to a cellulose or nylon support membrane following electrophoresis. A particular protein molecule can then be identified by probing the blot with a radiolabelled antibody which binds only the specific protein to which the antibody was prepared. Useful, for example, for measurement of levels of production of a specific protein in a particular tissue or at particular developmental stage. *The opposite word is* blot: northern blot; Southern blot.

Wet weight–The gross weight of a product with its full water content, or weight of fully hydrated tissue. *Search in* fresh weight.

Wetting agent–A substance that improves surface contact by reducing the surface tension of a liquid: *e.g.*, Triton X-10TM added to disinfecting solutions promotes the disinfestation process. *Search in* detergent; surfactant.

Wild type–An organism as found in nature; the dominant allele usually found in nature, and from which mutations produce other dominants or recessives alleles.

Wilt–Drooping of stems and foliage due to loss of water and decreased turgidity of cells. May be caused by water stress or by disease.

Wilting point (WP)–The moisture content of soil at which plants start to wilt, but not to the extent that they fail to recover when placed in a humid atmosphere. *The opposite word is.* permanent wilting point.

Wobble hypothesis–An explanation of how one tRNA may recognize more than one codon. The first two bases of the mRNA codon and anticodon pair properly, but the third base in the anticodon has some wobble that permits it to pair with more than one base.

WP–*Search in* wilting point.

x–The basic number of chromosomes in a polyploid series, monoploid = x; diploid = 2x; triploid = 3x; etc.

Xanthophyll (Gr. *xanthos*, yellowish brown + *phyllon*, leaf)–A yellow chloroplast pigment.

X-chromosome–A chromosome associated with sex determination. In most animals, the female has two, and the male has one X chromosome. The opposite occurs in birds, in which the equivalent is the Z chromosome.

Xenia–The immediate effect of pollen on some characters of the endosperm.

Xenobiotic–A chemical compound that is not produced by living organisms; a manufactured chemical compound.

Xenotransplantation–The transplantation of tissue from one species to another species, typically from non-human mammals to humans. This technology has become very important because of a worldwide shortage of human organs for humans requiring a new organ. The most popular non-human species involved in this technology is the pig.

Xerophyte (Gr. *xeros*, dry + *phyton*, a plant)–A plant very resistant to drought or that lives in very dry places.

X-linkage–*Search in* X-linked.

X-linked–The presence of a gene on the X chromosome.

X-linked disease–A genetic disease caused by an allelc at a locus on the X-chromosome. In X-linked recessive conditions, a normal female "carrier" passes on the allele on her X chromosome to an affected son.

X-ray crystallography–The deduction of crystal structure from analysis of the diffraction pattern of X-rays passing through a pure crystal of a substance.

Xylem (Gr. *xylon*, wood)–A complex tissue specialized for efficient conduction of water and mineral nutrients in solution. Xylem may also function as a supporting tissue, particularly secondary xylem.

Y-chromosome–The partner of the X-chromosome in the male of many animal species.

YAC–Yeast artificial chromosome, used as a vector system for cloning DNA fragments that can be hundreds of kilobases long. Linear cloning vectors constructed from essential elements of yeast chromosomes. They can accommodate foreign DNA inserts of 200 to 500 kb.

Yeast–A unicellular ascomycete fungus, commonly found as a contaminant in plant tissue culture.

Yeast artificial chromosome–*Search in* YAC.

Yeast episomal vector (YEp)–A cloning vector for the yeast *Saccaromyces cerevisiae* that uses 2?m plasmid as origin of replication and is maintained as an extrachromosomal nuclear DNA molecule.

Yeast extract–A mixture of substances from yeast. *The opposite word is* organic complex; undefined.

Yeast cloning vectors–The yeasts, and especially *Saccaromyces cerevisiae*, are favourite organisms in which to clone and express DNA. They are eukaryotes, and so can splice out introns, the non-coding sequences in the middle of many eukaryotic genes.

Z-DNA; zig-zag DNA–A form of DNA duplex in which the double helix is wound in a left-hand, instead of a right-hand, manner. DNA adopts the Z configuration when purines and pyrimidines alternate on a single strand, *e.g.*,

5´CGCGCGCG3´ or 5´CACACA CACA3´

3´GCGCGCGC5´ 3´GTGTGT GTGT5´

Zone of elongation–The section of the young root or shoot just behind the apical meristem, in which the cells are enlarging and elongating rapidly.

Zoo blot–Hybridization of cloned DNA from one species to DNA from other organisms to determine the extent to which the cloned DNA is evolutionarily conserved.

Zoo FISH–Fluorescent *in situ* hybridization of DNA from one species on metaphase chromosomes of another species. Typically, the hybridization is done separately for DNA libraries representing each chromosome. The result is a fascinating picture of the regions of chromosomal homology between species.

Zoospore–A spore that possesses flagella and is therefore motile. May be mitosporic or meiosporic zoospore.

Zygonema (adj: zygotene)–Stage in meiosis during which synapsis occurs; coming after the leptotene stage and before the pachytene stage in the meiotic prophase.

Zygospore (Gr. *zygon*, a yoke + spore)–A thick-walled resistant spore developing from a zygote resulting from the fusion of isogametes.

Zygote (Gr. *zygon*, a yoke)–A diploid cell formed by the fusion of two haploid gametes during fertilization in eukaryotic organisms with sexual reproduction. It is the first cell of the new individual.

Zymogen–Inactive enzyme precursor that after secretion is chemically altered to the active form of the enzyme. It is converted into an enzyme and a protein by the action of kinase or zymoexciton.

Chapter 5
Pesticide Terms

Abscission–It is a process by which a leaf or other part is separated from the plant.

Absorption–The process by which pesticides are taken into tissues, namely plants, by roots or foliage (stomata, cuticle, etc.) is called absorption.

Acaricide (miticide)–It is a particular agent that destroys mites and ticks.

Acetylcholine (ACh)–Chemical transmitter of nerve and nerve-muscle impulses in animals.

ACGIH–Abbreviation for the American Conference of Governmental Industrial Hygienists, a private organization of occupational safety and health professionals. The ACHIG recommends occupational exposure limits for numerous toxic substances, and it updates and revises its recommendations as more information becomes available. ACHIG limits are not legally enforceable. *See* Occupational Exposure Limlts.

Activator–Material added to a fungicide to increase toxicity.

Active ingredient (a.i. or AI)–Chemicals in a product that are responsible for the pesticidal effect.

Acute toxicity–The toxicity of a material determined at the end of 24 hours; to cause injury or death from a single dose or exposure.

Adjuvant–Adjuvent is an ingredient that improves the properties of a pesticide formulation. It may include various things like wetting agents, spreaders, emulsifiers, dispersing agents, foam suppressants, penetrants, and correctives.

Adsorption–Adsorption is a process of chemical and/or physical attraction of a substance to a surface. It refers to gases, dissolved substances, or liquids on the surface of solids or liquids.

Adulterated pesticide–It is a type of pesticide that does not conform to the professed standard or quality as documented on its label or labeling.

Aerosol–Colloidal suspension of solids or liquids in air is known as aerosols.

Agroecosystem–It is an agricultural area which is sufficiently large to permit long-term interactions of all the living organisms and their nonliving environment.

Algicide–Chemicals used to control algae and aquatic weeds are known as algicides.

Alkylating agent–Highly active compounds (chemosterilants) that replace hydrogen atoms with alkyl groups, usually in cells undergoing division.

Annual–Plant that completes its life cycle in one year, *i.e.*, germinates from seed, produces seed, and dies in the same season.

Antagonism–Decreased activity arising from the effect of one chemical or another (opposite of synergism).

Antibiotic–Chemical substance produced by a microorganism and that is toxic to other microorganisms.

Anticoagulant–A chemical that prevents normal blood clotting. The active ingredient in some rodenticides.

Antidote–A practical treatment, including first aid, used in the treatment of pesticide poisoning or some other poison in the body.

Antimetabolite–Chemicals that are structurally similar to biologically active metabolites, and that may take their place detrimentally in a biological rection.

Antitranspirant–A chemical applied directly to a plant that reducesthe rate of transpiration of water loss by the plant.

Apiculture–Pertaining to the care and culture of bees.

Aromatics–Solvents containing benzene or compounds derived from benzene.

Atropine (atropine sulfate)–An antidote used to treat organophosphate and carbamate poisoning.

Attractant, insect–A substance that lures insects to trap or poison-bait stations. Usually classed as food, oviposition, and sex attractants.

Auxin–Substance found in plants that stimulates cell growth in plant tissues.

Avicide–Lethal agent used to destroy birds but also refers to materials used for repelling birds.

Aziridine–Chemical classification of chemosterilants containing three-membered rings composed of one nitrogen and two carbon atoms.

Bactericide–Any bacteria-killing chemical.

Bacteriostat–Material used to prevent growth or multiplication of bacteria.

Band application–Application to a continuous restricted band such as in or along a crop row, rather than over the entire field area.

Biennial–Plant that completes its growth in 2 years. The first year it produces leaves and stores food; the second year it produces fruit and seeds.

Biological control agent–Any biological agent that adversely affects pest species.

Biomagnification–The increase in concentration of a pollutant in animals as related to their position in a food chain, usually referring to the persistent, organochlorine insecticides and their metabolites.

Biota–Animals and plants of a given habitat.

Biotic insecticide–Usually micro-organisms known as insect pathogens that are applied in the same manner as conventional insecticides to control pest species.

Biotype–Subgroup within a species differing in some respect from the species such as a subgroup that is capable of reproducing on a resistant variety.

Botanical pesticide–A pesticide produced from naturally occurring chemicals found in some plants. *e.g.* nicotine, pyrethrum, strychnine, and rotenone.

Brand–The name, number, or designation of a pesticide.

Broadcast application–Application over an entire area rather than only on rows, beds, or middles.

Broad-spectrum insecticide–Non-selective, having about the same toxicity to most insects.

Calibrate–To determine the amount of pesticide that will be applied to the target area.

Carbamate insecticide–One of a class of insecticides derived from carbamic acid.

Carcinogen–A substance that causes cancer in animal tissue.

Carcinogenic–Capable of causing cancer.

Carrier–An inert material that serves as a diluent or vehicle for the active ingredient or toxicant.

Causal organism–The organism (pathogen) that produces a given disease.

Ceiling Limit–The maximum amount of a toxic substance allowed to be in workroom air at any time during the day.

Certified applicator–Commercial or private person qualified to apply restricted-use pesticides as defined by the EPA.

Chelating agent–Certain organic chemicals (*i.e.*, ethylenediaminetetraacetic acid) that combine with metal to form soluble chelates and prevent conversion to insoluble compounds.

Chemical name–Scientific name of the active ingredient(s) found in the formulated product. The name is derived from the chemical structure of the active ingredient.

Chemosterilant–Chemical compounds that cause sterilization or prevent effective reproduction.

Chemotherapy–Treatment of a diseased organism, with chemicals to destroy or inactivate a pathogen without seriously affecting the host.

Chemtrec–A toll-free, long-distance, telephone service that provides 24-hour emergency pesticide information (800-424-9300).

Chlorosis–Loss of green color in foliage.

Cholinesterase (ChE)–An enzyme of the body necessary for proper nerve function that is inhibited or damaged by organophosphate or carbamate insecticides taken into the body by any route.

Chronic toxicity–The toxicity of a material determined beyond 24 hours and usually after several weeks of exposure.

Combustible–Able to catch on fire and burn.

Common pesticide name–A common chemical name given to a pesticide by a recognized committee on pesticide nomenclature. Many pesticides are known by a number of trade or brand names but have only one recognized common name. For example, the common name for Sevin insecticide is carbaryl.

Compatible/Compatibility–When two materials can be mixed together with neither affecting the action of the other.

Concentration–The amount of one substance in another substance.

Concentration–Content of a pesticide in a liquid or dust; for example, pounds/gallon or percent by weight.

Contact herbicide–Phytotoxin by contact with plant tissue rather than as a result of translocation contact.

Contamination–The presence of an unwanted pesticide or other material in or on a plant, animal, or their by-products; soil; water; air; structure; etc. (See Residue).

Cumulative pesticides–Those chemicals that tend to accumulate or build up in the tissues of animals or in the environment (soil, water).

Curative pesticide–A pesticide that can inhibit or eradicate a disease-causing organism after it has become established in the plant or animal.

Cutaneous toxicity–Same as dermal toxicity.

Cuticle–Outer covering of insects.

Deciduous–Plants that lose their leaves during the winter.

Decomposition–Breakdown of a chemical.

Decontaminate–The removal or breakdown of any pesticide chemical from any surface or piece of equipment.

Deflocculating agent–Material added to a spray preparation to prevent aggregation or sedimentation of the solid particles.

Defoliant–A chemical that initiates abscission.

Density–How much space a given weight of substance takes up. The density of a substance is usually compared to water, which has a density of 1. Substances more dense than water (which sink in water) have densities greater than 1; substances which float on water have densities less than 1.

Deposit–Quantity of a pesticide deposited on a unit area.

Dermal–By or through the skin.

Dermal toxicity–Toxicity of a material as tested on the skin, usually on the shaved belly of a rabbit; the property of a pesticide to poison an animal or human when absorbed through the skin.

Desiccant–A chemical that induces rapid desiccation of a leaf or plant part.

Desiccation–Accelerated drying of plant or plant parts.

Detoxify–To make an active ingredient in a pesticide or other poisonous chemical harmless and incapable of being toxic to plants and animals.

Diluent–Component of a dust or spray that dilutes the active ingredient.

Disinfectant–A chemical or other agent that kills or inactivates disease-producing microorganisms in animals, seeds, or other plant parts. Also commonly refers to chemicals used to clean or surface sterilize inanimate objects.

DNA–Deoxyribonucleic acid.

Dormant spray–Chemical applied in winter or very early spring before treated plants have started active growth.

Dose, dosage–Same as rate. The amount of toxicant given or applied per unit of plant, animal, or surface.

Drift, spray–Movement of airborne spray droplets from the spray nozzle beyond the intend contact area.

EC 50–The median effective concentration (ppm or ppb) of the toxicant in the environment (usually water) that produces a designated effect in 50 percent of the test organisms exposed.

Ecdysome–Hormone secreted by insects essential to the process of molting from one stage to the next.

Ecology–Derived from the Greek oikos, house or place to live A branch of biology concerned with organisms and their relation to the environment.

Economic level–The insect pest level at which additional management practices must be employed to prevent economic losses.

Ecosystem–The interacting system of all the living organisms of an area and their nonliving environment.

ED50–The median effective dose, expressed as mg/kg of body weight, which produces a designated effect in 50 percent of the test organisms exposed.

Emulsifiable concentrate–Concentrated pesticide formulation containing organic solvent and emulsifier to facilitate emulsification with water.

Emulsifier–Surface active substances used to stablize suspensions of one liquid in another; for example, oil in water.

Emulsion–Suspension of miniscule droplets of one liquid in another.

Environment–All the organic and inorganic features that surround and affect a particular organism or group of organisms.

Environmental Protection Agency (EPA)–The federal agency responsible for pesticide rules and regulations, and all pesticide registrations.

EPA Establishment Number–A number assigned to each pesticide production plant by EPA. The number indicates the plant at which the pesticide product was produced and must appear on all labels of that product.

EPA Registration Number–A number assigned to a pesticide product by EPA when the product is registered by the manufacturer or his designated agent. The number must appear on all labels for a particular product.

EPA–The Environmental Protection Agency.

Eradicant–Applies to fungicides in which a chemical is used to eliminate a pathogen from its host or environment.

Explosive Limits–The amounts of vapor in air which form explosive mixtures. Explosive limits are expressed as Lower Explosive Limits and Upper Explosive Limits; these give the range of vapor concentrations in air which will explode if heat is added. Explosive limits are expressed as percent of vapor in air.

Exterminate–Often used to imply the complete extinction of a species

Field scout–A person who samples fields for insect infestations.

Filler–Diluent in powder form.

Fixed coppers–Insoluble copper fungicides where the copper is in a combined form. Usually finely divided, relatively insoluble powders.

Flammable–Catches on fire easily and burns rapidly.

Flammable Limits–*See* Explosive Limits.

Flash Point–The lowest temperature at which the vapor of a substance will catch on fire, and then go out, if heat is applied. Provides an indication of how flammable a substance is. Not to be confused with Ignition Temperature.

Flowable–A type of pesticide formulation in which a very finely ground solid particle is mixed in a liquid carrier.

Foaming agent–A chemical that causes a pesticide preparation to produce a thick foam. This aids in reducing drift.

Fog treatment–The application of a pesticide as a fine mist for the control of pests.

Food chain–Sequence of species within a community, each member of which serves as food for the species next higher in the chain.

Formamidine insecticide–A new group with a new mode of action highly effective against insect eggs and mites.

Formulation–Way in which basic pesticide is prepared for practical use. Includes preparation as wettable powder, granular, emulsifiable concentrate, etc.

Full-coverage spray–Applied thoroughly over the crop to a point of runoff or drip.

Fumigant–A volatile material that forms vapors that destroy insects, pathogens, and other pests.

Fungicide–A chemical that kills fungi.

Fungistatic–Action of a chemical that inhibits the germination of fungus spores.

Gallonage–Number of gallons of finished spray mix applied per acre, tree, hectare, square mile, or other unit.

General-use pesticide–A pesticide that can be purchased and used by the general public without undue hazard to the applicator and environment as long as the instructions on the label are followed carefully. (See Restricted-use pesticide).

Gram–The unit of mass in the metric system. An ounce is about 28 grams, and a pound is approximately 450 grams. (A teaspoon of sugar weights about 8 grams.)

Growth regulator–Organic substance effective in minute amounts for controlling or modifying (plant or insect) growth processes.

Harvest intervals–Period between last application of a pesticide to a crop and the harvest as permitted by law.

Health Hazard–Anything which can have a harmful effect on health under the conditions in which it is used or produced. There can be both *Acute* and *Chronic* health hazards.

Hormone–A product of living cells that circulates in the animal or plant fluids and that produces a specific effect on cell activity remote from its point of origin.

Hydrolysis–Chemical process of (in this case) pesticide breakdown or decomposition involving a splitting of the molecule and addition of a water molecule.

Hyperplasia–Abnormal increase in the number of cells of a tissue.

Hypertrophy–Abnormal increase in the size of cells of a tissue.

Ignition Temperature–The lowest temperature at which a substance will catch on fire and continue to burn. The lower the ignition temperature, the more likely the substance is going to be a fire hazard.

Incompatible–Two or more materials that cannot be mixed or used together.

Inflammable–Same as Flammable.

Ingest–To eat or swallow.

Ingestion–Swallowing.

Ingredient statement–That portion of the label on a pesticide container that gives the name and amount of each active ingredient and the total amount of inert ingredients in the formulation.

Inhalation toxicity–To be poisonous to man or animals when breathed into the lungs.

Inhalation–Exposure of test animals either to vapor or dust for a predetermined time.

Insect pest management–The practical manipulation of insect (or mite) pest populations, using any or all control methods in a sound ecological manner.

Insect-growth regulator (IGR)–Chemical substance that disrupts the action of insect hormones controlling molting, maturity from pupal stage to adult, and others.

Integrated control–The integration of the chemical and biological control methods.

Integrated pest management–A management system that uses all suitable techniques and methods in as compatible a manner as possible to maintain pest populations at levels below those causing economic injury.

Intramuscular–Injected into the muscle.

Intraperitoneal–Injected into the viscera but not into the organs.

Intravenous–Injected into the vein.

Invert emulsion–One in which the water is dispersed in oil rather than oil in water. Usually a thick mixture like salad dressing results.

Kilogram–1000 grams. One kilogram equals about 2.2 pounds.

Label–All printed material attached to or part of the pesticide container.

Labeling–Supplemental pesticide information that complements the information on the label but is not necessarily attached to or part of the container.

LC 50–1.The concentration of a substance in air that causes death in 50 per cent of the animals exposed by inhalation. A measure of acute toxicity.

2.The median lethal concentration, the concentration that kills 50 percent of the test organisms, expressed as milligrams (mg) or cubic centimeters (cc, if liquid) per animal. It is also the concentration expressed as parts per million (ppm) or parts per billion (ppb) in the environment (usually water) that kills 50 percent of the test organisms exposed.

LD 50–1.The dose that causes death in 50 per cent of the animals exposed by swallowing a substance. A measure of acute toxicity.

2.A lethal dose for 50 percent of the test organisms. The dose of toxicant producing 50 percent mortality in a population. A value used in presenting mammalian toxicity, usually oral toxicity, expressed as milligrams of toxicant per kilogram of body weight (mg/kg).

Leaching–The movement of a pesticide chemical or other substance downward through soil as a result of water movement.

Liter–The unit of volume in the metric system. A liter is about the same as a quart.

Low–volume spray–Concentrate spray, applied to uniformly cover the crop, but not as a full coverage to the point of runoff.

M.L.D–Median lethal dose (LD 50).

Meter–The unit of length in the metric system. A meter is about 40 inches.

mg/kg (milligrams per kilogram)–1.Used to designate the amount of toxicant required per kilogram of body weight of test organism to produce a designated effect, usually the amount necessary to kill 50 percent of the test animals.

2.A way of expressing dose: milligrams (mg) of a substance per kilogram (kg) of body weight. Example: a 100 kg (220 pound) person given 10,000 mg (about 0.02 pounds) of a substance would be getting a dose of 100 mg/kg (10,000 mg/100 kg).

mg/m^3–A way of expressing the concentration of a substance in air: milligrams (mg) of substance per cubic meter (m^3) of air.

Microbial insecticide–A microorganism applied in the same way as conventional insecticides to control an existing pest population.

Mildew–Fungus growth on a surface.

Milligram–One one-thousandth of a gram.

Miscible liquids–Two or more liquids capable of being mixed in any proportions and of remaining mixed under normal conditions.

Molluscicide–A chemical used to kill or control snails and slugs.

Mutagen–Substance causing genes in an organism to mutate or change.

Mutagenic–Capable of changing cells in such a way that future cell generations are affected. Mutagenic substances are usually considered suspect carcinogens.

Mycoplasma–A microorganism intermeditae in size between viruses and bacteria possessing many virus-like properties and not visible with a light microscope.

Necrosis–Death of tissue, plant or animal.

Nematicide–Chemical used to kill nematodes.

Occupational Exposure Limits–Maximum allowable concentrations of toxic substances in workroom air to protect workers who are exposed to toxic substances over a working lifetime.

Odor Threshold–The lowest concentration of a substance's vapor, in air, that can be smelled. Odor thresholds are highly variable depending on the individual who breathes the substance and the nature of the substance.

Oncogenic–The property to produce tumors (not necessarily cancerous) in tissues. (See Carcinogenic.)

Oral toxicity–Toxicity of a compound when given by mouth. Usually expressed as number of milligrams of chemical per kilogram of body weight of animal (white rat) when given orally in a single dose that kills 50 percent of the animals. The smaller the number, the greater the toxicity.

Organochlorine insecticide–One of the many chlorinated insecticides, *e.g.*, DDT, dieldrin, chlordane, BHC, Lindane, etc.

Organophosphate–Class of insecticides (also one or two herbicides and fungicides) derived from phosphoric acid esters.

Ovicide–A chemical that destroys an organism's eggs.

Oxidizer–Any substance which reacts violently with oxygen or which gives off large amounts of energy in a chemical reaction.

Pathogen–Any disease-producing organism or virus.

Perennial–Plants that continue to live from year to year. Plants may be herbaceous or woody.

Persistence–The quality of an insecticide to persist as an effective residue due to its low volatility and chemical stability, *e.g.*, certain organochlorine insecticides.

Pesticide–An economic poison defined in most state and federal laws as any substance used for controlling, preventing, destroying, repelling, or mitigating any pest. Includes fungicides, herbicides, insecticides, nematicides, rodenticides, desiccants, defoliants, plant growth regulators, etc.

Physical selectivity–Refers to the use of broad-spectrum insecticides in such ways as to obtain selective action. This may be accomplished by timing, dosage, formulation, etc.

Physiological selectivity–Refers to insecticides that are inherently more toxic to some insects than to others.

Phytotoxic–Injurious to plants.

Piscicide–Chemicals used to kill fish.

Poison control center–Information sources for human poisoning cases, including pesticides, usually located at major hospitals.

Poison–Any chemical or agent that can cause illness or death when eaten, absorbed through the skin, or inhaled by humans or animals.

Polymerization–A chemical reaction in which individual molecules combine to form a single large chemicl molecule (a polymer). Usually involves the release of a lot of energy.

Postemergence–After emergence of the specified weed or crop.

Ppb–Parts per billion (parts in 10^6 parts) is the number of parts of toxicant per billion parts of the substance in question.

ppm–1.Parts per million. Generally used to express small concentrations of one substance in a mixture.

2.Parts per million (parts in 10^6 parts) is the number of parts of toxicant per million parts of the substance in question. They may include residues in soil, water, or whole animals.

Predacide–Chemicals used to poison predators.

Preplanting treatment–Made before the crop is planted.

Propellant–An inert ingredient in self-pressurized products that produces the force necessary to dispense the active ingredient from the container. (See Aerosol.)

Protectant–Fungicide applied to plant surface before pathogen attack to prevent penetration and subsequent infection.

Protective clothing–Clothing to be worn in pesticide-treated fields under certain conditions as required by federal law, *e.g.*, re-entry intervals.

Protopam chloride (2–pam)–An antidote for certain organophosphate pesticide poisoning, but not for carbamate poisoning.

Rate–Refers to the amount of pesticide active ingredient applied to a unit area regardless of percentage of chemical in the carrier (dilution).

Raw agricultural commodity–Any food in its raw and natural state, including fruits, vegetables, nuts, eggs, raw milk, and meats.

Reactivity–The ability of a substance to undergo change, usually by combining with another substance or by breaking down. Certain conditions, such as heat and light, may cause a substance to become more reactive. Highly reactive substances may explode.

Reentry (intervals)–Waiting interval required by federal law between application of certain hazardous pesticides to crops and the entrance of workers into those crops without protective clothing.

Registered pesticides–Pesticide products that have been approved by the Environmental Protection Agency for the uses listed on the label.

Repellent (insets)–Substance used to repel ticks, chiggers, gnats, flies, mosquitoes, and fleas.

Residual–Having a continued killing effect over a period of time.

Residue–Trace of a pesticide and its metabolites remaining on and in a crop, soil, or water.

Resistance (insecticide)–Natural or genetic ability of an organism to tolerate the poisonous effects of a toxicant.

Restricted–use pesticide–One of several pesticides, designated by the EPA, that can be applied only by certified applicators, because of their inherent toxicity or potential hazard to the environment.

RNA–Ribonucleic acid.

Rodenticide–Pesticide applied as a bait, dust, or fumigant to destroy or repel rodents and other animals, such as moles and rabbits.

Safener–Chemical that reduces the phytotoxicity of another chemical.

Secondary pest–A pest that usually does little if any damage but can become a serious pest under certain conditions, *e.g.*, when insecticide applications destroy a given insect's predators and parasites.

Selective insecticide–One that kills selected insects, but spares many or most of the other organisms, including beneficial species, either through different toxic action or the manner in which insecticide is used.

Selective pesticide–One that, while killing the pest individuals, spares much or most of the other fauna or flora, including beneficial species, either through differential toxic action or through the manner in which the pesticide is used (formulation, dosage, timing, placement, etc.).

Senescence–Process or state of growing old.

Sex lure–Synthetic chemical that acts as the natural lure (pheromone) for one sex of an insect species.

Signal word–A required word that appears on every pesticide label to denote the relative toxicity of the product. The signal words are either Danger-Poison for highly toxic compounds, Warning for moderately toxic, or Caution for slightly toxic.

Slimicide–Chemical used to prevent slimy growth, as in wood-pulping processes for manufacture of paper and paperboard.

Slurry–Thin, watery mixture, such as liquid mud, cement, etc. Fungicides and some insecticides are applied to seeds as slurries to produce thick coating and reduce dustiness.

Soil application–Application of pesticide made primarily to soil surface rather than to vegetation.

Soil persistence–Length of time that a pesticide application on or in soil remains effective.

Solubility–The amount of a substance that can be dissolved in a solvent, usually water.

Soluble powder–A finely ground, solid material that will dissolve in water or some other liquid carrier.

Solvent–Usually, a liquid in which other substances are dissolved. The most common solvent is water.

Spot treatment–Application to localized or restricted areas, as differentiated from overall, broadcast, or complete coverage.

Spreader–Ingredient added to spray mixture to improve contact between pesticide and plant surface.

Sticker–Ingredient added to spray or dust to improve its adherence to plants.

Stomach poison–A pesticide that must be eaten by an insect or other animal in order to kill or control the animal.

Structural pests–Pests that attack and destroy buildings and other structures, clothing, stored food, and manufactured and processed goods; *e.g.* termites, cockroaches, clothes moths, rats, and dry-rot fungi.

Stupefacient or soporific–Drug used as a pesticide to cause birds to enter a state of stupor so they can be captured and removed, or to frighten other birds away from the area.

Subcutaneous toxicity–The toxicity determined following its injection just below the skin.

Surfactant–Ingredient that aids or enhances the surface-modifying properties of a pesticide formulation (wetting agent, emulsifier, or spreader).

Suspect Carcinogen–A substance that might cause cancer in humans or animals but has not been proven to do so.

Suspension–Finely divided solid particles dispersed in a liquid.

Synergism–Increased activity resulting from the effect of one chemical on another.

Synthesize–Production of a compound by joining various elements or simpler compounds.

Systemic–Compound that is absorbed and translocated throughout the plant or animal.

Tank mix–Mixture of two or more pesticides in the spray tank at time of application. Such mixture must be cleared by EPA.

Target–The plants, animals, structures, areas, or pests to be treated with a pesticide application.

Temporary tolerance–A tolerance established on an agricultural commodity by EPA to permit a pesticide manufacturer or his agent time, usually one year, to collect additional residue data to support a petition for a permanent tolerance; in essence, an experimental tolerance. (*See* Tolerance.)

Teratogenic–Substance that causes physical birth defects in the offspring following exposure of the pregnant female.

Thermal–Involving heat.

TLV–Abbreviation for Threshold Limit Value. The average 8-hour occupational exposure limit. This means that the actual exposure level may sometimes be higher, sometimes lower, but the average must not exceed the TLV. TLVs are calculated to be safe exposures for a working lifetime.

Tolerance–Amount of pesticide residue permitted by federal regulation to remain on or in a crop. Expressed as parts per million (ppm).

Tolerant–Capable of withstanding effects.

Topical application–Treatment of a localized surface site such as a single leaf blade, on an insect, etc., as opposed to oral application.

Toxic Substance–Any substance which can cause acute or chronic injury to the human body, or which is suspected of being able to cause disease or injury under some conditions. Many toxic substances are chemicals or chemical mixtures, but there are other kinds of toxic substances as well (bacteria and viruses, for example).

Toxic–Poisonous to living organisms.

Toxicant–A poisonous substance such as the active ingredient in pesticide formulations that can injure or kill plants, animals, or microorganisms.

Toxin–A naturally occurring poison produced by plants, animals, or microorganisms; for example, the poison produced by the black widow spider, the venom produced by snakes, and the botulism toxin.

Trade name (trademark name, proprietary name, brand name)–Name given a product by its manufacturer or formulator, distinguishing it as being produced or sold exclusively by that company.

Translocation–Transfer of food or other materials such as herbicides from one plant part to another.

Trivial name–Name in general or commonplace usage; for example, nicotine.

Ultralow volume (ULV)–Sprays that are applied at 0.5 gallon or less per acre or sprays applied as the undiluted formulation.

Vapor Density–The density of the gas given off by a substance. It is usually compared with air, which has a vapor density set at 1. If the vapor is more dense than air (greater than 1), it will sink to the ground; if it is less dense than air (less than 1), it will rise.

Vapor–The gas given off by a solid or liquid substance at ordinary temperatures.

Vector–An organism, as an insect, that transmits pathogens to plants or animals.

Virustatic. Prevents the multiplication of a virus.

Viscosity–A relative measure of how slowly a substance pours or flows. Very viscous substances, like molasses, pour very slowly. Slightly viscous substances, like water, pour and splash easily.

Volatility–A measure of how quickly a substance forms vapor at ordinary temperatures.

Volatilize–To vaporize.

Weed–Plant growing where it is not desired regarded as using ground or hindering the growth of more desirable plants.

Wettable powder–Pesticide formulation of toxicant mixed with inert dust and

a wetting agent that mixes readily with water and forms a short-term suspension (requires tank agitation).

Wetting agent–Compound that causes spray solutions to contact plant surfaces more thoroughly.

Winter annual–Plant that starts germination in the fall, lives over winter, and completes its growth, including seed production, the following season.